Computer Systems for Occupational Safety and Health Management

OCCUPATIONAL SAFETY AND HEALTH

A Series of Reference Books and Textbooks
on Occupational Hazards • Safety • Health •
Fire Protection • Security • and Industrial Hygiene

Series Editor
ALAN L. KLING
Loss Prevention Consultant
Jamesburg, New Jersey

Other Volumes in Preparation

Computer Systems for Occupational Safety and Health Management

Charles W. Ross

WAPORA, Inc.
Madison, New Jersey

MARCEL DEKKER, INC.   New York and Basel

Library of Congress Cataloging in Publication Data

Ross, Charles, W. [date]
  Computer systems for occupational safety and
health management.

  (Occupational safety and health ; 10)
  Includes index.
  1. Industrial safety--Data processing. I. Title.
II. Series: Occupational safety and health (Marcel
Dekker, Inc.); v. 10 [DNLM: 1. Accidents, Occupational.
2. Computers. 3. Occupational Health Services--
organization & administration. W1 OC597M v.10 /
WA 412 R823c]
T55.R64 1984   363.1'1'02854   84-17055
ISBN 0-8247-7243-1

MARCEL DEKKER, INC.
270 Madison Avenue, New York, New York 10016

Current printing (last digit):
10 9 8 7 6 5 4 3 2

PRINTED IN THE UNITED STATES OF AMERICA

PREFACE

This book is written for safety professionals and other professionals
who have the responsibility for managing safety programs. It is
written for the professional who wants a safety data management system but who may not be familiar with computer systems.

In managing a safety program, the professional collects a variety
of reports. He or she must be able to analyze these reports and give
management a concise summary of what action is to be taken to minimize or eliminate the risks. In many situations the professional does
not have the time to collect the data from hundreds of case files, analyze the data, and report the results. For example, suppose a professional is asked to review last year's 350 accident cases to find the
four most prevalent causes, the five accidents, by type, that occurred
most frequently during last year, the three supervisors in whose departments most of the injuries occurred, and the total number of
recommendations still outstanding among all cases. It is an over-
whelming task to find and report that information. It can be done by
manually searching the files and sorting through the cases. It would
probably take a professional about 6 to 8 days of work to provide
such a report. This is assuming there are no interruptions, meetings,
phone calls, or other disruptions. The question then becomes, is
there time enough to do this kind of research? The answer is probably
not. It is frustrating to know that good information is available but
that there is simply no time to retrieve it manually.

I felt that, like myself, most of the professionals who manage
safety programs want to be able to search rapidly through their files
and compile specific information about a problem or problems. They
want to be able to provide management with information about accident

cases from which decisions can be made. They want information that
can document ways to cut accident costs. They want to be able to
advise management where potential high-risk conditions are located
and to recommend effective corrective action. They want to produce
all this information quickly and accurately. They want to be able to
change the parameters of the data, refine the information, and promptly
produce another report.

My interest in finding better ways to manage safety data started
in the early 1970s. It was not until 1976, however, when I heard
Bill Pope speak about the computerized system he developed at the
U.S. Department of the Interior, that I saw a way to cope with the
problem. Bill's description of his system really interested me. He
outlined a system that tracked dozens of items on an accident report
and he could get at the information easily. I wrote the Department
of the Interior for that system and when it arrived I studied it as
though I were preparing for a final exam. I was impressed with the
system, but I felt that I had no way to implement it.

Some months later I read about Bob Wright's safety data system
at Gulf Oil Canada, Ltd. His system was patterned after Bill Pope's.
I wrote Bob and he sent me a copy of his system. I saw that this system could be useful to me. The company I was working for at the time
encouraged me to propose a safety data system. I was successful in
designing a third generation to Bill Pope's system. In my work as a
consultant with WAPORA, Inc., I have been able to design other safety
data management systems for clients. This work has enabled me to
further improve the original system. These improvements are included
in the system described in this book.

The book is written as a step-by-step "how to" text that outlines
methods and techniques in developing a computerized system for managing safety data. I wrote it for the professional who may not be
familiar with computer systems. Thus, the book not only includes the
mechanics of creating a system to manipulate safety data but also discusses the steps necessary to persuade management that such a system
is a cost-effective and profitable venture. The effort and the preparation necessary to obtain approval of the system is as important as the
proper design of the computer system to manage the data. Both subjects are therefore treated with equal thoroughness. The book also
contains discussions on future applications that can be added as modules
to the basic system. The appendixes include several complete safety
data systems from different industries. Between the descriptions in
the text and the information in the appendixes, a professional will
be able to mix the systems and make one that will work for his or her
own application.

I wish to thank Bill Pope, founder of the National Safety Management Society, and Bob Wright, Safety Manager for Gulf Oil Canada, Ltd., for leading the way and developing their fine safety data systems on which mine is based. For her suggestions and ideas about Chapter 1, thanks to my daughter Amy. My love and thanks to my wife Betty for her support and her editorial comments in preparing the manuscript.

Charles W. Ross

CONTENTS

INTRODUCTION

In an era when businesses of all sizes are restructuring and adapting
to meet rapidly changing financial and technological demands, the
business leader must repeatedly reassess the viability of every component of the company. When profits are high and times are good,
marginal components are often tolerated; however, when inflation climbs
into double digits and interest rates on borrowed capital skyrocket,
frequent reevaluations must be made. The business leader must un-
hesitatingly prune the inefficient, the marginal, and the excess elements from the company. It is a necessary action if the business is
to survive.

During these times, those company work groups that survive
with few layoffs and minimal budget reductions are those that are
recognized as consistently operating as efficient, contributing profit
centers. Such groups are recognized by management to be effectively
reducing costs and continually developing new ways to increase profits. These work groups are the survivors because they have produced
results and, more importantly, they have consistently reported their
contributions to management. They are pragmatic in their recommendations, and they produce valid, believable dollars-and-cents results.
Over the years, these groups have repeatedly proved and justified
their worth to the organization.

The operating line work groups report their work through a series
of reports based on cost of raw materials, production rate, reject or
scrap rate, and value of goods moved into inventory or into transportation to customers. There are capital improvements and operations
improvements that can be shown to reduce costs and improve the product. Changes in production methods or materials can reduce scrap

rate and increase the number of goods ready for market. All of these
reports and special reports provide management with an easy measure
to assess the efficiency and effectiveness of a line work group.

In a staff group, a more sophisticated type of effort is required
to consistently show a dollars-and-cents contribution to economies
and to the profits of the company. In many cases, staff groups reflect their accomplishments in terms of numbers and the amounts of
services provided to the company. Their measurement techniques for
productivity usually are not the same as those of the line manager who
measures costs and productivity as functions of the business operation.
Thus, a staff group that uses numbers and not dollars to show accomplishment is in an awkward posture when presenting this information
to management.

The safety group usually is located in a staff function such as
personnel, labor relations, or engineering. From an organizational
point of view this is an acceptable position. It allows the safety
professional to act as an advisor to members of line management. The
safety group can produce unbiased inspection and accident analysis
reports and make recommendations for action. All of the data in the
reports prepared by the safety group are usually developed manually.
In many cases, the safety group produces accident data in terms of
number of cases, frequency rates, and severity rates. Some safety
groups use cost of cases and compensation costs to show dollars
saved. This is the exception, not the rule. The safety group is
effective in reducing accidents and saving many dollars for the company, and safety is regarded by management as helpful in complying
with federal and state safety requirements. For this reason, management accepts the added expense of a safety staff.

However, unless the safety staff group has devised effective and
verifiable ways to demonstrate consistent contributions to cost savings
and to profits, it is among the first to suffer budget and personnel
reductions during a recession. In order for the safety professional
and the safety staff to remain intact, they must be recognized by
management as making a contribution to the efficiency and the profits
of the business. The cost savings derived from the safety function
and their contribution to profits must be reported in an accurate and
timely manner. Safety's high level of contribution to the business
must be made obvious to top management. Safety must be regarded
by management as an essential element that consistently contributes
to the profit line. Only when this condition has been achieved will
the safety staff be counted among the essential groups in the company.

A positive way that the safety group can report its contributions
is through the use of a computerized safety data system. This system can put the safety manager in a position to provide timely, effective information to perform three functions: (1) efficiently pinpoint

those areas that have the potential to cause harm, property damage
or both; (2) provide safety reports from which a manager can make
precise decisions and avoid costly business interruptions due to injuries and damage; and (3) report dollars-and-cents savings and contributions to profits on a consistent basis. Due to the computer's
high speed and random search capability, it can accomplish the three
functions faster and more accurately than can be done manually.

In addition, savings accomplished from the computer's speed are
twofold: (1) the use of the computer reduces manpower requirements
to produce reports from large accident files; and (2) more importantly,
because of the computer's speed, more data can be accessed and analyzed in a brief time period. Thus, better and more timely reports can
be available to direct management's efforts toward areas where the
savings potential are greatest.

A computerized safety data system vastly expands the capabilities
of the safety group. A data system enables the safety manager to
easily search and manage the data formerly only maintained in manual
files. The computer reports allow the safety department to demonstrate to management better contributions to the profit line. The
timely and efficient reports allow accurate decision making. With
precise data about causes of accidents and their costs, a line manager
can accurately apply recommendations to reduce injury levels significantly. The additional cost savings are easily tracked since it is all
on computer. This kind of efficient reporting raises the awareness
of the management group concerning the value of the safety department. Thus, through the use of a computerized safety data system
the safety professional can consistently show a contribution to savings
and to profits. Such a data system can enable the safety group to
join the ranks of those recognized as essential to the company's success.

PART I

PREPARATION AND ANALYSIS: WHERE ARE YOU NOW?

1

COMPUTERIZATION: A BETTER WAY

There exist limitless opportunities in every industry. Where there is an open mind, there will always be a frontier.

Charles F. Kettering

INFORMATION PROCESSING: THE REDUCED-STATURE SYNDROME

All around us, in every field of human endeavor, there is a seemingly endlessly increasing mass of information. Experts and practitioners in every profession are generating more and more data as they conduct research to help others do a better job in their fields. Every day, studies, reports, research findings, government regulations, conference papers, and speeches are published concerning every dimension of business and industrial activity. "Today the United States government alone generates 100,000 reports each year plus 450,000 articles, books and pamphlets. On a worldwide basis scientific and technical literature mounts at a rate of 60,000,000 pages a year."[1] It is no wonder that in the safety profession, which is relatively small, there has been a corresponding increase in the studies, statistics, commentary, and other material published each year.

Those of us working in the safety profession face this flood of useful information each day. There are government reports to read and industry publications to review plus the regular company reports, such as injury and inspection reports, to read and analyze. Our reading and reviewing must be sandwiched between meetings, more inspections, visits, investigations, and consultations that go on every day.

To deal with all this safety information efficiently and to make effective use of such data, ways have to be found to allow us, as safety

professionals, to use our time effectively. The difficult part seems
to be the sorting and culling through of all the various pieces of
information and reports to find useful data. Once found, it is necessary to provide an analysis for our management or our client. Often,
the pace and the volume of work are so great that there simply is not
enough time to read and research all the information that might be
useful to our work in safety. On the other hand, there are occasions
when some time is available, but not enough time to extract and syn-
thesize all the bits of information filed away in many separate places.

In both of these cases, the level of performance and the reputation
of the safety professional can be diminished if the information that
can be assembled and presented to management is not sufficient or is
not compelling as that presented by other departments. The result
is that other departments get more attention and more time with manag-
ment. Consequently, they may obtain larger shares of the budget
and may continue to have management's ear for other projects. This
process can become repetitive, reducing the stature and effectiveness
of the safety professional. This reduced-stature syndrome may continue to the point that safety professionals come to be viewed as not
really contributing to productivity and business efficiency, as compared
with other staff and line groups. This can place the safety professional at a distinct disadvantage. As a result of the reduced-stature
syndrome, the safety function itself may not be valued highly, and
thus the salary for that job may not be comparable to the salaries of
others at the same peer level. When cuts are made in manpower, the
safety staff may suffer a higher percentage of loss than other collateral
or supporting groups such as labor relations, personnel, recruiting,
security, and training. When competing for time with management,
safety may be viewed as less important than other departments that,
in the eyes of managers, contribute directly to the profit line.

The safety professional needs an efficient way to counteract the
reduced-stature syndrome. To do this, the safety professional must
find a way to effectively get at and produce the kind of information
that will attract attention and be recognized by management as a significant contribution to the efficiency and the profitability of the
enterprise. That is a tall order. But it is what must be done if the
safety professional is to be taken seriously.

Let us step back a minute and review a typical injury case to see
how the safety information is handled. From this we shall draw some
conclusions. Following that, we shall discuss the sources and history
of the statistical systems we now use. The intent is to direct attention
to the way we handle information now, with an eye toward finding a
better way to handle it in the future. Ultimately, the goal is to find
a way to cure the reduced-stature syndrome and increase the stature
of the safety professional.

THE CASE OF BILL SMITH, MAINTENANCE ELECTRICIAN

The Accident

It was a bright, sunny spring morning. The electrical shop foreman,
Tom Elder, had just assigned Bill Smith the job of removing the guy
wires from the six utility poles that carried power into the plant. As
he walked out of the shop, Bill felt that the job was not really challenging. Frustration built inside him. He considered it beneath his
skill level to do such work. After all, it was only undoing a few bolts
and taking the wire-rope guy from the pole to allow the contractor to
replace the existing poles.

He had worked at the Townsville plant for six years. In that time
he had advanced from laborer to journeyman electrician. He was good
at what he did, and he liked to do the tough jobs. Still, Bill admitted,
the job of removing those guy wires, by union rules, called for an
electrician of at least a journeyman level; these lines did carry the main
13,000-volt power for the whole plant. In his mind, Bill finally grudg-
ingly agreed that it was a proper assignment.

Bill drove his blue pickup truck out to the job site. The new poles
were already being readied by the contractor's crew to replace the old
ones. He took his aluminum extension ladder from the rack on the back
of the pickup and carried it to the first pole. What a simple job thought
Bill: loosen the turnbuckle at the ground stanchion, climb the ladder
and remove the nut holding the eyebolt to the pole, pull the eyebolt
loose from the pole, and let the guy wire fall.

The first two poles were no problem. On each one, the eyebolt nut
backed off with only a little pressure on the wrench. The sun beat
down. It was going to be one of those hot summery days, he thought.
He was really starting to sweat. At the third pole, the turnbuckle
was rusty. It took a lot of strength to break the turnbuckle loose and
turn it. After several minutes, Bill got the slack he needed on the
guy. He climbed the ladder and got his wrench on the eyebolt nut.
It would not budge. Bill wished that he had thought to bring the
penetrating oil to loosen the rust. He tried again. The nut was frozen.

Bill looked at the situation. All he really had to do was get the guy
wire down. The eye of the eyebolt gave him an idea. It was opened
slightly wider than the other two had been. He climbed down and went
to his truck for his large pair of lock-grip pliers. He went back up
the ladder, got a good grip on the eye, and began to bend the eye
farther apart. Slowly the metal gave, and the eye bent open. It
was looking more like a hook all the time. Finally, the eye was bent
into an L shape. By this time Bill could feel the sweat running all
over his body. Beads of sweat were pouring into his eyes. He wiped
his face with the back of his sleeve. With a sigh of relief, Bill reached
over and lifted the guy wire off the L of the opened eyebolt. The
opened eye of the bolt was facing straight up. Bill raised the guy

wire up to clear the hook, but it snagged at the tip of the hook. With
a jerk, Bill snatched the wire clear. That sudden move threw the
guy wire and Bill's arm high above the hook. As his arm reached the
top of its swing, Bill could see his hand and the wire passing just
beneath the high-voltage line.

The electric arc cracked like a rifle shot. Workers along the line
snapped their heads around to see what had happened. They saw
Bill falling to the ground, the guy wire still clutched in his hand.

From up and down the line of poles men scrambled to help. Bill lay
quivering and jerking in the grass beneath the pole. His breath was
labored. The weeds at the base of the guy-wire stanchion had been
burned and were smoldering. One man raced to his truck and radioed
for the emergency squad. Another man threw a sweater over Bill.
Others just looked on, helpless.

Within minutes the siren of the emergency ambulance could be heard.
Other cars soon arrived, raising a cloud of dust. Tom Elder was one
of the first to get there. Tom quickly assessed the situation. He
went over to Bill and knelt down. Bill's eyes were open, but his
breathing was erratic. The emergency medical team arrived seconds
later. There was a flurry of activity. Quickly, oxygen was administered, and Bill, wrapped in blankets and on a stretcher, with sirens
screaming, was taken to City General Hospital. Tom stood there looking
at the ladder and the length of guy wire lying limp on the ground next
to Bill's yellow hard hat. Tom took out the small notebook he always
carried in his shirt pocket and started making notes on what he saw.

As Tom was writing and gathering information from the men who had
been near the scene, Ernie Fletcher, the plant safety manager, arrived
in his pickup. Together, the two men continued to investigate the
accident. During this time, the plant manager, maintenance manager,
and others in management arrived and left.

About an hour and a half later a doctor from the hospital called Tom
at his office. The doctor said he was sorry, but Bill Smith never re-
gained consciousness and died in the emergency room. Tom told Ernie,
who was in the office at the time. Ernie said that he had better call
the local office of the Occupational Safety and Health Administration
(OSHA); he hurried off, saying he would be back in a few minutes.
Tom continued to look over his investigation notes for a few more
minutes, then left his office and drove to meet with the plant manager
and the maintenance manager. Together they would decide how Bill's
wife was to be told. Tom had never before had a man die on the job,
at least not one who worked directly for him. It was not going to be
easy to forget this day.

The Investigation

Both a compliance officer from OSHA and other plant personnel investigated the accident of Bill Smith. The OSHA compliance officer remarked

at the closing conference that citations would probably be issued on at least two standard violations: one, using an aluminum ladder near energized high-voltage power lines; two, working on energized equipment without proper protective gear, such as gloves and leather protectors. As he left, the compliance officer said the citations would be issued in several weeks.

ACCIDENT ANALYSIS REPORT

DATE OF REPORT		
MO	DAY	YR
04	22	92

GENERAL		
ACCIDENT NO.– SAFETY **4467**	PLANT **Townsville**	ESTIMATED COST
DEPARTMENT **Plant Maintenance**	ACCIDENT OCCURRED — MO **04** DAY **21** YR **82**	☒ L.T.A. ☐ S.I. ☐ OTHER
MACHINE OR AREA **West Plant Road at 3rd line pole**	TIME **10:45 AM**	

EMPLOYEE

NAME OF EMPLOYEE	JOB TITLE	NATURE OF INJURY
William L. Smith	Electrician Class II	Electrocution – Fatal

ACCIDENT DESCRIPTION

Bill Smith was assigned the job of removing guy wires from the incoming power poles carrying 13 Kva. He had removed two poles. On the third pole he had trouble removing the guy from the pole and opened the eye. When he lifted the guy from the hook he came too close to the power line and received an arc. He was standing on an aluminum ladder and the other end of the guy was attached to a stantion in the ground. The arc passed through his body to the ladder and then to ground.

ACCIDENT CAUSE

UNSAFE ACT	UNSAFE CONDITION
Lifting the guy wire up instead of allowing it to fall down.	Using an aluminum ladder and not using rubber protection equipment.

WHY ACT WAS COMMITTED	WHY CONDITION EXISTED
Poor technique possibly unaware of the arcing hazard of high voltage equipment.	Aluminum ladders have been permitted for use on non-electrical work for many years. Removing the guy wires was termed non-electrical work.

HAZARD CONTROL

UNSAFE ACT CONTROL	UNSAFE CONDITION CONTROL
① Training session will be provided concerning the hazards of high voltage equipment. ② Sessions on various electrical repair techniques will be started for all electrical trades.	① Electrician will only use wooden ladders. ② Supervisors will check all jobs above 440 volts prior to start of the job. ③ High voltage work above 4K vac will be done by a licensed contractor.

FOLLOW-UP

RESPONSIBILITY FOR CONTROL	RESPONSIBILITY FOR CONTROL
Maintenance Shop manager.	Maintenance Shop Manager
EST. COMPLETION DATE – 07/30	EST. COMPLETION DATE – 06/30

	REPORTING SUPERVISOR
	PLANT MANAGER	SAFETY

Exhibit 1-1

The accident report,(Exhibit 1-1) written by Tom Elder and Ernie
Fletcher, explained the facts concerning the work situation and the
accident. Delivered to the plant manager by Tom and Ernie, it was
reviewed and the necessary corrective, preventive measures were
instituted. The citations from OSHA were answered with the same care.
Maintenance supervisors made the necessary checks on ladders and job
starts. Refresher training for the electrical workers was scheduled
as a regular monthly program in the maintenance department. Management personnel at the plant felt confident that they had done a good
job of correcting the problem.

Adequacy of Monthly Statistics

When Ernie Fletcher, the safety manager, published the accident figures
for the month, they were neatly typed and presented at a morning staff
meeting (Exhibit 1-2). The plant manager and his department managers
were there. It was discussed and reviewed in the meeting. Both the
maintenance and warehouse managers agreed to work with the safety
manager to reduce the number of injuries occurring in their departments. The plant manager said he wanted to be kept informed of the
actions taken and the progress made on the recommendations. Several
of the other department managers said they believed that more work
needed to be done in safety. The plant manager agreed and asked the
safety manager to develop a plan to meet that request. The production
manager reminded the group that production goals were 21 percent
behind target for the month. He said that quality had to be improved
and back orders had to be reduced. He stressed that production had
to be given first priority if the plant was to meet its production goals
for the month. He recommended that each manager examine each activity that cut into the production time of their supervisors and make
some hard decisions. The discussion continued.

Epilogue

Have you heard of a story like this? Many? Have you lived this kind
of situation? Do you find it typical? Do you believe that more could
have been done? Would you feel frustrated trying to cope with the
crude analysis methods depicted here?
The safety professional uses the contemporary statistical structures
illustrated by this sample case. These structures are the norm for the
National Safety Council and the OSHA. These formulas produce statistics that can give an indication of the levels at which the events
(errors?) are occurring. These statistics can be used to compare one
plant with another or to compare a plant against the industry levels.
The frequency-rate method used at the Townsville plant is based on
the OSHA incidence-rate system.

May 3, 1984

    To: Staff
  From: Ernie Fletcher
    Re: MONTHLY CUMULATIVE ACCIDENT ANALYSIS (January-April 1984)

	Cases			Lost Work-days	OSHA Incidence Rates		
	Medical-treatment	Lost-time	Total		Total Rate	Lost-Workday Case Rate	Lost-workday Rate
April 1984	3	2*	5	14	4.3	2.0	11.9
March 1984	8	0	8	0	6.8	0.0	0.0
April 1983	6	0	6	0	6.5	0.0	0.0
January-April 1984	18	6	24		5.9	1.5	14.5
January-April 1983	24	1	25		5.8	0.2	5.3

Analysis:
   The following is a breakdown of the locations and types of injuries that have
occurred over the past 4 months. Please note that the largest number of injuries happened in the warehouse and maintenance departments (45 percent of
the total injuries). This number includes the two lost-time injuries that occurred
this month, as well as the others that happened in the previous months of this
year.

Locations of accidents

Maintenance	7 (4)*		Electrocution	1
Production	3		Fall, different level	1
Finishing	2		Laceration	4
Warehouse	4 (2)		Fall, same level	2
Power plant	2		Back injury	3
Administration	1		Burn	1
Stock preparation	2		Cut	6
Cleanup crew	2		Eye irritation	3
Q-C laboratory	1		Puncture	1
			Sprain	2

*Parentheses indicates the number of lost-time cases.

Recommendations: Review the training procedures in both the maintenance and
warehouse departments. Have the safety committees make additional safety surveys of those areas. Increase safety awareness through safety meetings and
posters.

Exhibit 1-2

   The following sections develop a frame of reference for the OSHA
incidence-rate system and compare it with the method of calculating injury rates used by the American National Standards Institute (ANSI
Z16.1).

CURRENT METHODS OF REPORTING INJURY EXPERIENCES

OSHA Incidence-Rate System

The statistical frequency-rate calculation used by OSHA and the U.S.
Bureau of Labor Statistics is called the incidence rate. This rate system is based on 100 workers working 40 hours per week for 50 weeks
per year. This amounts to 200,000 annual worker-hours. The 100
workers are the average workforce for a plant in the United States.
The rates are related to that figure through a formula. The calculation uses those injuries that are recordable on the OSHA Log of Occupational Injuries and Illnesses. This includes both lost-workday
cases and medical-treatment injuries. The definition of a lost-workday
case (sometimes referred to as a lost-time case) is one in which the
worker cannot return to work on the next regularly scheduled shift.
The injury is counted as one case. The case of Bill Smith, for example,
would count as one case. No lost workdays would be charged, because
Bill died the same day. In this system, the only lost days that are
recorded are those that are regular workdays. Therefore, if a plant
operates only on weekdays and is closed Saturday and Sunday, then
those days would not be counted. For example, there is no lost time
recorded for a case in which a worker is injured on Friday afternoon,
goes to the doctor and is sent home, but returns to his regular shift
on Monday. However, depending on the circumstances surrounding
the injury, the case may be recordable as a medical-treatment.
   A medical-treatment case is one in which a medical professional performs a procedure that cannot be done by a non-professional trained
only in first aid. Diagnostic procedures are excluded from consideration. This medical treatment can include such procedures as setting
a bone, giving a shot, and suturing a laceration. Prescribing special
medication for an illness caused by the workplace, such as exposure to
toxic substances, can also be classed as medical treatment. This
classification applies to a patient who returns to work on the next
regularly scheduled shift and does not lose a workday; otherwise it
will be charged as a lost-workday case.
   Both of these types of cases are counted in calculating the total
case incidence rate. This formula is as follows:

$$\text{Total case incidence rate} = \frac{(\text{No. of lost-workday cases} + \text{No. of medical-treatment cases}) \times 200{,}000}{\text{No. of worker-hours for the period}}$$

   Worker-hours can be calculated in either of two ways. One way
is by approximation. The other method uses computer runs of time
cards. The computer runs are quite accurate if all employees, both
management and employees, are included in the system. The approxi-

mation method is effective if computerized data are not readily available. This method uses the average number of employees for the month and multiplies that number by the number of workdays in the period, times eight for the 8-hour shift. The resulting answer will provide a close approximation of the actual worker-hours. It is considered a close approximation, because worker absences and time off will just about balance out against overtime hours.

Besides the total case incidence rate, two other rates can be determined from this formula. These are the lost-workday case rate and the lost-workday incidence rate. The rates are calculated in the same way, but either lost-workday cases or total workdays for the period are substituted in place of the total number of reportable cases. Table 1-1 shows us how these different rates are computed.

ANSI Z16.1 Standard

The accepted norm for recording injury statistics prior to introduction of the OSHA incidence-rate system was the ANSI Z16.1 standard. The ANSI standard was developed from work started in 1920. The U.S. Bureau of Labor Statistics published Bulletin 276, which recommended the use of a standard formula to calculate rates of injuries. This rate calculation came to be used in both government and private sectors. In 1926 the bulletin was revised by a committee set up by the American Standards Association (now called the American National Standards Institute). The first edition of the standard was published around 1937. The standard has been revised several times between that date and the present.[2]

Standards such as ANSI Z16.1 are produced as consensus documents. This means that a committee formed of interested parties from industry, labor, government, and the public at large work on a proposed standard until unanimous agreement is reached concerning the standard's contents. Often this process takes years. However, such a process usually produces a standard that is acceptable to the groups represented and, more important, one that will be used.

The ANSI Z16.1 standard, "USA Standard Method of Recording and Measuring Work Injury Experience," differs from the OSHA incidence-rate system in several significant ways. ANSI counts only lost-time injuries, whereas OSHA records lost-time cases and another category called medical-treatment cases. ANSI records all days lost. Under ANSI, neither the day of the injury nor the day of return is counted, but all the days in between are recorded as lost days. OSHA does not count the day of the injury or the day of return either, but it differs by counting only lost workdays (that is, days that the employee would normally have worked). ANSI uses 1,000,000 man-hours as a base, whereas OSHA uses 200,000 worker-hours. ANSI also has a schedule of charges in lost days for various disabling injuries, depending on

Table 1-1 Determination of Incidence Rates: An Example

Plant population average for period:	1500
Number of workdays in the period:	21
Number of shifts:	1

Cases for the period: Lost workdays   2   (14 lost workdays)
                        Medical treatment   9
                        First-aid cases   23[a]
              Total reportable cases   11

Worker-hours = 1500 X 8 X 21 = 252,000 (approximation method)

$$\text{Total case incidence rate} \quad = \frac{11 \times 200,000}{252,000} = 8.73$$

$$\text{Lost-workday case rate} \quad = \frac{2 \times 200,000}{252,000} = 1.59$$

$$\text{Lost-workday incidence rate} = \frac{14 \times 200,000}{252,000} = 11.11$$

---

[a]Not reportable.

severity. For example, a fatality merits a charge of 6000 lost days.
Loss of an eye receives a charge of 1800 lost days, and a hand lost
at the wrist is recorded as 3000 lost days. All parts of the body that
can be lost or damaged in an injury are accounted for in the schedule
of charges. These lost-day charges are included in the calculation of
the severity rate. OSHA does not make any charge other than for the
actual days lost due to an injury. The calculations for injury rates
within the ANSI system are as shown in Table 1-2. The plant situation is the same as the OSHA example in Table 1-1.

A Critical Look at Both Rate Systems

Both of these rate systems generate statistical information. The rates
provide a statistical basis to describe the number of events that are
occurring in a given period of time. Each rate system produces frequency and severity information, severity meaning the number of days
lost per time period per number of worker-hours. The data generated
in each system can be quite accurate, depending on the quality of the
investigation and the way the worker-hours are developed. Thus,
both formulas will produce accurate frequency and severity rates.
     Our logic tells us that frequency means that this is a rate at which
events are happening. Further, it is assumed the events will continue

Table 1-2 ANSI Z16.1 Injury-Rate Determination

Plant population 1500
One-shift operation

Cases : Lost workdays — 2

Case 1 : 4 days (injured Thursday the 5th, back Thursday the 26th)

Case 2 : 16 days (injured Tuesday the 10th, back Thursday the 26th)

Total lost days: 20

Worker-hours: 252,000 (approximation method, Table 1-1)

Total cases — 2^a Frequency rate $= \dfrac{\text{No. cases} \times 1,000,000}{\text{Worker-hours}}$

Lost days — 20 Severity rate $= \dfrac{\text{No. days lost} \times 1,000,000}{\text{Worker-hours}}$

Frequency rate $= \dfrac{2 \times 1,000,000}{252,000} = 7.94$

Severity rate $= \dfrac{20 \times 1,000,000}{252,000} = 79.36$

---

[a] Had one of these two lost-time cases resulted in a fatality, the schedule charge of 6,000 days would be added to the days lost. This is a theoretical number charge based on the amount of time a 30-year-old worker would work during the rest of his career.

to occur at the same rate unless something is done to change the rate. The information is presented to management as an indication of the rate at which future injuries will occur. Recommendations are made to reduce this predicted rate of events. Management follows the recommendations. Sometimes the rate goes down. Everyone is pleased. But sometimes more injuries happen, and the rate goes up. The succeeding recommendation to management is to try harder and follow the recommendations closer. Perhaps management did not follow the recommendations completely. Then, too, other factors may be involved that are not evidenced in the current rate information, factors such as the number of rush orders, special one-time operations, the start of a second-shift operation, worker malpractice, marginally hazardous conditions, extra overtime, equipment abuse, and many others. All of these factors can be operating in a plant, but they do not become significant to the reporting system until they cause an injury, death,

or property damage. This is not to say the safety professional does
not try to find the problems before they cause injuries, just that the
rate systems do not record these factors until they produce injuries.

Another point is that injury events, which are recorded as cases,
are random events. In other words, injuries do not occur at a standard rate; they pop up, as it were, from within a stream of other events.
This means that it is faulty logic to believe that frequency rates predict that injury events will occur at a standard rate. Random events
do not follow any set pattern.

If frequency rates actually track an unrelated string of random
events, how did the misconception get started that allowed us to believe
otherwise? In the early 1930s an insurance investigator, H.W. Heinrich,
published a study of the types of injuries that occurred at an industrial plant. He said there was a ratio between major, or disabling,
injuries and those injuries of a lesser degree of severity. His study
showed that in that plant 1 disabling injury occurred for every 29
minor injuries, and for every 29 minor injuries, about 300 noninjury
events occurred. This ratio of 1:29:300 was published in a book;
it came to be endorsed and accepted as fact by the safety profession.
That ratio has been included in many books, manuals, and other safety
publications, and it was not challenged by research until 1969.

In 1969, the Director of Engineering Services for the Insurance
Company of North America (INA) made an analysis of 1,750,000 accidents reported by 300 corporations representing 21 different industries. This group of companies had a total annual employment of over
1,750,000. This employee population had an exposure of over 3
billion man-hours during the study. The study results verified that
Heinrich was right in believing there was a relatively constant ratio
between different types of injury events. The INA study found a
ratio of 1:10:30:600. This stands for 1 serious injury per 10 minor
injuries per 30 property-damage cases per 600 near-miss cases. Near-
miss cases are those events that cause neither injury nor property
damage but could have if the circumstances had been different. It
is interesting to note that only 53 percent of the companies in that
study investigated and reported property-damage cases. In addition,
88 percent investigated only disabling-injury cases.

There was significant work done by INA in the area of incident
recall during the study. This technique involves interviewing employees and encouraging them to remember events that resulted in neither
injury nor property damage but that would have been capable of either
under different circumstances. Chapter 12 has a complete review of
this technique.

The incident-recall technique used by INA provided the extra information needed to establish and to verify the ratio. This is especially true for the near-miss events in this ratio. Even considering
that there may be some statistical deviation in the INA study, their
mass of data does indicate that a ratio can exist between different types
of injury events.[3]

We may choose to believe Heinrich, or we may rely on INA's report of their ratio. Others may wish to develop a study of their own and find a different ratio between injury events. The fact remains that there is some relationship between injury events. The key point, however, is that a ratio may or may not be accurate in predicting future events, because all of the events are occurring in random order. Where injuries occur this month may not be the same place they will occur next month. An if, as the INA study shows, most companies investigate and report only lost-time injuries, an analysis of these cases probably will not generate a sufficient amount of data to be interesting or persuasive to management. Controls cannot be accurately recommended for prudent management decisions based on so few random events.

What INA and Heinrich proved is that there is a pyramid of events, with a lost-time event at the top. Thus, if safety professionals today want to attract management's attention, they must examine more cases than just lost-time injuries. If strong arguments are to be developed to persuade management to take decisive action, then as many cases as possible must be investigated and reported. This effort must include the near-miss category.

The designers of the OSHA incidence-rate system recognized the need to look at greater numbers of cases. Under that system, both lost-workday and medical-treatment cases are reported and included in the rate calculations. This increases the numbers of cases significantly. It provides much more data than can be provided through the ANSI system, which tracks only lost-time cases. The most important point to recognize is not that more cases are reported, but what is done with the information once it is collected.

THE INFORMATION WITHIN INJURY CASES

Heinrich: Unsafe Acts and Unsafe Conditions

For a moment, let us step back and take a historical view of where our knowledge of the accident sequence originated. In a previous section we discussed the accident rates developed by Heinrich. Let us expand on that knowledge. In his book, *Industrial Accident Prevention,* Heinrich listed 10 axioms of industrial safety. The first axiom stated that "[the occurrence] of an injury invariably results from a completed sequence of factors, the last one being the accident itself. The accident in turn is invariably caused or permitted directly by the unsafe act of a person and/or a mechanical or physical hazard."[4] The second axiom stated that "the unsafe acts of persons are responsible for a majority of accidents."[5]

Since his book was published, these axioms have provided much of the foundation for the teaching of safety. These 10 axioms have been

quoted and used by safety specialists everywhere. The first two
axioms especially directed safety workers to concentrate on finding the
immediate causes of accidents. The emphasis was heavily placed on
finding the unsafe act and the unsafe condition that caused the injury.
Thus, it came to be that only two causes would be singled out and
listed on the investigation report form. Investigators looked for the
unsafe act and the unsafe condition. Once found, these two causes
were entered on the form. Corrective action centered around developing ways to prevent unsafe acts and conditions. This highly concentrated effort brought results in the years after the book was published, and these successes cemented the technique into the profession.
    Accidents, however, kept occurring. Others in the profession
believed that what Heinrich had alluded to in the first axiom needed
to be reviewed again, specifically the statement "...an injury invariably results from a completed sequence of factors,..."[6] It was becoming apparent that there was a set of other contributing factors
that needed investigation.

The Injury Sequence: How Things Happen

In investigating accidents, the investigator searches for causes that
contributed to the event that resulted in an injury or property damage
or both. Many times the investigation ends when the immediate causes
of the injury have been found. This approach is based on Heinrich's
theory. It is true that these immediate causes that triggered off the
injury event are important, but there were errors and conditions that
preceded the event in space and time. These events and errors are
causes that also contributed to that injury. Unless corrected, these
underlying causes will continue to contribute to future injuries.
    In industry, the majority of accidents are caused by errors. Er-
rors are the result of many factors. They are not confined to worker
errors alone. All of these factors lie in a time continuum behind the
accident event. Other error sources are supervisory misdirection,
inadequate procedures, upper management policies, lack of operator
knowledge due to inadequate training, and so on. Figure 1-1 illustrates some of the other error source factors that can contribute to
the accident-event sequence. This model depicts the conditions, policies, and operating climate that contribute to an accident. Singly,
these conditions, decisions, policies, etc., probably would not cause
injury, but if some combination is together in the right time sequence
an accident may result. The more factors, the greater the probability.
Once the accident sequence begins, the outcome is subject to the whims
of fortune or chance. The final result can lie anywhere in the entire
spectrum of outcomes, from a no-injury, no-property-damage near-
miss to a catastrophic event involving loss of life and severe property
damage.

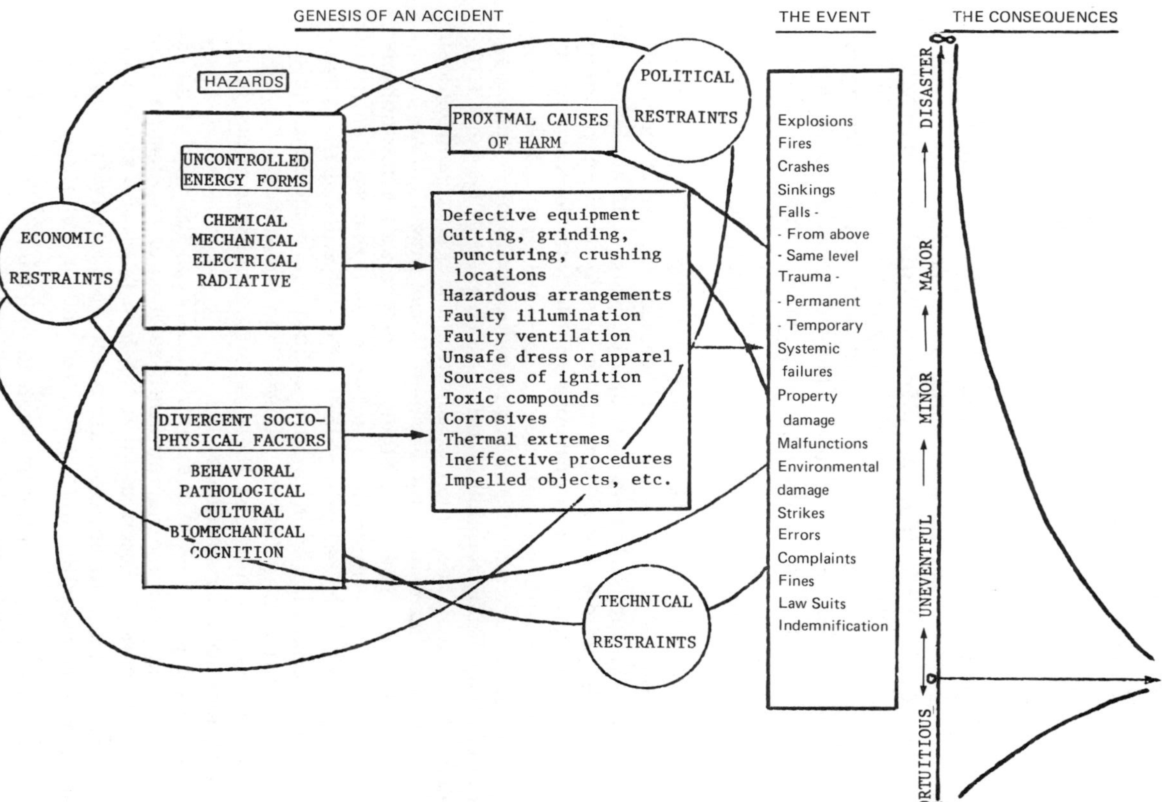

Figure 1-1 Adapted from lectures by Dr. John V. Grimaldi, New York Univ., 1974-75.

The Theory of Multiple Causation: Beyond Heinrich

Simply stated, there are multiple events that occur on a time continuum
during an operation. If event sequences line up and uncontrolled
physical forces are in proximity, an accident can occur. Let us see
how this relates to the domino theory Heinrich used:

Most safety people have preached this theory many times.
Many of us have actually used dominoes to demonstrate it. As
the first domino tips, it knocks down the other four dominoes
unless at some point a domino has been removed to stop the
sequence. Obviously, the easiest and most effective domino to
remove is the center one—the one labeled "unsafe act or condition." This theory is quite clear; it is also quite practical
and pragmatic as an approach to loss control. Simply stated,
"If you are to prevent loss, remove the unsafe act or the unsafe condition."

We use this theory in two fundamental areas today: in accident
investigation and in inspection. In accident investigation, almost
invariably the forms that we use, or that we give to our supervisors
to use, ask that one unsafe act and/or unsafe condition be identified and removed. This, of course, seems very logical, considering the statements and principles expressed by the domino theory.
It is in fact a very practical and pragmatic approach. Perhaps,
however, our interpretation of this domino theory has been too
narrow. For instance, when we identify a single act and/or a single condition that caused the accident in the investigation procedures of today, how many other causes are we leaving unmen-
tioned? When we remove the unsafe condition that we identify in
our inspection, have we really dealt with the cause of the potential accident? Today we know that behind every accident there
lie many contributing factors, causes, and subcauses. The theory
of multiple causation is that these factors combine together in random fashion causing accidents.

Let us briefly look at the contrast between the multiple causation theory and our too-narrow interpretation of the domino theory.
We shall look at a common accident: A man falls off a stepladder.
If we investigate this accident using our present investigation
forms, we are asked to identify one act and/or condition:

The unsafe act: climbing a defective ladder.
The unsafe condition: a defective ladder.
The correction: getting rid of the defective ladder.

This would be typical of a supervisor's investigation of this
accident under the domino theory.

Let us look at the same accident in terms of multiple causation.
Multiple causation asks what are some of the contributing factors
surrounding this incident. We might ask:

1. Why was the defective ladder not found in normal inspec-
    tions?
2. Why did the supervisor allow its use?
3. Did the injured employee know he should not use it?
4. Was he properly trained?
5. Was he reminded?
6. Did supervision examine the job first?

The answers to these and other questions would lead to the following kinds of corrections:

1. An improved inspection procedure.
2. Improved training.
3. A better definition of responsibilities.
4. Prejob planning by supervisors.

With this accident, as with any accident, we must find some
fundamental root causes and remove them if we hope to prevent
a recurrence. Defining an unsafe act of "climbing a defective
ladder" and an unsafe condition of "defective ladder" has not led
us very far toward any meaningful safety accomplishments. When
we are looking at the act and the condition, we are looking only
at symptoms, not at causes. Too often our narrow interpretation
of the domino theory has led us only to accident symptoms. If
we deal only at the symptomatic level, we end up removing symptoms but allowing root causes to remain to cause another accident
or some other type of operational error.

Root causes often relate to the management system. They
may be due to management's policies and procedures, supervision
and its effectiveness, training, etc. Root causes are those which
would effect permanent results when corrected. They are those
weaknesses which not only affect the single accident being investigated, but also might affect many other future accidents and
operational problems.[7]

An important part of the mission of the safety professional today
is directed toward predictions and the correction of potential sources
of harm. That is why inspections are made. That is why thorough
accident investigations are performed. That is why we make analyses
of accident cases. Identifying errors and omissions is the goal. With
the methods and tools currently available, finding these sources is a
difficult and time-consuming task. Our investigations, predicated
on the principle of preventing a recurrence, may detect part of the
error sources. Let us say our investigations are thorough, and 20

to 25 percent of the errors are found by this method. Safety inspections may find another 20 to 25 percent of the sources of harm. Together, these two methods account for a 40 to 50 percent identification
of potential sources of harm. That means that half of the sources of
harm will be under analysis for correction at any one time. Why?
Because the industrial operation is always growing and changing, and
thus new hazards and problems are moving into positions where they,
too, can cause harm. It is a perplexing problem. Part of the answer
in finding the other 50 percent of the future sources of harm lies in
handling the data we now obtain more efficiently. Another part lies
in using techniques designed to predict future sources of harm.

SOURCES OF HARM

Analysis of Safety Data

The monthly accident report that Ernie Fletcher, the safety manager
in the Bill Smith case, presented to his plant management is an example of a monthly report (Exhibit 1-2). Many of us have seen or have
prepared similar reports at one time or another. This report format
can be expanded to analyze other parameters, such as the following:

1. The part of the body injured.
2. Injury cases by location.
3. Cases by time of day.
4. Cases by day of the week.
5. Cases by job being performed.
6. Length of service.
7. Age of employee.
8. Number of years in that job.

Usually, each accident case that happens during a month is listed
in the report. Each case has a brief description of what happened,
the corrective action recommended, and what will be done. Trends
are identified in the first part of the report with the frequency rates.
The usual discussion of the report by management centers around
what action is being taken in each case, and then there is a brief review of the trends. In other words, there is a case-by-case examination of what is being done, or what should be done, to prevent a recurrence. This approach addresses each case squarely and usually
is successful in either eliminating or reducing the recurrence possibilities for that particular problem. This work on the accident problem
is effective as far as it goes. It simply does not go far enough.
There is more information to be extracted from each accident case
than we are currently using. Consider this: There are 20 or more

bits of information contained in the report on Bill Smith (See Table
1-3). Suppose an analysis is made of 100 cases. It probably will uncover some trends and subtle points to be checked out that are not
evident when looking at one case at a time. If this is done on a regular basis, it probably will improve the safety professional's ability to
predict future sources of harm. Working through case files is a time-
consuming task. If it were possible to somehow do the sorting job
faster, that would be a major improvement.

Reviewing Safety Data Files

Occasionally a line manager will ask the safety professional about a
specific injury cause or a potential problem. For example, suppose
Bob Davis, the maintenance manager from our Bill Smith case, asks
Ernie Fletcher how many steam burns have occurred at the Townsville
plant in the last year. Also, Bob wants to know in each case where the
incident happened, the kind of process or operation involved, and what
part of the body was injured. Ernie first goes to his monthly report
with its listing of cases and identifies all the cases of steam burns.
He then pulls the case files, extracts the information, analyzes the
data, and prepares a report. Time spent? Probably an entire day
from start to the completed typed report.

When Ernie reviews the report with Bob, he is pleased. They discuss the information in detail. Bob asks Ernie if all the burns were
caused by steam hoses and how may of the injuries happened to new
or inexperienced employees. He asks Ernie to provide just a little
more information. This means Ernie must go back to the files, repeat the process, and deliver another report. Time-consuming? Very
much so. Important to their working relationship. Yes.

Any review of injury case records in a fairly large plant or company can be a time-consuming job. For example, suppose at the
Townsville plant there are, on the average, 35 injury cases reported
each month. This figure includes all kinds of cases, including first
aid, medical treatment, and lost time. Over a 3-year period this will
amount to more than 1260 cases on file. To review the case files for
the pertinent cases, even with a file log or index, as Ernie Fletcher
did, can take many hours. Analyzing the information and preparing
one such report probably will take a day or more of a safety professional's time. A second analysis of the files, as in the foregoing example, will require pulling some of the same case files and performing
a further study of the data. Of course, a secretary can gather the
case files, but the safety professional, like Ernie Fletcher, still must
conduct the review. The second analysis also must be written up and
prepared in polished form to be presented to the manager.

Reviewing files manually is a time-consuming job, one that is not
done too often because of the time and expense involved. However,

Table 1-3 The Case of Bill Smith: Accident Analysis Report Information

---

The accident report contains the following:

1. Report number	12. Nature of injury
2. Plant identification	13. Description of accident
3. Type of case	14. Unsafe act
4. Date of the report	15. Unsafe condition
5. Estimated cost	16. Why act was committed
6. Department	17. Why condition existed
7. Location of the injury	18. Unsafe-act control
8. Date of the accident	19. Unsafe-condition control
9. Time of the accident	20. Responsibility for control (act)
10. Name of the employee	21. Responsibility for control
11. Job title	(condition)
	22. Reporting supervisor

---

a search will be performed if a good client, such as an important manager, asks for the data. Usually it will turn out to be not just one
search either, but two or three, because the initial report, if it is
well done, will generate interest in further information.

Follow-up: An Aggravating Problem

For the safety professional at any level in an organization, follow-up
on recommendations is a never-ending task. However, it is something
that must be done. Someone has to verify that those recommendations
are carried out. For example, a supervisor may believe that the recommended changes have been accomplished and may relay that message
to safety. On visiting the department, the safety professional may
find that one of several things has happened: One, the recommendations have been carried out as specified. Two, the supervisor has
misinterpreted the instructions and has done something different.
Three, the recommended changes have been only partially accomplished
because of misperception of what needed to be done.
    Follow-up, therefore, is a two fold project. First, the safety de-
partment must have a feedback system in order to be informed that
the recommendation has been carried out. There is an assumption
here. The assumption is that at the beginning of the process the
recommendation was originally transmitted to those who were to take
action. Second, once receiving the feedback, those in safety must
then take the time to verify the results of the action that was taken
to assure that what was accomplished was what was intended to be
done.

On the surface, this appears to be a rather simple and straight
forward task. It consists of making recommendations, receiving feedback about them, and verifying that the changes have been accomplished. But, like many tasks that appear simple, this can be quite
complex and time-consuming. First, there are many sources from
which recommendations are generated. These include in-house safety
surveys and inspections, injury and property-damage case reports,
OSHA inspections and citations, insurance reports, reviews of new
and renovation construction plans, and new and revised process reviews, as well as general recommendations for changes in the way
things are done. Within each of the written reports there will be at
least one recommended action, and usually more. In an operation of
any size, this reporting activity will keep as many as 100 or more recommendations in process at all times. And that number includes just
the written recommendations. Beyond that are the verbal recommendations that the average safety professional passes out over lunch and
at other times.

Ideally, each recommended change should be documented and verified for completion. This is a time-consuming and complex undertaking, because the status of the recommendation is always changing.
Parts are ordered; then they are back-ordered. Delays and changes
of all kinds must be considered a given, not a variable. The recommendation, if it is extensive, may require that plans be drawn up by
engineering, and therefore the safety department must check with
engineering to assure that the plans correctly describe the details of
the recommended change. Management may delay the recommendation
for lack of funds until the next fiscal year, which may be only a month
or so away. Perhaps a guard to go between two machines must be
fabricated, but must wait until the machines are relocated to a new
production line. Of course, the new line must also be inspected for
safety. This inspection may generate more recommendations. The list
of possible variations in the changing status of a recommendation is
endless, and this is considering only those recommendations that are
accepted by management.

The process of following up, which is quite important, must be
performed as conscientiously as possible in the time the safety professional can spare. In many instances, the time simply is not available to follow up on all the outstanding items. Follow-up may be performed once on a recommendation, but then time passes. It may be
a month or more before an opportunity is again available to check on
the recommended change. By that time, more recommendations will
have entered the system. Thus, the process is slow and tedious.
Another factor is the problem of remembering the status of each recommendation and when to check again. A method can be devised to
track each recommendation, but it must be updated as each action is
taken. If the follow-up system is operated manually, it will generate

file drawers full of papers that must be reviewed again and again to
discover where the recommendation stands. This also requires further
administrative controls to ensure that those in safety will know when
the next follow-up is due.

Use of a Computer System in Safety Work

The discussion in this chapter has concentrated on the traditional ways
a safety professional goes about certain phases of the job, specifically,
how information is developed, stored, retrieved, and then presented
to someone else. All of us in safety receive and store great amounts
of information on injuries, damage, surveys, inspections, reviews,
and meetings. Most of us use some manual means for retrieving and
culling through our stored data.

In most cases a safety professional cannot use all the data that
are stored. Why? Because there simply is not enough time to manually retrieve, extract, analyze, and present the information. Also,
usually there are not many requests. Why? Because most of us do
not encourage them. Again the problem is time.

As in the example of Bob Davis, the maintenance manager of the
Townsville plant, how one handles a request for specific information
from injury case files can be important in maintaining good client relationships. Again we emphasize that this work is time-consuming.
One report that is well done and provides good data can easily generate more requests. The time it takes to perform such tasks manually
is enormous. This poses a dilemma. Suppose a safety professional is
asked to develop data for a manager (client). The cost of this project
will be significant in terms of time, but if the information is not provided, the client may think that the safety professional is not interested in providing a service. Consequently, the manager may be less
inclined to ask for help again. Worse, the manager may not want to
cooperate in the next new safety program.

There is a parallel time problem in following up on recommendations.
This job is necessary, but it also is quite time-consuming. In many
cases, if the follow-up is not performed, nothing will happen. The
department head may or may not take the recommended action. If it
is done, the action may not be complete or may be inadequate to control the exposure. In other words, if follow-up is to be done, the
safety professional must do it.

A computer system can be useful in improving the efficiency of
the safety professional. The three processes just discussed can be
made more effective and efficient through the use of a computer system. For injury cases, a computer system can accept 50 or more
items of information for each case. The information will be sorted and
calculated, and a report will be quickly generated in a format that can
be easily analyzed. For example, the Bill Smith case could be entered

into the computer system and compared with other cases that occurred that month, the previous month, or the previous year. Comparisons can be made for 5 or 10 separate items of information for each case. For example, the computer could list the information for comparison by type of case, department, location of injury, job function of the injured, management factors, human errors, and cost. Such reports can be generated in seconds by a computer. It eliminates the manual labor of searching the files and allows the safety professional to make the analysis. From the analysis, relationships probably will appear that were too subtle or obscure to show up when reviewing the files case by case.

Follow-up procedures on recommended changes are also easy to track on a computer. Reports on follow-up activities can be produced on a regular basis, providing regular reminders of those recommended changes that are in progress and those that have been completed.

One real advantage of computerization of safety data is that by this means the safety professional is using the same information storage and handling methods that are used by other parts of the business operation. This can be an important factor in raising the status of safety information management and raising the stature of the safety professional in the eyes of management. These are two of the more important advantages to using a computer system: First, more effective and more efficient use of the available safety information formerly stored away in file cabinets. Second, management is more apt to take notice and listen when presented with computer-generated safety reports. Why? Because they are in the same format as the other reports that management regularly sees. They are computer-generated!

SUMMARY

Because of increases in the amount of information available, there is a need for a better method to sort and cull through this data. Our traditional manual methods and statistical structures take too much time and do not provide impressive information. This results in reduced stature for the safety professional in the eyes of management. Both the OSHA incidence-rate system and the ANSI Z16.1 standard are based on frequency rates. These formulas presuppose that the current frequency rates will continue into the future. The trouble with this notion is that injury cases are random events and thus may not repeat at a standard rate. Random events work against the safety professional who is attempting to predict future trends and provide recommendations for action.

The concentration on lost time and disabling injuries had its origins in the book *Industrial Accident Prevention* by Heinrich. That book outlined the theory of the unsafe act and the unsafe condition as the main causes of injuries. The ratio of 1:29:300 among disabling injuries,

minor injuries, and non-injury events was proposed as the ratio
describing how events happen. The ratio idea was confirmed by the
INA study in 1969, but a different ratio was found. Still the emphasis
was on prevention of disabling injuries. Later, the idea arose that
there could be more than two contributing causes in an injury event.
This is the theory of multiple causation, which broadens the ideas of
Heinrich to embrace the notion that there can be a wide range of
contributing causes in any injury.

The goal of the safety professional is prediction of accident trends
and elimination of future sources of harm. Achievement of this goal
requires analysis of injury case records, follow-up on safety recommendations, and performance of inspections. All of these activities
take time. Much of the safety professional's time is spent culling and
sorting through records to compile persuasive reports that will convince management to take action. A computer system can greatly reduce the time spent finding and collating data that are in storage.
At amazing speed the computer can analyze injury case files, report
the follow-up status of recommended changes, and provide statistical
summaries and lists of information for easy analysis.

The computer is a basic tool of management. Those who use computers in compiling the reports they submit to management tend to be
regarded as working efficiently. The safety professional who uses
a computer for safety data gains in two ways: One, by gaining better
control over the safety information and working at greater efficiency,
the safety professional presents more persuasive analysis and more
impressive data. Two, if computer-generated reports are presented
in the accepted format, management will recognize that the safety
professional is contributing to the efficiency and profits of the business.

NOTES

1.   Alvin Toffler, *Future Shock*. Random House, New York, 1970,
     p. 31.
2.   American National Standards Institute, *USA Standard Method for
     Recording and Measuring Work Injury Experience*. Z16.1. ANSI,
     New York, 1976.
3.   Harold E. O'Shell, *Modern Principles of Loss Prevention and Con-
     trol*. International Safety Academy, Houston, 1972.
4.   H. W. Heinrich, *Industrial Accident Prevention*. McGraw-Hill,
     New York, 1st ed., 1931, 2nd ed., 1959, p. 69.
5.   Ibid., p. 69.
6.   Ibid., p. 69.
7.   Dan Petersen, *Safety Management—A Human Approach*. Aloray,
     Huntington, N.Y., 1975. p. 17-18.

2

THE CURRENT REPORTING SYSTEM: A LOOK AT WHAT IS HAPPENING NOW

It is always better to proceed on the basis of a recognition of what is, rather than what ought to be.

Stewart Alsop

When I started to develop my first safety data system, I quickly discovered a very important fact: When a large sum of company money is to be spent on a project and the project requires the services of other departments, the justification for that project must be well presented. To prepare a proposal, the first thing I did was to examine the current accident reporting system. This analysis solved two problems. First, it developed data for comparison with the new operation using the computer system. Second, if, for whatever reason, I was not successful in getting permission to develop a computerized safety data system, I could use the analysis to make improvements in the current system. Fortunately, I was able to marshal enough facts to convince management to support the development of a computerized safety data system.[1]

In this chapter we shall discuss ways to analyze an accident reporting system. Through this technique and others in later chapters, I shall show you the methods I used to persuade management to agree to the creation of a computerized system.

Devising a computerized safety data system is a relatively straightforward task. The problems arise when permission is required to build such a system. Persuading others to accept your ideas is a most challenging and rewarding effort. In many cases, the contributions of others will further improve the original idea. Thus, when proposing a new project, it is wise to examine in detail what is now being done.

Every aspect of the current system is important. Each part needs to be examined in detail. Among these aspects are the types of summary reports that are sent to management, the investigation and reporting cycle, the nature and the uses of the current injury case file, and how accident costs are developed. All of these subjects are significant parts of the current reporting system. The more that is known about how the system acutally works, the better you, the safety professional, will be equipped to discuss the subject with management. Knowledge of the present system will enable you to intelligently review the advantages and disadvantages of converting to a new system.

SAFETY REPORTS SENT TO MANAGEMENT

Content and Format

In a reporting system there is usually an end product that is regularly sent to management. This end product is called by many names, but it is a summary of the accident events that happened during a specified time period. In Chapter 1, the summary report prepared by the safety manager in the Bill Smith case was a monthly report (see Exhibit 2-1).

In an examination of the current reporting system, the place to begin is the end product. The reason for the survey reports and the way in which they are viewed by management will provide avenues of inquiry into the rest of the system. As an analysis of these documents proceeds to uncover perceptions and historical facts about the origins of the reports, the investigator will begin to understand why the system operates as it does. To begin the examination of the summary reports, here are some sample questions that can be used:

1. How is the information about numbers of cases presented in the report? Are only raw numbers given, or is there a frequency rate computed as well? Is there an analysis of the meaning of the rate or raw numbers?
2. Is there a comparison of frequency rates used? In other words, is this month's rate compared with that of last month or a year ago? If this is done, what does it show to management? What can a manager do with the information?
3. From these statistics, can a prediction be made as to where future accidents will happen? Can the sources of serious harm be identified? Try this. Take a report prepared 2 years ago, and try to predict future injury sources and trends. Confirm your findings with the current reports.
4. In the report, what kinds of cases are reported? Does the list include lost time, medical treatment, first aid, near-miss, property damage, and other incidents? All of them or just some? If it includes some and not others, find out why.

May 3, 1984

    To: Staff
  From: Ernie Fletcher
    Re: MONTHLY CUMULATIVE ACCIDENT ANALYSIS (January-April 1984)

| | Cases | | | Lost Work-days | OSHA Incidence Rates | | |
	Medical-treatment	Lost-time	Total		Total Rate	Lost-Workday Case Rate	Lost-workday Rate
April 1984	3	2*	5	14	4.3	2.0	11.9
March 1984	8	0	8	0	6.8	0.0	0.0
April 1983	6	0	6	0	6.5	0.0	0.0
January-April 1984	18	6	24		5.9	1.5	14.5
January-April 1983	24	1	25		5.8	0.2	5.3

Analysis:
    The following is a breakdown of the locations and types of injuries that have occurred over the past 4 months. Please note that the largest number of injuries happened in the warehouse and maintenance departments (45 percent of the total injuries). This number includes the two lost-time injuries that occurred this month, as well as the others that happened in the previous months of this year.

Locations of accidents

Maintenance	7 (4)*	Electrocution	1
Production	3	Fall, different level	1
Finishing	2	Laceration	4
Warehouse	4 (2)	Fall, same level	2
Power plant	2	Back injury	3
Administration	1	Burn	1
Stock preparation	2	Cut	6
Cleanup crew	2	Eye irritation	3
Q-C laboratory	1	Puncture	1
		Sprain	2

*Parentheses indicates the number of lost-time cases.

Recommendations: Review the training procedures in both the maintenance and warehouse departments. Have the safety committees make additional safety surveys of those areas. Increase safety awareness through safety meetings and posters.

Exhibit 2-1

5. What is the history of these reports? How did they come into
    being? Who developed them? If possible, find out why these
    particular reports are used. If you were the originator of this
    set of reports, describe the process of development and the
    decisions that were made during the beginning period. For
    example, during the development period, were any compromises
    made? If so, these should be explained and included in the
    examination.
6. Are costs included in the report? How are these costs devel-
    oped? What is the degree of accuracy of the costs? (Are
    they estimates or actual?) Are both direct and indirect costs
    included in the report? If not, why?

This list of questions is aimed at finding the real purpose for making
a summary report. If the reason is that it has always been done or
that it is expected, then a harder look is necessary. To be useful,
the report has to contribute to the business enterprise. A report is
an effective tool if it can provide solid information on which to make
decisions. The purpose of good safety information is to reduce downtime and work stoppages by controlling injuries. The review of this
report should be directed toward this purpose. The optimum would
be a report that provides information about where there are sources
of harm that have yet to cause an injury. This optimum report would
also indicate malpractices and causes of injuries that were too subtle
to be discovered by reading a single case. The analysis of the current
report should uncover every scrap of information about what the report does and does not do as an effective decision-making tool.

Routing Reports to Management

The quality of a report's content can be an important factor in its
distribution, or rather its redistribution. A report is initially distributed by the originator. Thus, the originator has control over who
will receive the report. This is distribution. Redistribution is where
the first reader passes along the document to others for their review.
Rerouting is usually done when the first reader believes that the
report is important enough to be shown to others on the staff.
    In the case of the safety report, what happens is an important
aspect in learning if the content of the report is recognized as being
worth reading and passing around. What kind of impact does it have?
Ask the readers. If it is given a courteous reading and then filed
away, the report may be perceived as a nice report, but not one on
which decisions can be made. This is a point for careful consideration. The purpose of the report and the reader's perception of it
have to coincide; otherwise the document will not be used as intended.

This short discussion on the routing of a report is intended to raise some questions about that activity. It is also intended to help you expand your examination of the reporting system. In concert with describing the purpose of the report and its distribution, the kind of feedback that is received is also important.

Feedback Concerning Reports

Connected closely with the routing of a report to management is the type of feedback that is received. This kind of communication can be received in two forms: verbal and written. If it is written, it can be formal, as in a memo, or informal, as in a note written right on the report. These memos or comments can run the entire scale, from questions, to praise, to criticism, to formal requests for assistance on a problem. These are typical types of written responses.

Verbal feedback is the other form of reply concerning reports. Remarks about the report can come from a discussion at a formal meeting or from a casual conversation over lunch or in the workplace. This verbal feedback can sometimes be more informative than what is written. This is especially true in informal discussions, where the atmosphere is less constrained. Feedback in terms of feelings and suggestions in a relaxed climate can provide substantial insight into ways of improvement. The key point, though, is that there *is* feedback of some kind.

Comments and criticism about one's work complete the two-way portion of the communication system. Replies and feedback are especially important aspects for all parts of the safety program. They provide the safety professional with an understanding of how the program is working, or not working.

In this particular instance, we are discussing comments and perceptions of the regular safety reports sent to management. For example, if there is no comment, this may indicate a lack of concern or may show a general apathy about the content of the report. Conversely, an absence of comments about the regular reports may mean that they do not warrant comment. Consider this: Suppose at a certain plant the safety department had generated reports month after month. Each month the safety manager made a presentation about the report. For a considerable length of time, no feedback or comments pro or con were received. If you were this safety manager, would you not want to find out if anyone was listening? I believe you would. So would I.

What about compliments? This is the "thanks for a job well done, pat on the back" we all enjoy. Appreciation and compliments for work performed are scarce commodities, and when they are received, they are always welcome. Positive feedback can be quite useful; it can be a means to develop a good give-and-take dialogue.

The favorable comment about the report is the entrée. Return the compliment by showing your appreciation to that person for taking the time to make the comment. Feedback followed by a sincere expression of appreciation can lead to a good discussion of the report, and possibly one or more excellent suggestions may come out of the meeting. Real unsolicited feedback, good or bad, is difficult to elicit on a regular basis. When it happens, it is important to grab the opportunity. During such a chance meeting it may be worth the effort to ask the manager about others who also might want to discuss the report. This can be a lead to others who might want to contribute their ideas to improving the safety program, but who, for various reasons, have not come forth.

Sometimes we are uncomfortable encouraging feedback even when it is understood that this process is important in improving a report's usefulness. So often we are reluctant to ask someone else how our work is perceived. We would rather not hear what other people, especially those in management, have to say about our work. Why? It is a risk. A risk of hearing a negative comment.

In an examination of the current system, it is important to enlist feedback and comments about the system's condition. It is also important to review the quantity of feedback and try to determine whether or not it reveals how the reports are received. The risk is small, and the rewards in ideas are great.

Use of Cost Data

Evaluating how cost data are used in the current reports is another part of the examination of the reporting system. It can reveal many things about the system. Each aspect of the handling of accident cost information should receive careful attention.

The use of injury cost data has always been a subject of controversy because of the varying viewpoints about how these costs should be handled. Views and opinions differ depending whether or not the person believes the subject of injury costs is significant. For example, someone in accounting may believe that the subject of injury costs is adequately addressed through worker compensation payments and medical payments. These expenses are tracked through the normal accounting channels; therefore, any other tracking is extra work.

Line operations managers may agree with accounting and may add that damage and actions taken to correct a situation created by an injury are recorded as a cost through maintenance expense; therefore, everything is covered. The safety professional may or may not agree. The safety professional may believe that other costs are important and should be included as part of the accident picture. The

safety professional may have tried unsuccessfully to persuade others
that additional cost categories are important.

How injury and property-damage cost are kept is the real subject
of the review. The answer may be that cost levels, both direct and
indirect, are being recorded and are used to their best advantage as
part of the safety reports sent to management. On the other hand,
direct costs may be all that is reported. In any case, all the information that is found will be useful in the development of an overall
analysis of the current reporting system.

THE INVESTIGATIVE SYSTEM AND ITS FORMS

Events That Get Investigated

In the previous sections we looked at four aspects of the reports
that are prepared and sent to management: the content and format
of the reports, the receiver of the reports, the feedback that is received, and whether or not injury cost data were included in any of
the reports. In this section we shall look at the source events that
provide the basis for the reports.

The reason for the reporting system is that events occur that require explanation through a report. In this case, a safety reporting
system gathers, at a minimum, data on injury events. The depth and
breadth of information will depend on what events are investigated.
In reviewing the current reporting system, the types of accidents
that are formally investigated and reported will indicate several factors. One, the more types of cases that are investigated, the greater
will be the amount of information. Two, the broader the scope of the
investigation, the more opportunity there is for finding more sources
of harm.

The larger the number of injury cases of different severity levels
that are investigated and reported, the broader will be one's understanding about what is actually happening in the workplace. As mentioned in Chapter 1, human errors and malpractices occur quite frequently. Errors that sequence together to produce an injury are
relatively rare events. Thus, if only lost-time injuries are reported,
the amount of information accumulated will be small. Conversely,
if first-aid, medical-treatment, and lost-time cases are investigated
and reported, then a greater mass of information about errors and
losses will be available. Statistically, the greater the number of
events that are recorded per time period, the more reliable will be
the resulting data.

Table 2-1 shows the kinds of accident and nonaccident events that
can be investigated. Compare this list with those events that are
actually being investigated in your organization to determine the amount
of information that is being generated by the system.

Table 2-1 Events That Can Be Investigated

Injuries	Property damage	Near-miss (no injury, no property damage)
Fatalities	Fires	Error by operator
Lost-time cases	Damage to equipment	Hazardous condition
Medical-treat- ment cases	Damage to buildings and facilities	
First-aid cases	Abuse to equipment	
	Severe weather damage	

Let us digress for a moment and consider the term *to investigate*. This term, for our discussion, means that someone, preferably from management, took the time to go to the scene, talk with witnesses, examine the evidence, and prepare a report. This report states the findings, lists the conclusions as to causes, and makes recommendations for action. The inquiry into the accident does not have to be exhaustive and minutely thorough, but it should be performed with reasonable care. Usually, the degree of thoroughness is directly proportional to the seriousness of the accident.

Overall, an evaluation of the types of cases that are investigated can show us which events are only noted as happening and which events events are investigated and reported. The thoroughness of the investigation is also of concern. How detailed are the investigations? The entire subject of investigation, if probed fully, will contribute to the picture of the reporting system as it is currently functioning.

Development of the Current Reporting System

Closely tied to the question of the types of events that are investigated is an inquiry into the history of the current system. How and why the current reporting system was developed have direct bearing on what is now investigated, as well as what investigation forms are currently in use. Old policies and management styles may have shaped the current system; thus, those who were in management at the time should be identified.

Perhaps those who developed the current system have retired or left the company. It might be worth the effort to make a telephone call or two to find the answers to the origins of the system. This may also provide an understanding into the forces that shaped the system and the company. These policies, management styles, and concepts probably have extended beyond the philosophy of the company. Compiling these notions will also be useful later in our review of the company.

If you were the architect of the current system, then you already
know what assumptions and concepts were used. However, it may be
prudent to look again at the previous system and, if you have not done
so, review how it came into being. Many times when we change a
system this is not a point of thorough inquiry. It should be. Knowledge of who participated in and influenced the initial development can
be a source of enlightenment. Changing anything requires skill if it
is to be a lasting change; therefore, the more that we know of the
background of the system and of former systems, the more useful
material we will have for designing future changes

Training Investigators

A factor entering directly into the quality of the injury investigation
report concerns how adept the investigator is at finding and evaluating
the facts of the case. Effective training in investigation techniques
can help develop better information, leading to reports of better quality than would ordinarily be possible without such training. When we
review investigative training, the historical perspective should be
taken into account. The examination should concentrate on the type
of training in investigation techniques that is now provided. It should
also check out the kind of training provided 1, 2, 5, or even 8 years
ago. Why? Because those trained and those who received no training
several years ago may still be in the group performing today's investigations. If training records are available, this is a good place to
start.

Investigative training is carried out in many different ways in
various companies. This type of training is sometimes combined with
other topics in a safety course for supervisors. In other cases this
training is given as a single course. In our review, it is important
to look at the educational structure of each course. This will include
examining factors such as the amount of course time devoted to investigative techniques. The types of examples and the structure of
the course are other factors to consider. What specific topics did the
course cover? Here is a list of subjects that could be used in a course
on investigation techniques:

1. Interviewing witnesses.
2. Surveying the scene.
3. Evaluating evidence.
4. Researching facts.
5. Developing findings.
6. Preparing conclusions.
7. Writing recommendations.

Trained investigators usually turn out investigative reports of uniformly better quality than do those who have had to learn what they can from watching others. In making a review of this part of the system, it is well to document the findings especially from the historical perspective.

Investigation Reports: Content and Format

The way an investigation form is laid out and the information it asks for will help determine its effectiveness in acquiring the needed information. Poorly designed forms may elicit misleading data that can cause those who analyze, compile, and report the data to reach ineffective conclusions. For example, a form that requests only one unsafe act and one unsafe condition overlooks the fact that management oversights may be contributing to many accidents; therefore, concluding that there were only two causes and that they were the main cause of the accident might be an ineffective conclusion.

An investigation form can provide flexibility to include a wide range of facts and conclusions, or it can be very rigid. For example, a flexible form can ask for the development of as many causes for each accident as possible. It can ask that several recommendations be developed. On the other hand, a rigid form may ask for one unsafe act and one unsafe condition. Under recommendations, the rigid form may ask the investigator to develop recommendations to prevent a recurrence of that particular injury. Both forms have their uses. However, the flexible form will help develop more information and better recommendations than the more rigid form. The chances of prediction and prevention improve when there is more usable information.

The investigation form should be thoroughly reviewed to see what information it will yield. If the safety department ordinarily does not perform investigations but leaves it to other management personnel to fill out these forms, it may be a good exercise to complete several forms to see how well the investigative information fits into the form. Performing this task several times will highlight the good points and the flaws in the form.

The following is a list of subjects that a flexible investigation form should address:

1. Case identification: Within this category is such information as location, time, date, the type of case (whether it was property damage, injury, fire, near-miss or other), the contaminants or chemicals involved, and type of operation being performed.
2. People: If a person was involved in the case, the information will include name, employee number, supervisor's name, injury

    classification, workdays lost or days of restricted activity,
    and job title.
3. Description of the occurrence: In some detail, the facts of
    the event should be listed, as if telling a story.
4. Findings: These are facts discovered during the investiga-
    tion that support conclusions as to the causes of the accident.
5. Analysis of the causes: In any case there are usually several
    main causes and many minor contributing causes. All of these
    should be listed.
6. Recommendations for action: Sometimes the recommendations
    can be limited to the specific case. However, it is better if
    they can be expanded to suggest looking at a class of similar
    work practices, management policies, or conditions.

These six categories should be addressed to some degree in all investigations. A well-designed reporting form can facilitate the collection
of such information. The easier the form is to use, the more efficient
will be the process of collecting meaningful information.

Routing Investigation Reports

The routing of the investigation report is important. Routing should
be selected on the basis of how other company reports are distributed.
Once a report is put into the company or plant distribution system,
it becomes part of the mass of mail each person receives. Therefore,
the attention a report receives will depend on how well it competes
with all the other reports each person is receiving, as well as that
person's perception of its worth.

Usually the routing of an investigation report is fairly straightforward and simple, depending on who begins the sequence. The
originator of the report may be the supervisor who did the investigation, the plant nurse who treated the worker, or the safety professional. This is assuming that the injury or property-damage event is
all that is reported. If the system includes the reporting of a wider
scope of events, such as hazardous conditions and near-misses, then
the group of potential originators will be larger. Near-miss reporting
usually is originated by employees who have been encouraged to report such incidents. On the other end of the reporting spectrum
would be the investigation and reporting of only those cases required
by law. This would restrict reporting to lost-time, medical treatment,
and occupational illness cases. In this situation the originator could
be the nurse or the immediate supervisor. Routing would, if it began with the supervisor, be either to the next higher level of management or to safety, or both, depending on the plant's past practice.

Adding more different types of events to the list of things that are
investigated will add more variety to the possible routings from the

originator. An important factor in examining the routing system is
to learn what is done with the report after it is prepared. Who gets
the report? Is there a check to ensure that the completed report
reaches the safety department? Does the routing system have a good
sequence? Where does the corrective action on the recommendation
begin? That is, at what level of management does the action begin?
In addition, when does the action begin? As you continue to examine
the routing you probably will find other important considerations.
Examination of the routing system also can reveal perceptions of safety
within the organization, because it will show how high up in the organization accident events are reported, and in what detail. In the
right context every scrap of information can be significant.

Insurance Forms in the Reporting System

Even though there are company investigation forms, the insurance
carrier for workers compensation usually requires the submission of
its form as well. For those companies that are self-insured, there
probably are state forms that must be completed. Each of these may
require some data different from the data entered on the company
investigation form.

These insurance forms may be completed by the safety group, the
medical department, or personnel, depending on how the company is
structured. It is useful to review these forms and compare their
content with that of the current company forms. One should contact
the insurance carrier and ask about the type of accident summary
data they generate regarding injuries and their causes. This information may be useful for the development of cost information, because
medical expenses and workers compensation payments are reported
and tabulated by the carrier.

Quality of the Current System

When quality is mentioned, all too frequently there is a defensive reaction. In some cases that reaction may be well-founded. In many
companies the work of the safety professional is not regarded as
having the same importance as the work of the manager in traffic,
purchasing, production, or maintenance, Therefore, the reporting
of injuries may or may not be considered as important as other aspects
of a manager's or supervisor's job, and this can affect the quality of
the end product—the report.

Thus, the quality of the accident reporting system reflects atti-
tudes about several aspects of the system: the importance of the re-

port in the minds of those doing the reporting; how well investigators
have been trained and supported in making good investigations and
preparing thorough reports. With these possible limitations on quality
in mind, here are several parameters that can be used to measure the
quality of investigation reports:

1. Timeliness: Was the report received within 72 hours, or what-
    ever is considered prompt?
2. Completeness: Is the report complete? Are all the sections
    filled in thoroughly?
3. Description: Is the story of what happened accurate and
    factual?
4. Cause development: Does the report identify causes in terms
    of (a) worker malpractices or errors, (b) management system
    factors, and (c) hazardous conditions?
5. Recommendations: Does each recommendation have an expected
    completion date? Do some recommendations apply to situations
    that could exist companywide, or at least in other similar de-
    partment locations? Do the recommendations extend preventive
    measures to other oversights, malpractices, and conditions
    that have not yet produced injuries?
6. Follow-up: Is there a plan whereby management and safety
    will follow up and report on the compliance with these recom-
    mendations?

These six parameters for measuring report quality can be quantified. A point system can be developed to compare various degrees
of compliance in the different categories: There are six categories;
suppose we allot 10 points per section. For example, for the category
of completeness, the report will be considered excellent and will receive all 10 points if it is complete in every section. If the report
has one section left blank, then 9 points will be awarded, and so on
down to poor or unacceptable, where four categories are left blank.
This type of quantified rating system can give management a more
precise picture of the quality of the reports being received as compared with what is expected.

Quality is the element in the system that can make the difference
between well-prepared, accurate information and mediocre data. If
those in authority will insist on high standards for report quality
and will measure compliance against such a standard, then the system's workers will provide that high quality: "What gets measured
is what gets done."[1]

CASE FILES: A BANK OF IMPORTANT INFORMATION

Nature of the Current Filing System

Files on injury cases can be kept in a number of ways. They can be
filed by date, by month, by serial number, by department, or by
location. Cases can also be filed in a way that will combine several
of these sorting methods. Whichever system is used, it should be
examined to determine several factors: One, why are the cases filed
as they are? Two, how many years are readily available on file, and
how many years of files are retained in storage elsewhere? Three,
is there a synopsis of each case available for quick and ready reference? Four, how easy is it to locate an individual case? Five, how
complete is the average case? Six, how accessible are the case records in records storage? These are a few of the points to review in
evaluating the current filing system. The better the detail in which
the current reporting system is known, the more effectively this background information can be used in justifying a change to computerization.

Processing Requests for Information

Part of the evaluation of the current system concerns the handling
of requests for information about cases. Basically, this is an administrative procedure. However, it can be useful to learn how often
the accident case files are used: What type of record is kept concerning requests for information? How is this record kept? Who
keeps the record, and what is done with it? Is any report or summary
made about the numbers and kinds of requests that are generated?
Many other questions can arise concerning the processing of requests
for information about cases.

This type of request can also be tied to an overall system for
monitoring safety activity. A log such as that illustrated in Figure
2-1 may be used. In such a case, the analysis of how requests are
handled will center around the effectiveness of the request system.
Such questions as whether or not verbal requests get recorded should
be answered as well. If there is no system to track requests for
assistance, then the reasons should be documented.

Searches and Special Reports

The file of case histories is a source of information about what has
occurred over the years so far as losses and injuries are concerned,
and one should tabulate the frequency at which these files are used
each month to generate special reports. The log system described

earlier will track this kind of activity. If a log is not kept, then perhaps there are special reports on file that will indicate how often special searches of the files are performed.

Another type of information that will be quite useful concerns the mechanics of searching the current files. If this is not done routinely,

SAFETY SERVICES ACTIVITY REPORT

```
M   -  MEMO                                  NAME:_______________________
P   -  PROJECT
PH  -  PHONE CALL                            DATE PREPARED:______________
L   -  LTR
V   -  VERBAL
```

PROJECTS/REQUESTS

REQ. LOG #	ACTIVITY DESCRIPTION	IN PROCESS	COMPLETED	DEFERRED	REMARKS & DATES OF COMPLETION
	TOTALS				

Figure 2-1

it may be necessary to carry out a few test searches to see how much
time it takes. These could include the following types of file searches:

1. A search for single injury type and cause over the past 3 years;
    for example, a search for the number of back injuries over the
    period. Find the most frequent cause by ranking the causes
    of these injuries from the most frequent to the least.
2. A search for multiple causative factors; for example, over the
    last 5 years, the number of hand injuries that occurred. What
    two departments had the highest numbers and what were the
    three most frequently listed causes in those departments? Are
    these causes different from the most frequent cause for the
    plant or the company as a whole?

During this exercise, the following information should be recorded:

1. Total time to complete each of the searches.
2. Time required to prepare a report of the results of each search.
3. The difficulties and problems encountered in performing the
    searches.
4. The numbers of errors and false starts during each search
    exercise.

There are several ways to gather this information. The most difficult method, but also the one that will give the best information, is
to perform each search completely. Perhaps the search can be done
on a current problem, rather than on the examples given, to make it
time well spent. That way, all the information can be verified with
actual data. A second technique would be to search the case information for the preceding year and then make an estimate of the time
it would take for a complete search. A variation of this technique
is to calculate the time it takes to extract the data for one case, then
multiply by the number of cases for the number of years included in
the search. This last method can introduce a large deviation from
the actual time factors, but if that is understood, then the work
can proceed.
   If case file searches are made as a routine activity of the safety
office, then the foregoing exercise is unnecessary, because preparation time will already be known. The reader in that situation can
move on to the next section after documenting the time cost for such
special searches. Of course, if this type of searching is a rare occurrence, then it will be important to obtain these data, because
this will be a key point used to help promote the computerized safety
data system (CSDS). Such a system can be used to perform special

searches and deliver a completed report in a matter of minutes. That
is why it is important to be as accurate as possible about the time
cost in performing manual searches. Management is quite sensitive
to labor savings on routine tasks through the use of a computer.

RECORDING THE COST OF ACCIDENTS

Types of Costs That Are Generated

The title of this section is virtually self-explanatory. The big question: Does the current reporting system include the collection and
recording of cost information? Does this include recording direct
costs, such as the following?

1. Medical costs.
2. Workers compensation costs.
3. Damage and repair costs.
4. Equipment replacement costs.
5. Insurance premiums.

Do the data also include the following indirect costs?

1. Supervisor's time cost to investigate and prepare the accident
    report.
2. The time cost of management review of the report and the costs
    of actions taken to correct.
3. The time cost of follow-up action by management.
4. Employee time costs incurred while observing the aftermath of
    the accident, and waiting time while machines are down.
5. Overtime costs due to machine downtime or replacement of the
    injured worker or both.

For the moment we shall assume that the current reporting system
does record accident costs. The thrust of the review then will be to
understand the methods by which these costs are developed. Direct
costs are generated from several sources and are easily identified.
As mentioned earlier, the insurance carrier usually provides the information on workers compensation costs. The personnel department will
know the medical costs. Plant maintenance will have the damage estimates for equipment as well as the invoices for actual repairs. These
departments and the insurance carrier are the sources for direct cost
information. How well and how accurately the information is developed
for each case will come out in the review.

Data on indirect costs, as listed earlier, are more difficult to acquire and are subject to controversy. Many companies simply do not keep records on these costs because of the problems involved. Because we are assuming that the current system does keep records on costs, direct as well as indirect, we shall consider what is important about these costs. Indirect costs of accidents derive from loss of time from production and management activities. Because these are time costs, they must be developed as estimates. There is always some degree of error involved in any estimate. The amount of that error is the subject of the review.

Each type of indirect cost can generate a different level of error, depending how it is estimated. One should select several cases and find out how the costs were generated. The estimating method should be clear enough to calculate the percentage of error. After doing several cases, the overall accuracy can be estimated. This information is valuable in three ways: One, it will be a useful part of the review that can be transferred to generate costs in the CSDS. Two, the time it takes for manual tabulation of cost information can be included to justify the CSDS. Three, the information can be used to propose changes in the existing system to improve estimating accuracy.

Current Use of Cost Data

The manner in which cost information is presented to management will have a direct bearing to the impact and interest generated by that information. Accident cost information can be an effective tool for the safety professional if it is used in a way that will show how safety operations contribute to the operation of the business. Unless this relationship is properly developed, the information will arouse only passing interest. For example, the cost data can be related to numbers of units produced, to earnings per share, or to percentage of profit. All of these relationships can be used to demonstrate to management the value of the recommendations for action, as well as to enhance management's perception of the safety group. The review should include an analysis of how the data are handled when presented to management.

SUMMARY

Successfully convincing management to spend money and commit manpower from the resources of other departments for a new project requires thorough analysis and a compelling argument. The justification for a new project must convince management that the change is sound business and will improve the profitability of the company. A starting point to build such a convincing argument is information gathering.

Because we are interested in changing the current accident reporting system, it is appropriate to evaluate this system thoroughly. This evaluation has several aspects: One, the format, distribution, and feedback concerning the accident summary reports which must be sent to management. Two, the evaluation must consider the investigation, the accident report form, the history of the form, the investigator training courses, the quality of the reports, and the report routing. Three, it must consider the case file system in regard to how files are kept and the ease with which they can be used. Four, it must consider the types of costs that are developed on each case and how this information is used. By compiling carefully documented data concerning the condition and efficiency of the current system, two things are possible: One, this information will be useful as part of the justification for the CSDS. Two, the data can be effectively used to improve the current system to make it more efficient and responsive.

Appendix A presents a sample analysis questionnaire based on this chapter.

NOTES

1. John V. Grimaldi, and Rollin H. Simonds, *Safety Management,* 3rd ed. Richard D. Irwin, Homewood, Ill. 1975.

3

THE COMPANY STRUCTURE

> *To get something done involving several departments, divisions*
> *or organizations, keep quiet about it. Get the available facts,*
> *marshall your allies, think through the opponent's defenses, and*
> *then go.*
>
> *Robert C. Townsend*[1]

Any group of individuals who call themselves the management of a
small company have a "collective personality." The ways in which a
company's management conducts its business, interacts with its employees, and deals with its customers depend on the concepts and
ideas held by this collective personality regarding how a business
should be run. In hiring and promoting, those in management tend
to recommend people with personalities similar to their own as candidates for managerial positions. As the company grows, the collective
personal'ty may broaden at the company base, but it may stay relatively stable at the top.

The manner in which management deals with its employees in terms
of communication, promotion, hiring, and firing is based on how those
in management believe such dealings should be carried out. In many
cases these perceptions are derived from the past experiences of those
in management and are modeled on prior situations. As the organization matures, these concepts can be modified by pressures from government, from organized labor, or from changing local conditions.
Still, the collective personality is there with its ideas and past experiences, and it will try to hold fast to as many of its own concepts as
possible in spite of outside pressures.

There is a parallel with respect to management's image of safety.
How this part of the business is perceived and judged depends on two
factors. First, it depends on management's individual perception of
safety as an idea. Through past experiences, from as far back as
childhood, each manager has formed a picture of what is safe and what
is an acceptable risk. In many cases, management evaluates the recommendations of the safety professional and the industrial safety

program's contribution to the business based on these preconceived
notions of what is an acceptable risk. The second factor has to do
with changes in the first: how successful the safety professional has
been in cultivating various managers and enriching their personal notions of safety.

The first of these two factors is infused into the collective personality and becomes a part of management policies. Learning about these
two factors and other related concepts is an important facet in the
collection of background information.

REVIEWING THE CURRENT MANAGEMENT STRUCTURE

To prepare to propose a new and different manner in which safety
will collect data and present its output reports to management requires
that the safety professional know as much as possible about the management structure and the way the company operates. Because support is necessary to promote the idea of a computerized safety data
system (CSDS), it is important to know where that support can be
generated and what efforts will make the project successful.

In the development phase of the project, this information will be
useful in the design of the output documents. If the new document
has the backing of influential managers, the format will gain easier
acceptance by others.

The review should concentrate on both the formal and informal
lines of communication, as well as the power structure. Each of these
aspects will be useful in the presentation and development of the CSDS
concept. The formal lines of communication are evident in an organization chart. However, the identified and block-diagrammed reporting
relationships may or may not reveal what actually happens in the operation of the business. The informal lines of communication and the
actual relationships must also be known.

Management: The Authority Structure

Within the company or corporate organization there are two structures
that function sometimes together and at other times in disharmony with
one another: one, the recognized and official authority structure of
the organization; two, the informal power structure. The second,
the power structure, may or may not match the line authority structure.

The authority structure is the one that appears on an organization chart and depicts the recognized levels of control within the organization. During an analysis of this structure, it will be useful to
know the following:

1. The budget level of each department. Learn about the past, present, and proposed budget levels.
2. The number of people in each department. Learn whether the department is growing or shrinking.
3. The number of management people and the numbers in different levels of management. What changes are planned?
4. The span of control of each department head, i.e., how many people report directly to that vice-president. Is control too broad or too narrow? Can you learn why?
5. Background information on each key person in management. Include details about education, birthplace and age, career progression, outside activities and organizations, licenses (CPA, MD, etc.), and close associates.
6. How each department interacts with other departments.

Most safety professionals are well aware of their company organizational structures. Most have committed those structures to memory. This is a good start. The more specifics that are known about the inner workings of the structure, the better. This analysis should be as complete and as detailed as possible, because these data have a variety of uses. For example, they will be used to compare the authority structure and the power structure. Background data are useful in planning meetings and training sessions for the CSDS (see Chapter 5). This information will help in making up report distribution lists. It will be useful to determine areas of support.

Profit Orientation and Profitability

This portion of the inquiry relates to the way the company management views the business operation. Is there a high profit motivation to the exclusion of other considerations? Is everything rated secondary to production and its goals? If this is true, then only if there is a direct impact on profit and production will a new proposal receive attention. For example, in safety, the profit motive will be evidenced by the case of a worker who is caught in a machine and injured. This event stops the production line for 2 hours. Management hears of this and is deeply concerned. Action is swift. Supervisors and managers are required to explain why guards were left off and what the injury cost in production time. Prior to this incident the safety professional had visited the department many times. Management had been reminded of the danger of leaving guards off. The department manager agreed to comply. However, he saw a loss in production time because of the delay in removing and replacing guards. Only later, after the injury has cost a 2-hour delay in production, does management take the recommended action. The injury cost more production time than would

have the occasional removing and replacing of a guard. The reaction
was not to prevent an injury but to keep production at a high level.

Profitability, or the level of profits of the company, is directly
tied to the level of efficient production. If efficiency is low compared
with the amount of effort required to produce goods, then company
management will press for higher production levels and greater efficiency. Conversely, if profit margins are high compared with the production level and effort required to produce the product, then management will consider that the operation is running efficiently. This
concept will pervade the entire organization and affect almost every
phase of the business.

An example of this is the business that has an extremely narrow
profit margin and must severely control expenses even in good times.
It places controls on all budgets and requires multiple approvals for
even small expenditures. This kind of control may be necessary because of the nature of the industry, but it also can be an attempt to
make an inefficient operation profitable. For safety, this situation
of high or low ratios of profitability to cost of goods should be thoroughly reviewed and documented. In fact, it will be worth the effort
to thoroughly analyze the source and background of such a severe
cost-control system to determine if it is caused by the profit situation
or by another business consideration. The reason for the careful inquiry is obvious. The more severe the cost-cutting measures, the
stronger must be the justification for a safety data system.

The Unwritten Power Structure: The Company Politic

When compared with the organization-chart authority structure, the
power structure may be vastly different. The power structure evolves
from alliances formed within the organization, from family ties, from
school friendships, and from the need to protect those in the alliance
from others. It seems that this is more prevalent in the upper levels
of management, but power struggles can also be found in the lower
levels of management.

The safety professional can determine the authority structure from
the organization chart, but the unwritten power structure is not so
easy to determine. Nevertheless, it is there, and over a period of
time this informal chain of command can be defined. There are several
interesting properties of this structure. It is transient. Changes
in alliances and positions on issues occur on a regular basis. Because
the power lines constitute an unwritten organizational structure, it
changes with circumstances. The safety practitioner must know the
personalities and backgrounds of the players in order to understand
the power game. It is important to continue one's research into the
backgrounds of these people. Find out, for example, who are neigh-
bors, who are members of the same bowling team or country club, who

play bridge, tennis, or golf together, who are graduates of the same
schools, and who are related by marriage. These background points
are a few examples of power connections. Find as many as you can.
This will be quite useful in documenting the connections within the
power structure.

   Because it is not a formal organization, the power structure must
be identified in ways other than by looking at the organization chart.
Developing background information on as many people in management
as possible is the first step. It provides clues to the next step, which
is observation. Observation is the quickest way to identify power
lines once backgrounds are known. You have background information, such as home address, schools attended, former departments
and working relationships, and so on. Now observe the managers'
working patterns. With whom do certain managers eat lunch? Is there
a special table, not designated as such, where certain executives always eat together? Who else joins this group either regularly or occasionally? Are there gathering spots in halls or courtyards where
certain managers always congregate? Other than work, is there a connection between these individuals? In casual conversations with different managers you can confirm the background information and collect more, such as sports and recreation activities. Each of these bits
of data will form connections or possible relationships to observe at
work. In this manner a schematic of where the power lines develop
and flow can be documented.

The Communications System

In reviewing the communications system, there are several points to
consider. First, how is information transmitted? There seem to be
many obvious answers to this question, but on a second examination
it may turn out to be a more complex subject. As the saying goes,
"Nothing is simple." The immediate answers would be telephone calls,
memoranda, letters, one-to-one meetings, and informal group get-
togethers. That would be true if we were trying to learn only general
information. On a second look at this question of how information is
passed, there appears to be a need for more specific answers. For
example, the safety information concerning injuries and other events
is routed and distributed through normal channels. Our review has
already documented this part of the communications network. Are
other similar reports and information vital to the operation transmitted
in like manner? How are changes in production and other manufacturing information relayed? How much information is passed around during meetings? There are several types of meetings. Of the formal
types, there are the occasional, the regular, and the staff meetings.
Other meetings are informal in nature but are connected with the unwritten power structure. Meetings of any kind can take on great
significance when viewed from the aspect of the exercise of power.

All power is ritual and myth, as it always has been, and those
who seek power must be prepared to enact the rituals of power
and take their place in the local mythology. Certain events have
totemic significance. A meeting called by one person for a specific purpose exists for its own sake . . . or for his. But fixed
meetings, whether of committees that meet at regularly appointed
times, or board of directors' meetings, meetings which do not occur at one person's whim but take place by schedule, automatically become invested with magic significance. On one level, they
represent meetings of the tribal elders, whether real or self-ap-
pointed; but on another, deeper level they symbolize the soul and
the continuity of the tribe, which is why it is so hard to change
their timing or their place. It is not necessary that such meetings
be productive, or even that substantive questions be discussed;
it is only necessary for them to *take place* so that the rhythm of
tribal life is maintained in an unbroken pattern. Without such
routines, life would seem chatoic, unorganized, and there would
be no calendar of events to give form to work.[2]

It is useful to be aware of the significance to those in power of
the meeting structure. The formal, regularly scheduled meeting is
the most powerful type of gathering; thus, most executives who play
the power game desire to chair such gatherings. This does not apply
to many regularly scheduled meetings where the attendance is not
held to the highest level and is limited by invitation of the chairman.
The meeting where any number can attend and where replacements
are often sent loses its significance as a gathering where important
things happen.

Memos and letters, as well as all other written communications,
enter into the two structures (authority and power). The way data
are passed around the organization will depend on these two structures. It is a good check on the unwritten power structure to see
who gets the information but was not on the regular distribution. It
is important to be fully knowledgeable about how management uses
the telephone system for communications. Is more information passed
over the telephone or verbally than in written form? With tact, this
kind of information can be added to the review of the company structure.

Reports: Routing and Format

It is important to collect information concerning how other reports are
prepared and presented to management. This piece of research will
be valuable when developing a usable format to computerize the safety
reports. The inquiry should include a list of the report titles that

are normally sent to management from various departments or locations. Especially important are reports sent in from production, quality control, maintenance, and warehousing. These departments are
the main clients of safety. If they are accustomed to seeing and preparing reports in a certain format, then one of similar format and appearance may have a better chance of acceptance.

The reports that other departments submit to management present
another avenue of inquiry. The routing of such reports will reveal
how other departments disseminate information. Included in the review of the routing system should be a little history about how each
particular report was developed, who started it, and why it began
that way. These data will provide the safety professional with a more
complete idea of what reports are being seen and by whom.

While checking on the reports from other departments it may be
good to learn how the data for the report are developed. If forms
are used to get the original data, a copy of each should be obtained.
Learning about how the information is generated can provide ideas
and possibly new formats with which to develop safety information of
all types.

The Company Data System

The company systems group will be the main supporting department
for the CSDS. Thus, that department and the data system should
be well researched. Because the safety professional has learned
about the major reports that are submitted to management, it is also
known which are computer-generated, as well as what format is used.

The development of a CSDS is facilitated by having a personnel
data system already in place; much of the information about employees
can be used by both systems. For example, the name, date of birth,
employee number, job title, and location and department will already
be on the computer file in the personnel data base. This will facilitate
development of the CSDS.

What if there is no personnel data system? What can be done?
Two options are available: One, help the personnel department to
develop a company personnel data system. Two, include the necessary
personnel information on the investigation form. Input the information
into the CSDS for each case. Both techniques will achieve the desired
result. Having a personnel system on the computer in the first place
does save time. Do not be discouraged; a safety data system will work
quite well without a personnel data system.

In reviewing the company data system, information such as how
that department is organized to perform its job will show efficiencies
and potential problems. For example, is the data-processing department an independent group? If not, does it report to or through another group such as finance? If it is independent, the data-processing

department is more apt to serve the needs of a wider group of clients than if it is reporting to a special-interest group such as finance or accounting. After all, whose work is going to get a higher priority and be done first — yours or their own?

The type of equipment, the machine languages that are used, and how processing is done are all good things to know. In fact, the more that is known about the computer system, the easier it will be to work with the data-processing people. If you can talk their language and converse in their jargon and understand what the equipment will do, it will be easier to find solutions to problems that arise.

MANAGEMENT AND SAFETY: THE INTERFACE

Safety's Position in the Management Structure

Although many in safety feel that their position and reporting level in the management scheme of things is never quite high enough, each of us is usually quite aware of where we are within the structure. No matter where it is in the organization, the position of the safety department should be reviewed. This review should include the evolutionary history of the department, how the reporting relationships developed, and the position and power of the person to whom safety reports. How many levels of management down from the location general manager or president of the company is the safety department? How large is the department? What size is the budget? Are recommendations from safety acted on in a timely fashion? Are the priorities for safety recommendations equal to those given to other staff groups? These are just some of the questions that should be answered. More can be added to this review list. The main point here is to document the position, power, and influence the safety department has within the organization.

Perception of the Safety Function: Support from Management

It is sometimes quite difficult to determine how a department is perceived by others in the organization. Sometimes, however, that perception can be determined through the kind of support that is received from management. Support can consist of backing during a meeting discussion or it can be an endorsement of a safety proposal. These kinds of actions identify support lines and people. The supportive people can be contacted. A discussion of their feelings can be a start toward an understanding about how safety is perceived by others in the organization. Most safety professionals have a good idea of who

in their organization truly believes in safety, who appears to give
some support but does so only when it is politically expedient, and
who puts up with safety measures only when directed to do so. These
kinds of relationships are part of any job, but they are more evident
in safety than in other lines of work.

Going through this exercise of identifying how safety as a department is perceived and supported by others is well worth the effort.
This information will be useful when assistance is needed to persuade
management to accept the idea of developing a safety data system.
Such support will help influence those who are undecided and will help
sway those who really do not care because the proposal will not affect
them. Therefore, knowing where this support lies will help later on
in the process by facilitating discussion of ways to win more support.

The Management Accountability System: Appraisals

In some corporations the appraisal system is quite formal and well documented, whereas in others it is freer, with the individual supervisor
being given little direction. Appraising people for their work performance, their quality, their objectives, and their accomplishments is a
difficult task. Most appraisal systems are imperfect. Many apparent-
ly good, well-documented formal systems can be quite disappointing
when analyzed. Other systems that appear to be quite imperfect seem
to inspire and promote excellence. The safety professional within an
appraisal system needs to understand how that company's system
works and to what degree it is effective.

The system usually has an appraisal form, with an accompanying
document providing directions for completion of the form. If there
has been a recent review and analysis of the system, see if a copy
of the report can be obtained for review. Discuss the effectiveness
and usefulness of the system with those who administer it. Ask about
the appraisal system from others in management who use the system.

The appraisal system is important to the safety professional in
a number of ways. If safety is included in an individual manager's
objective and goals, then it may be included in the appraisal system.
There may even be a section for safety on the form. The instructions
may direct that an appraisal of safety performance be part of the system. Conversely, there may be no mention of safety in the appraisal
system at all. The safety professional must objectively review the
company's system as it is. During the review, the safety professional
must not try to judge whether it is good or bad or somewhere in between. Just collect as many facts about the present system as practical for future use.

Individual Safety Performance Measurement

The title of this section might indicate that it should rightly be in the
previous section, because it could be related to the appraisal. The
subject of individual performance can also be related to the appraisal
system known as management by objectives (MBO). What is really
needed is to look again at performance measurement and see how workers and management are measured for individual and group performance in safety. What are the criteria that are applied to determine
if a worker, supervisor, or manager is operating with an adequate
degree of safety? These are objective questions that require objective
answers. Suppose that numbers of injuries and dollar amounts of property damage and production losses from injuries are used as the only
measurements of safety performance; it must be noted how this is
done. Our interest is only to determine what is happening, not to
appraise. This is not the point in our examination to determine or
judge if what is being done can be improved. By sticking to gathering facts we can learn as much as possible about how management now
appraises safety performance. Check with several managers to see
how they evaluate their people for safety performance. Also, ask how
safety is put into the MBO program for individual managers and supervisors. Get copies of the safety goals to see how they are written
and what they cover. This will give a good idea of some of management's perceptions of safety and will add to the accumulated body of
knowledge.

PHYSICAL COMPANY STRUCTURE

Physical Layout

In developing a safety data system for a company, another point to
consider is where everything is located physically. Where plants are
located and how many are to be in the system is important information.
The logistics involved in information transfer must be part of the evaluation and plan. Telephone communications are excellent within the
United States, but if overseas locations are to be included in the system, then other data-transmission factors must be evaluated. Getting
the data in on time has to be included in the plan. The numbers of
employees at each plant, the plans to expand, company acquisitions,
and plant closings are all points to include in this part of the review.
    If the company has a computer system, how are those data being
transferred? There are several ways in which information can be sent
between locations, such as teletype and video terminal through telephone lines or to a printer. Other methods that are available are the
telecopier, the mail, and courier services. Most companies will use

a combination of all of these data-transfer methods, depending on the volume of material, its urgency, and its destination. The world of data transmission is changing so rapidly that the most modern methods used today may seem laboriously slow by next year's standards. Therefore, thorough discussions with administrative services and the information systems people will reveal how they do it and will give advance notice of any improvements that are in the works.

Plant Control and Autonomy

The exercise of control over the plants, regions, areas, divisions, or groups within the company and the degree of autonomy that each is afforded can be larger considerations than the physical locations of the various units. Autonomy or central control will vary from corporation to corporation. Some companies foster a parental, military-like central control. This type of authority management will give little latitude to the individual manager below a certain level. The degree of autonomy or central control is exhibited in many ways that are relatively easy to spot. Financial limits are a good indicator. For example, what level of approval is required to spend money? This includes approvals for capital-improvement funds, purchase requisitions, budget limits for expenses, and approvals for travel. Another, more subtle indicator of control involves approval of MBO programs, authority to hire, and approval of a new position.

All of these considerations are tied back to the unwritten power structure in many ways, because approval levels may vary widely from one division to another. For example, in one division there may be tight central control; in another, individual managers may have a greater degree of autonomy, but within the corporate limits.

The degree of autonomy or central control will have a direct effect on the type of data system you may want to recommend, and it can be an influencing factor in the successful development and acceptance of the system. This information will tie back to both the authority structure and the power structure.

Interaction Between Units

The degree of cross-communication between plants is an interesting aspect of a company's operating philosophy. The amount of formal reporting to top management and the types of reports that are passed between plants will provide a picture of the degree of interaction between plant managements. Meetings and forums that bring together managers from various locations can accelerate the information exchange. Association meetings and conferences can bring together managers and their counterparts from other companies. The numbers

and types of all these points of contact, both formal and informal, can
be indicative of management's willingness to exchange ideas. On the
other side of the equation is upper management's degree of encouragement and support for interaction between various management groups.
   An atmosphere that encourages free exchange of ideas is indicative of a management system in which there are open, honest relationships. Conversely, if there is little interaction below the formal-re-
port level, this may imply an authoritarian management style in which
informal exchange of ideas between management groups is inhibited.
Interaction between company units will tie back into both the authority structure and the power structure as a check on how the company
operates and relates to information flow. Each piece of knowledge
about the company's information systems, both formal and informal,
is quite useful.

SUMMARY

The management of a company is composed of individuals who together
exhibit a collective personality. As the company grows, management
tends to hire individuals who will fit into that personality. Within
management, there are two structures: the authority structure, which
is depicted on the organization chart, and the power structure, which
consists in informal lines of communication, direction, and natural
leadership that may or may not follow the authority structure.
   The safety professional must understand and be thoroughly familiar with both structures. Such factors as report routing and format,
the information systems department, the computer data system, the
physical plant layout, the locations of plants around the world, plant
management's authority and autonomy, and interactions between plants
are all important aspects of the inquiry into how the company is organized. This part of the review should also examine safety's position
in the structure, management's perception of the safety function, the
way safety performance is measured, and management's accountability
system with regard to safety performance. All of these bits of information, when combined with the other parts of the analysis, will be useful in designing acceptable report formats, preparing persuasive presentations to management, distributing information about the system,
and keeping the system efficient once it is operating.

NOTES

1. Robert C. Townsend, *Up the Organization*. Alfred A. Knopf,
    New York, 1970, p. 55.
2. Michael Korda, *Power*. Ballantine, New York, 1975, p. 177-178.

4

SELLING COMPUTERIZATION TO MANAGEMENT

Conviction is a flame that must burn itself out — in trying an idea or fighting for a chance to try it. If bottled up inside it will eat a man's heart away.

Robert C. Townsend[1]

SAFETY WORK REQUIRES SALESMANSHIP

When we think of selling, we may remember selling magazines in college or selling merchandise in a department store, or we may think of the professional salesman who works for a company and markets that firm's products or services. From time to time, we, as safety professionals, may think of some parts of our work as requiring selling skills. In fact, selling is very much a part of safety work. Consider, for example, the use of personal protective equipment (PPE). To promote a recommendation for protective gear to a manager and to employees requires the use of proven selling techniques. First, there must be a need for the equipment. Second, both management and the wearer of protective gear must recognize the need for the equipment; otherwise it will not be worn. Of course, the wearing of protective gear can be ordered by supervisors. However, unless the wearer believes in the equipment, this will create an extra management burden to constantly monitor workers and order them to wear their equipment. Conversely, if employees understand the benefits of wearing the PPE, they will be more likely to do so. Understanding the benefits translates into acceptance.

The art of persuasion, or selling, involves sets of principles that are much the same in all contexts. The basic selling principles are simple: discover a need, suggest a solution, and provide a product or service that will achieve that solution. This is the basic formula for selling any product, and it is the means for persuading management to approve a new project. If people are to accept a new idea, change an old routine, or modify a traditional way of doing something, they must first understand the need. They must believe that the change will benefit them. They must believe that the solution offered will be a benefit, one to which they will contribute. These three points are integral parts of the selling process, and they are equally important in promoting new projects.

In promoting a project to management, one must make a detailed analysis of the situation, just as in a selling situation. First, managers' operational problems and situations must be understood. A thorough understanding of these problems will enable the safety professional to advance persuasive reasons why the proposed change will fit into the operation and provide a solid foundation for help and support. Second, through individual discussions, uncover every major obstacle, and bring unvoiced doubts into the open. Third, crystallize the idea of the need, and confirm all areas of agreement. Fourth, resolve any objections through continued exploration and forthright discussion.[2] These are the principles for persuading the prospect, the manager, to support the data system proposal.

Computerization of a safety data system is a relatively new idea. Although computers have been in use for many years, safety is an area in which computers have not been used to a great extent. Thus, if management is to approve a computerized safety data system (CSDS), then the need for such a system must be convincingly demonstrated. In addition, the specific benefits of such a system must be demonstrated to each manager who will use the system.

Persuading a number of managers to accept such a system would appear to require a lot of effort. It will. Selling in this context will require the same principles as selling a product. The safety professional must develop finely tuned selling skills to complement other techniques used on a day-to-day basis.

THE OPPOSITION: POSSIBILITIES

No reasonable person in a company will openly oppose a project that will improve safety conditions for workers. No one will be heard to say that safety measures hamper production; at least no one will make a pronouncement to that effect in a general meeting. There is never open opposition to protecting workers from injury. However, there are conflicting views of what should be done to make the workplace safe. And this is where those who believe that "the place is already safe enough" will try to prevail. This attitude derives in part from a basic resistance to any change and in part from a belief by some that too many safety measures will hurt production.

Fear of the unknown is another source of opposition to computerization. This can be manifested as uncertainty about the exact nature of the changes that will be involved, or it can reflect the underlying

discomfort some people feel regarding the power of computers. At
the bottom of these fearful ideas lies insecurity. This fear can show
up in seemingly rational reasons for opposition. Here are some reac-
tions to watch for:

1. It probably will increase my work load.
2. Will it change my relationships with my peers?
3. I am not comfortable working with large numbers.
4. What can it do for me?
5. It will keep me at my desk and keep me from making my usual
    rounds.
6. Top management will not appreciate the time I will have to
    take away from production work.
7. I am not sure safety needs computer time. They got along
    without it before. Why do they need it now?
8. The computer will affect line positions (job insecurity).

Fear and a basic mistrust of safety programs are possible sources
of opposition. Another source can be executives who fear that their
power will be eroded if they allow safety to have a computer capability. They may fear that sharing such a capability with another department will reduce their power. Each of the several possible sources of opposition described in this overview must be addressed. Op-
ponents must be persuaded into agreement or neutralized in order to
achieve the widest possible basis of support for the CSDS.

INFORMATION AND ITS EFFECTIVE USE

Cost of Preparing Safety Reports

In the previous two chapters we examined the types of background
information and facts that will be useful in developing a persuasive
case for adopting a CSDS. The high cost of manual preparation of
safety reports is another important piece of information. It can be
a convincing selling point. Management is always concerned with profit erosion due to creeping operations costs. Therefore, cost comparisons between the manual method and the computer method of preparing reports will be helpful in demonstrating the benefits of the CSDS.
There are two sets of cost figures concerning the current system
that should be generated: (1) the first cost of developing and trans-
mitting safety data from field or department units; and (2) the cost
of producing and distributing the report at the main office and back
to the field units. These can be calculated as follows:

Cost of Developing Monthly Safety Data by Field or Department Units

Variables:

(a) Time required to assemble and complete the required forms
(b) Number of people at each location who work on the report
(c) If reviewed, the amount of time required
(d) The hourly rate for each person who works on the report
(e) Percentage of the hourly rate attributed to the cost of company benefits

Calculation: Time to prepare × number of people × (hourly rate + % of salary for benefits) + time to review × number of people × (hourly rate + % of salary for benefits) = cost of single location × number of locations (this can be calculated for each location or department if desired)

Costs of Producing the Safety Report at the Main Office

Variables:

(a) Time required to calculate and prepare a draft of the report
(b) Time required to review, analyze, and prepare a trend summary of the data
(c) Time required to prepare the final report
(d) Time required for final review
(e) Time and cost of reproduction
(f) Time required to distribute the report
(g) Hourly rate for each person involved in items a through f
(h) Percentage of the hourly rate required for company benefits
(i) Number of employees involved in each item a through f

Calculation: a × number of people × (hourly rate + benefit cost) + b × number of people × (hourly rate + benefit cost) + c × number of people × (hourly rate + benefit cost) + d × number of people × (hourly rate + benefit cost) + e × number of people × (hourly rate + benefit cost) + f × number of people × (hourly rate + benefit cost) = cost of preparation and distribution.

By adding both sets of costs together, we get the total cost of preparing the report each month by the traditional system.

The other part of the cost analysis is an estimate of what it will cost to prepare the reports generated by the CSDS. Precisely how the output documents are generated from the system will determine how much cost is involved. For example, in the system that is used as an example in this book, there are two output report documents generated each month. One lists all the cases for a department; the other makes an analysis of those cases for the period and compares it with the same period in the previous year. The initial analysis is then completed by the computer. Only when further special searches are needed is an analysis required. The savings from such a system can be quite substantial. The return on investment can be a very persuasive factor if only the report savings are used; when improvements in safety results are achieved by use of the system, enthusiasm for the CSDS can be quite impressive.

Uses of Management-System and Reporting-System Analyses

By this time, you probably have thought of several excellent uses for the analysis suggested in Chapters 2 and 3. In any case, here are some of the ways I have used these analyses:

Management-System Analyses

1. Information about the power structure was important in deciding the most effective ways to approach different segments of management.
2. In structuring interviews, knowledge of each executive's background was valuable because it allowed me to emphasize the benefits that would appeal to each person.
3. Dissemination of information is more efficient when the information is routed to those who believe it will be beneficial. Knowledge of the power structure and its subtle relationships was advantageous in assuring that no one was overlooked.

Safety Reporting-System Analyses

1. The weaknesses and inefficiencies of the current reporting system, which cannot be easily corrected, can be effectively compared with the way the CSDS will efficiently process the same information in less time and at lower cost.
2. The strengths of the present system, which will be retained in the CSDS, can be used to demonstrate continuity from the current system to the new system.
3. Certain changes and improvements in the current system that can be made easily and will remain in place with the CSDS can be used to show immediate payback from the introduction of the new system.
4. The operating-cost differential between the current system and the CSDS is a practical way to demonstrate a return on investment for the start-up costs as well as the long-term savings. A graph can depict the savings in cents per share or as reduction in line costs for units produced.

These are a few ideas on how effective analyses can be used to promote the system and persuade management to agree to the development of a CSDS. You may wish to add to this list as you think about your specific situation.

Support for a Safety Data System

In any plant or company, those who work in safety are aware of particular managers who truly believe in safety. They know that those managers believe that a safe operation leads to better efficiency and cost savings. These are the managers who ask for safety's help in improving their safety programs and who promote the cause of safety in management meetings. These managers form the core group from which to build support for the CSDS.

It is this core group who also serve as the management link to information about how management is operating. This group can provide insight into subtle changes in the management power structure, new directions in the way management is managing, and other vital points of which safety must be aware. This group also can provide truthful feedback about how a program is operating, as well as guidance on what a proposed program element will do for them.

Overall, this group of advocates will vary in their levels of backing, depending on the particular situation. For some in management, their involvement will depend on forces within the management structure. For others, their support is consistent, because they understand that an efficient safety program is always beneficial to them. Thus, backing from the not-so-dedicated segment of management will remain constant so long as safety continues to provide benefits that are both useful in their operation and politically advantageous.

The Opposition and Why It Is There

In the beginning of this chapter we mentioned that there may be opposition to safety. Not opposition to keeping workers from injury, but in the methods used to achieve that objective. Opposition can arise unexpectedly from any source. For example, it can even come from a person who has been considered to be friendly to safety, one who, we learn, has an underlying fear that others might find out too much about their operations with a safety data system in place. Once brought out into the open, his fears about computers can be dealt with by detailed explanations. After the sessions were over and the safety data system was on line and providing good information, this person liked the benefits of the CSDS. He was impressed by how much it facilitated the process of decision making. He became an ardent advocate of the CSDS.

Fear of the unknown and of complicated machines can cause enormous problems in introducing a CSDS. To gather support when introducting a change, the proper groundwork must be done to present good information about the change: exactly what it entails and what the benefits will be. The other ingredient that is needed is feedback from those who will be using it. Fears and misunderstandings must be explored, fully discussed, and resolved.

Other possible sources of opposition are those who believe that computerization of reports and the use of the company computer confers status. These individuals tend to believe that functions such as safety, personnel, and other staff groups are nice to have but are not essential; therefore, these peripheral groups do not need access to the computer. They believe that these groups can adequately perform their functions manually. This kind of opposition will be need-oriented. Safety, they believe, does not need a computer to do its job. Only when a detailed cost-benefit analysis shows the payoff in graphic terms will these persons be persuaded to support a CSDS.

A variation on this theme of opposition on the basis of need is the "turf principle." Some executives believe that to have one's reports computerized is a symbol of status. Therefore, these executives feel that the fewer groups who have their reports prepared by the computer the more status there is for those who do. Ridiculous? Yes. But a person who holds this opinion is an opponent who must be won over. These people are experts at political games and the use of power. If they cannot veto an undesired proposal, they will recommend having a committee review it. They will suggest gathering more supporting data for good decision making. They will recommend that the proposal be shelved or tabled for 6 months because business conditions are not right.

All of these tactics are available to a determined opposition. The safety professional must be aware of these maneuvers and be prepared to deal with them. The key is careful preparation.

Other Companies' Systems

What are other companies doing? How many times have we heard this question? It is a valid one for many in management who are excellent businessmen. The conservative executive will want to know what other companies are doing. As part of our preparation, this question must be addressed. Otherwise, according to Murphy's law, it will be asked when we are least prepared to answer.

Checking with other companies can be a good source of information and ideas, and therefore a benefit to the development of a CSDS. When asking another company about their system, there are two possible answers: Yes, they have a system. No, they do not have a system. In either situation it is wise to inquire if they know of another company that does have or is developing a system. This may

provide good leads that can be followed up. Finding a company that
has such a system can provide a contact for ideas and someone with
whom to discuss the problems of systems development.

The list of companies to be called initially should include those
within your industry that are the same size as your company or larger. Also, the list should include several companies outside your industry that are large enough to have a full-time safety professional
who may have developed a CSDS. When the review of other companies
is completed, 25 to 35 firms will have been contacted, and perhaps
half will have data systems. A questionnaire that can be used to record these contacts with other firms is often helpful (Exhibit 4-1).

SYSTEMS QUESTIONNAIRE – OTHER COMPANIES

Company ________________________   Contact Date ________________

Location ________________________   Phone Number ________________

________________________________   Persons Contacted ___________

____

1. Does your company have its injury or loss data on computer?

    Yes_____ No_____ (If answer is no, go directly to Question 7.)

2. What kind of data (e.g., injuries, property damage, fires, near-

    misses, and hazardous conditions) are recorded on your system?

    ____

3. How long have you had your system in operation? _______________

4. What kind of reports does the system produce for you? _________

    ____

5. Can you describe how your system works, i.e., how the accident

    data gets into the system? _____________________________________

    ____

6. Is it possible for me to visit your facility and discuss your system?

    Yes _____ No _____

7. Do you know of other companies that have developed safety data

    systems? Yes _____ No _____

8. Whom would I contact in those companies? ______________________

    ____

Exhibit 4-1

Locating other companies that have CSD systems will accomplish
two things. First, as already discussed, it will provide a source of
information and support for the development of your own system.
Second, it will provide a base of statistical information about what
others are doing. These data can be used to show that your company
is not a lonely pioneer in the development of such a system. The cau-
tious executive may feel uncomfortable about being first; therefore,
knowing that your company is among a group of companies that are
developing or have developed such a system may well be an aid in persuading this person to back the proposal. This will continue to demonstrate to management that there has been thorough preparation
and research concerning the feasibility of the proposed system.

METHODS OF PRESENTATION

Major Presentation and Report Techniques

When the analysis has been completed, the cost data collected, other
companies contacted, and preliminary designs for the output documents from the system prepared, it is time to consider how to present
a proposal to management. There are several choices to be considered.
These choices are the major presentation, the slow sell through a pilot
study, and the creation of a groundswell. We shall discuss the merits
of each approach.

The major presentation and report to management can be risky,
but this approach can be rewarding if the proper preparations are
made and the timing is correct. Because this is a single presentation
for a decision, it can get the proposal approved and the system running in the shortest amount of time. This is an advantage over the
other types of presentations.

In making a single major presentation to management, the decision
makers must be present during the entire session. While the session
is going on, everything required for a decision must be provided.
This means that every question must be anticipated. The visual aids
used to enhance and support the presentation must be readily understandable and compelling. The written material should include an executive summary, the completed proposal, background material on computers, and an approval letter. The letter is provided so that the
decision makers have only to sign when the presentation is completed.

A crucial ingredient in the major presentation is timing. A perfect
presentation made at the wrong time can result in the project being
shelved without decision or rejected entirely. Therefore, the safety
professional who decides to use the major presentation method to seek
approval must be acutely aware of the political and business climate.
If there are political undercurrents operating that can lead to rejec-

tion, the presentation should be delayed until a more advantageous
time. For example, sluggish sales due to recessionary pressures can
cause the proposal to be rejected, delayed, or held without decision.

In summary, the major presentation technique is a fast way to gain
access to decision makers and possible approval. To be successful,
however, the presentation must be crystal clear, persuasive, and well
timed.

A Slow Sell Through a Pilot Study

In many companies, a new idea or a new way of doing something will
not be approved for use throughout the company without first under-
going a period of testing on a small scale. A conservative management
will want proof that the method or idea will work in their company.
As confirmation, a conservative manager usually will be receptive to
a pilot study of sufficient size to show whether or not the new method
or idea is feasible for use by the entire company.

The pilot study method has several advantages. One, because
it is not a complete changeover for the company, many of those who
might object can adopt a wait-and-see attitude. This means that they
probably will not vigorously oppose the pilot project. Two, the pilot
study will give the project developers a chance to find any flaws in
the system and make changes and adjustments. Three, a pilot study
will get the project moving through the development stage. Four,
with the pilot study data, information sessions can be run to educate
other managers about the benefits of the new CSDS. Five, the commitment by management is limited; this shows care in decision making,
and the commitment can be reversed if the pilot project does not provide the expected results.

Depending on company size, a pilot study may be the accepted
means of testing new methods. However, the proposal must be as well
presented as if a major company changeover were involved. In a
sense, this is what you are after for the long term; therefore, your
preparation for the presentation must be thorough and rigorous. In
fact, any time a proposal is made, the preparation should be exhaustive. Never risk having an excellent idea turned down because the
presentation was poorly done.

Creating a Groundswell: Selling an Idea

In introducing a change from a current method of doing something,
those involved with the current method will be more receptive and co-
operative if they are fully informed of the change and have an opportunity to offer suggestions and ideas. The opening session for the
users should answer questions like these: What exactly is the change?

How will it affect those who use the current system? What benefits
will those involved gain as a result of the change? How much work
will it be to use the new system? Who is to provide training? Who
will lose their jobs as a result? How long will be the period of disruption until the change is complete? All of these questions must be addressed if a change is to be successfully put in place.

To promote acceptance of such a change, a well-planned series
of meetings should take place on both a one-to-one basis and a group
basis. These meetings should be designed to educate and inform, but
also to enlist suggestions and ideas about the system from the potential users. These are some of the points to be discussed: the problems with the current reports and the current reporting system; ideas
on how to make the present system better; the concept of computerization and what computers can do.

In all of these meetings a major emphasis should be the encouragement of participation. All participants should be made to feel that
they are contributing to the overall development effort. All of the
potential users must understand that the system is not a proprietary
safety data system owned by the safety department. Rather, it is
a company system owned and developed by members of the company
for their benefit. If this kind of participation and sharing of ideas
can be encouraged successfully, then each person will feel a real
sense of contribution, of authorship, and will want the project to succeed. This can create enthusiasm for the project, a groundswell of
support.

The work of creating this type of involvement is time-consuming.
It requires checking and rechecking with many individuals for their
comments and ideas, making changes and compromises to aid one person or one group. It is difficult to continually review and discuss
a proposal until there is agreement. But this type of activity can produce excellent rewards, lowering the level of opposition and developing a broad base of advocates. It has an excellent rate of success.
The time required just may be worth it after all.

Combinations of Techniques

Rather than use a single approach to convince management to approve
the CSDS proposal, it may be well to use several techniques. For example, if it is customary in a particular firm to make a single major
presentation to management, it might be a good idea to lay sufficient
groundwork through information sessions, group meetings, and one-
to-one sessions to create a strong groundswell. This can provide an
informed management group; then, when the time comes for the major
presentation, most everyone will already be convinced. Also, in the
major presentation itself, several alternatives can be presented: a
pilot study in a department, a test in a division, or a complete change-

over of the company. By presenting the advantages and disadvantages of each, and then recommending the one that seems best for the
political and business climate at the time, you demonstrate your knowledge of the business and of the decision-making process. This kind
of approach, using a combination of techniques, can be more successful than using a single method. "Reaching a decision as to which alternative will be accepted and incorporated into the plan is not the
end of the planning process. The details of the plan have to be
worked out and presented in a form that is understandable to all concerned."[3]

AFTER THE SALE THE REAL SELLING BEGINS

Reinforcement: Confirmation of a Good Decision

In business, in many meetings and sessions, proposals are made and
approved. Those persons who receive their approvals move on with
their projects. Many do not contact the group from which the approval was given until the next time another approval is needed. The
technique is acceptable. All of us tend to avoid "bothering" management unless it is really necessary. Is there a risk involved in avoiding contact with management after receiving their concurrence? I believe there is, and it needs to be addressed.
     Suppose that every time you met a certain friend you asked for
something. What do you think will eventually happen? Sooner or later
the friend will begin to avoid you, fearing you will ask for another
favor. The same is true in business. Why? Because the person who
approves the project, or grants the favor, is not receiving anything
in return. I am not speaking of money or gifts, but of ego satisfac-
tion. For example, you make a presentation at a meeting and sell your
idea to the group. The group makes small changes in your proposal,
but essentially it is your work. The group approves of the project,
and you are free to go ahead. Now comes the feedback step. By acknowledging the group's contributions, individually and collectively,
you feed both the individual ego and the group ego. As time passes
and the project proves successful, it is necessary to compliment the
original management group for their good judgment and their business
sense. Compliments, not flattery, will cement their support for the
project and encourage their receptivity and cooperation the next time.
     The same is true of upper management. You need to provide con-
firmation of a good decision. After approval of the CSDS, it is appropriate to show that the decision was sound by demonstrating the actual
cost reductions, efficiency levels, and more effective reports. This
is a tribute to the decision maker. It is a compliment, based on fact.
Periodic reports on progress will create good feelings toward you and
your project.

The Opposition: Their Next Move

One must always prepare for the possibility that there will be opposition to the proposal for a CSDS; there may even be opposition after the project has been approved. Therefore, it is good business to continue with the process of periodically keeping everyone informed about the progress of the proposal, as was outlined in the previous section. Also, consideration should be given to the various ways an opponent might slow down the project or cause it to be unsuccessful.

It is never pleasant to realize that there are those who will oppose a safety project, but there are some. The assignment then is to decide how and in what form the opposition can materialize. Prior to the project's approval, the opposition can try to play down the savings benefits, create fear about misuse of the computer system, or recommend an exhaustive committee study. All of these tactics can slow down or even destroy a budding project.

After the project is approved, and during the initial pilot study, for example, the opposition can still recommend committee review or a tabling of the work at midpoint because of "cost" restrictions. Another method is to wait for the final report of the pilot study; then, during the presentation of the study results, they may recommend a more thorough review of the results. They may suggest that a special, independent group research the findings to determine if the system should be expanded.

All of these maneuvers can destroy a project unless there is efficient planning to meet these possible challenges. The best way is to anticipate these tactics and be in position to control the recommendation. If it is a study committee, have one already picked out, and make the recommendation first. If it is a further study or a broader report, try to see that your present report covers all the points to prove exhaustively that this system works. It will take foresight to anticipate problems from the opposition, but it is so much easier to plan in advance than to try to pick up the pieces after being turned down or shunted over to a hostile committee for review.

Keeping the Project Alive and Well

We have discussed the methods of presenting the proposal and of gaining approval. After the project is approved, we have talked about techniques for reinforcing those people who provided backing. We have considered ways of anticipating and defusing the possible tactics of the opposition. The part yet to speak of, now that the project is underway, is what should be done to keep it alive and well. Should anything more be considered? After all, the project was approved and is now in development. Are you not home free, so to speak? In light of the previous sections, I would have to say no, you are not there yet.

To make sure that the management support you have developed up to this point remains strong and, if possible, improves and broadens as the project moves to completion, it is necessary to keep a steady flow of information and requests for suggestions going to those who are your supporters. The greatest compliment one person can give another is to use the advice that is given. Therefore, while the project is moving along, it is wise to routinely meet with each backer, and even those who were neutral. Keep them up to date on the progress of the project. Ask for their suggestions, comments, and ideas on parts of the system's development. Use as many of their suggestions as possible. Thank them for their ideas in front of others.

By providing periodic progress information to the management group, especially to those who supported the project, you will keep the project alive in their minds. By asking for ideas and suggestions and using them, you will create pride of authorship within the group. This is a positive force that is contagious. Support for the project will expand as a natural event.

SUMMARY

Selling and persuasion are not limited to products and services sold to customers; they are part of the work of those who want to promote a project or idea. The CSDS is an idea and a project that must be carefully presented to persuade management. Opposition to the CSDS can originate from several sources: a vague fear of computers, mistrust of "too much safety," and a basic resistance to change.

One of the arguments that supports conversion to a computerized system for safety reports is the high cost of manual preparation of the current reports. A cost analysis will demonstrate that the CSDS will provide better information, and at lower cost. Add to this argument an analysis of the inefficiencies of the present reporting system and there are two powerful reasons for conversion. In addition to providing selling points for conversion, an analysis of the management system will be useful in structuring interviews, approaching various members of management, and efficiently distributing reports.

Generating support for the idea of a CSDS is all-important. There are those in management who are always positive about safety. This group, and those who are usually neutral, should be kept informed about the benefits of the system. At the same time, the opposition must be acknowledged and their position understood as clearly as possible. The opposing forces are expert at political games and tactics to delay, cancel, or table a project. One of the requests the opposition will make is for data about CSDS development in other companies.

Therefore, it is wise to gather such information before the question is asked.

There are several ways to present the CSDS proposal to management: (1) the major presentation for an immediate decision, (2) the recommendation of a pilot study, or (3) the creation of a groundswell. All of these techniques can be used separately or in combination to ensure that the proposed CSD system is approved. In all cases, however, all levels of management should be kept informed and consulted about the system. Their involvement will develop a sense of authorship and pride and can increase the overall support level. Feedback about the project can be used to reinforce management's original decision to go ahead. This can be accomplished by reviewing the benefits and savings that are realized from the project once it is on line.

Even after approval for the system to go into development, those in opposition can still attempt to kill the project. They can recommend delays or budget cuts or request exhaustive amounts of justifying information. The only way to succeed is to anticipate the opposition's requests and carefully prepare arguments to overcome their tactics. Much of the opposition to the system can be overcome through the same techniques used to generate continuing support for it. Informing, consulting and giving recognition for good ideas will keep the system's development moving along. Without frequent reminders that the system is alive, backers may forget or lose interest.

NOTES

1. Robert C. Townsend, *Up the Organization*, Alfred A. Knopf, New York, 1970, p. 33.
2. Joseph Rothman, *The Surest Way to a Sale*, Macmillan, New York, 1964, p. 163.
3. Dale Carnegie and associates, *Managing Through People*, Simon & Schuster, New York, 1975, p. 90.

PART II

DEVELOPING A SYSTEM

5

HOW TO BEGIN

The very essence of all power to influence lies in getting the other person to participate.

Harry D. Overstreet

In the first part of this book we talked about the preliminary work before actual system development. This work includes gathering information about the company, about the present accident reporting system, and about what other companies are doing in CSDS development. This information is collected for one purpose: to marshal strong facts to make a successful proposal to management. Various presentation techniques were discussed. Following the successful presentation and approval, the reinforcement technique was reviewed where safety continues to confirm management's decision in favor of developing the CSD system.

Beyond these preliminary moves, the next step is to discuss safety data system development. This part of the process is divided into four phases: (1) developing the output reports, (2) writing a user's manual that contains the files and the codes of the system, (3) designing an investigation form for the CSD system, and (4) finally putting all three pieces together and bringing the system on line. In this part, each of these subjects will be addressed in a separate chapter.

OUTPUT DOCUMENTS: MANAGEMENT INPUT

Trial Designs for Management

When I started to develop my first safety data system, I began with the investigation form and the user's manual. For several weeks I worked on designing a form and getting the outline of a user's manual. Then I met with the information systems people. During that meeting

we discussed what progress I was making on the system and how the
data base should be set up. They inquired what kind of information
the system would produce as output. I explained that once I had
completed writing the user's manual and designing the investigation
form, I could make up the output documents. They suggested that
possibly I had been working from the wrong direction. Information
produced as output from a data system, they explained, must be in
the data base files. To be in designated files, information must be
listed in the user's manual. For the user's manual to be useful, basic
information must be generated by a properly designed investigation
report form. Therefore, to avoid double and triple work backtracking
and rechecking, the output documents, especially the regularly re-
curring reports, must be developed first. Based on the requirements
of these documents and the special report needs, the user's manual
and the investigation form can be developed to support the output
reports.

The logic, once considered, is quite precise concerning what to
develop first. Because the entire system is designed to produce information, the primary consideration is what information will be the
product. Once the regular output documents have been designed,
the other supporting parts of the system will fall into place. This
avoids a lot of rechecking to see if any piece of data needed for the
output documents has been overlooked.

Management is the client and the recipient of the reports. The
reports are designed and produced for their use and benefit; therefore, they should have a large part in deciding what format the output reports should take.

To speed this process, several basic parameters for report design
should be decided upon. For example, you may wish to consider the
following list and add to it, based on your needs. Reports should:

1. Be easily understood by the user.
2. Provide a good understanding of the types of accidents and
    their cases.
3. Be of similar format to other reports produced by other de-
    partments.
4. Be designed to provide comparisons and totals where needed.

So that management can consider alternatives, several trial designs should be developed. Ideas for the trial designs can come from
a variety of sources. For example, I used such sources as other departments' report formats, other companies' formats, and one format
I designed. These trial designs are shown in Exhibit 5-1.

The purpose is to provide format ideas from which to get a consensus of what management would like to see. A broad selection of

INCIDENT TYPES

January 1, 19XX - January 31, 19XX

Type of Incident	Incidents per 200,000 Wkr Hrs				Estimated Cost				Year-to-Date Incidents per 200,000 Wkr Hrs				Year-to-Date Estimated Cost			
	FA	MT	LT	Total	FA	MT	LT	Total	FA	MT	LT	Total	FA	MT	LT	Total
Caught in or Between	XX	XX	XX	XXX	$XX	$XX	$XX	$XXX	XX	XX	XX	XXX	$XX	$XX	$XX	$XXX
Fall Same Level	XX	XX	XX	XXX	$XX	$XX	$XX	$XXX	XX	XX	XX	XXX	$XX	$XX	$XX	$XXX
Muscular Overexertion	XX	XX	XX		$XX	$XX	$XX		XX	XX	XX		$XX	$XX	$XX	
Spills	XX				$XX	$XX			XX	XX			$XX	$XX		
Electrical Shock	XX				$XX				XX	XX			$XX			
Laceration	XX				$XX	$XX	$XX	$XXX	XX				$XX			
All Types	XX	XX	XX	XXX	$XX	$XX	$XX	$XXX	XX	XX	XX	XXX	$XX	$XX	$XX	$XXX

Exhibit 5-1A

Report Date _______________

INCIDENT/INJURY REPORT

January 1, 19XX – January 31, 19XX

Organization Code 1234

Case #	Date/Time	Case Type	# Days Lost	Case Cost	Potential		Immediate Causes			Underlying Causes	
					Freq.	Severe	Human Error	Condi- tion Defect	Mgmt. Over- sight	Primary	Secondary
XXXX	XX/XXXX	XXXX	XX	$XXX	XX	XX	XXXX	XXXX	XXXX	XXXX	XXXX
XXXX	XX/XXXX	XXXX	XX	$XXX	XX	XX	XXXX	XXXX	XXXX	XXXX	XXXX
XXXX	XX/XXXX	XXXX	XX	$XXX	XX	XX	XXXX	XXXX			
XXXX	XX/XXXX	XXXX	XX	$XXX							
XXXX	XX/XXXX	XXXX									
XXXX	XX/XXXX										
XXXX											

Exhibit 5-1B

INCIDENT TYPES

January 1, 19XX – January 31, 19XX

Type of Incident	Incidents per 200,000 Wkr Hrs				Estimated Cost				Year-to-Date Incidents per 200,000 Wkr Hrs				Year-to-Date Estimated Cost			
	FA	MT	LT	Total	FA	MT	LT	Total	FA	MT	LT	Total	FA	MT	LT	Total
Packaging A	XX	XX	XX	XXX	$XX	$XX	$XX	$XXX	XX	XX	XX	XXX	$XX	$XX	$XX	$XXX
Chemical Mfg.	XX	XX	XX	XXX	$XX	$XX	$XX	$XXX	XX	XX	XX	XXX	$XX	$XX	$XX	$XXX
Packaging B	XX	XX	XX		$XX	$XX	$XX		XX	XX	XX		$XX	$XX	$XX	
Warehouse	XX	XX	XX		$XX	$XX	$XX		XX	XX	XX		$XX	$XX	$XX	
Maintenance	XX	XX			$XX	$XX			XX	XX			$XX	$XX		
Compounding	XX	XX			$XX	$XX			XX	XX			$XX	$XX		
Admin. A	XX				$XX				XX	XX			$XX			
Research	XX	XX	XX	XXX	$XX	$XX	$XX	$XXX	XX	XX	XX	XXX	$XX	$XX	$XX	$XXX
Divisional Total	XX	XX	XX	XXX	$XX	$XX	$XX	$XXX	XX	XX	XX	XXX	$XX	$XX	$XX	$XXX

Exhibit 5-1C

EXPENSE CENTER XXXX
INJURY/ILLNESS REPORT
January 1, 19XX - January 31, 19XX

	First Aid				Medical Treatment				Lost Time				Total			
	Number		--Cost--		Number		--Cost--		Number		--Cost--		Number		--Cost--	
	Mo.	YTD	Mo.	YTD	Mo.	YTD	Mo.	YTD	Mo.	YTD	Mo.	YTD	Mo.	YTD	Mo.	YTD
Nu Cost	X	XX	$XX	$XXX	X	XX	$XX	$XXX	X	XX	$XX	$XXX	X	XX	$XX	$XXX
Bel Celo	X	XX	$XX		X	XX	$XX		X	XX	$XX		X	XX	$XX	
O-Shine	X	XX			X	XX			X	XX			X	XX		
Too-Good	X	XX	$XX	$XXX	X	XX	$XX	$XXX	X	XX	$XX	$XXX	X	XX	$XX	$XXX
	X	XX	$XX	$XXX	X	XX	$XX	$XXX	X	XX	$XX	$XXX	X	XX	$XX	$XXX

Exhibit 5-1D

choices for a format can stimulate discussion. The discussion can lead
to thoughts and feelings about what each manager would like to see
in a report. Obviously, this discussion of formats will be carried out
in a series of one-to-one meetings with members of management.

Objectives of the Management Interview

The main objective in meeting with various members of management
is to show them the sample report formats and to obtain their ideas
for a best format for the output documents. In addition to the main
purpose, there are several other subsidiary objectives that also should
be addressed during each session:

1. Understanding computer systems in general and the CSD system specifically. Does the manager know what benefits will
   be derived from the change to the CSDS?
2. Quality requirements of the new system, i.e., investigation
   report and input information quality.
3. The availability of the special search report and other reports
   that the CSD system will produce.
4. Other feelings and ideas about the CSD system.

All of these, both the main objective and the subobjectives, can
be talked about in the sessions with members of management quite
easily because they are interrelated. The overall goal of the session,
as with all contacts with management members, is to encourage them
to feel comfortable with the CSD system. They need to consider it
as an additional information source designed with them for their benefit. If this philosophy is followed carefully and faithfully, the discussions and presentations will stress those points necessary to accomplish this overall goal.

As the great salesman Frank Beltcher said, "Apply this rule: Try
to find out what people want, and then help them get it."[1]

Planning the Interview Structure

The objectives of the interview are to review the prospective report
formats and to discuss the system, its properties, and its benefits.
The structure of the interview should center around these objectives
and their accomplishment.

It is important to structure the interview as an informal visit. In
an informal type of setting there can be an easy exchange of ideas.
The reason to establish a structure is to provide an organized way
to ensure that important points will not be overlooked. A structure

also gives standardization to each interview. This will generate comparable information, ideas, and feedback about the formats and the system.

From outward appearance, the meeting is a casual, informal one-to-one get-together for you to elicit this person's ideas about the system. This is not a trick, but a carefully planned approach to obtain a lot of information. The safety professional's part is structured to take advantage of several techniques. First, using memorized outline and carefully prepared questions, the interviewer gains confidence and control of the meeting. The structure also allows the interviewer to concentrate on answers, because the question sequence is known in advance. This eliminates groping for the next topic. Second, this type of organization will avoid overlooking points because of nervousness or emotion during the session. Third, a structure provides a uniform quality of information from the interviews that can be used as a basis of decisions about the report formats. Table 5-1 presents a sample interview structure that can be used as a guide to develop your own structure.

Testing the Interview Design

It can generate confidence to practice the structured interview several times. This can be done with people who do not have direct interest in the CSDS. This presents a nonthreatening setting and an open channel for feedback about the quality of the interview structure and the lead-in questions. In the past, I have used secretaries and laboratory technicians as substitutes for testing interview structures. There is a good level of sharing and involvement in this arrangement. It provides those people with a break in their routine and a sense of department participation.

During these sessions it is good to include an observer to take notes and give feedback. The observer, usually a safety professional from the staff, has the outline and the questions for the interview, plus a list of the important points to watch for during the practice interview. This provides a guide for the observer to catch the high-interest areas and give good feedback.

The stand-in interviewee is provided with all the material about the system that has been given to management. In addition, the observer and the interviewee make up some questions to ask during the interview. This kind of preparation contributes to the realism of the role playing. It gives the stand-in a part to play and can provide excellent practice for the actual interviews.

Usually the number of practice interviews can be limited to three or four. This number allows for two or three adjustments to polish both the structure and the questions, and that is usually sufficient preparation for a smooth confident interview style. Conversely, this

Table 5-1 Sample Interview Structure

I. Introduction

    A. Background of why the system is a benefit

    B. Purpose of the meeting

        1. To review proposed report formats for the CSDS.

        2. To discuss how the CSDS can improve management control of safety performance.

II. Overview of points to be covered

    A. Comparison between the current safety reports and the proposed CSDS formats

    B. Discussion of the special report capability

III. Questions to be used as leads into discussion

    A. Current reports and CSDS comparison

        1. Looking at these current reports, what problems do you have with them?

        2. In these prospective CSDS reports, how would you like the information presented? Would you want more, less, or different information than what is presented here in the sample formats?

        3. Using the proposed CSDS reports, which do you feel would provide the kind of information you would want to see? Would you want to combine ideas from several reports?

        4. With the new reports as you like them, how would you measure your department's safety performance?

        5. How frequently would you want the reports?

        6. What other things would you like to see in the system?

    B. Special report capability

        1. Under what circumstances would you use the special search?

        2. What other occasions would there be when you would want a special search?

        3. What type of format would be helpful in reviewing a special search report?

        4. Who in management should be permitted to request a special search?

    C. General discussion

        This might feature a small diagram of the information flow into the system and a list of data that could be part of the output to stimulate these ideas and suggestions.

---

small number of rehearsals may not cover all possible contingencies,
but most, if not all, of the managers to be contacted are known be-
forehand, and this reduces the odds of a really terrible result.[2]

Final Review of the Interview Design

The final review begins during the last practice interview and continues after it. Feedback from each practice interview is used to make
adjustments and polish the interview presentation. When the final
practice interview is completed, a review is made of the format. Present during this review should be everyone who participated in the
practice interviews. They can discuss the final product and contribute their ideas as to any other adjustments to consider.

The final check builds confidence that the interview has been
thoroughly practiced under conditions approximating reality. A study
meeting such as this also serves to promote a sense of involvement
and participation within the staff group. Involvement creates its own
enthusiasm.

MANAGEMENT INTERVIEWS: A TWO-WAY STREET

Using the Trial Designs for Participation

After the pleasantries of introduction and small talk are over, the report formats discussed earlier in this chapter are the materials to use
in the initial phase of the interview. These trial designs are excellent
talking points to portray the breadth and depth of information that
can be reported by the CSDS. It is dramatic to list the current report
information on one side of a page and the information from one proposed CSDS format on the other. Another point to mention is the
amount of information that is entered into the system on each file.
This is a list that is different from the proposed reports and will show
the additional information contained on the data base.

Each of these points can create interest and generate discussion
of how the system can best serve an individual manager. Because
you are asking for advice and suggestions about how the system can
best serve that manager, this will create a sense of obligation to contribute something. Ideas and comments, no matter how small, from
a number of people will provide a good overall consensus of their feelings and opinions as a group.

Session Taping

This is a very touchy subject that should be carefully considered before you decide to use a tape recorder in your interview sessions. A

taped interview is an excellent aid. It allows the interviewer time,
after the session, to replay the conversation and reconsider all the
points of the discussion. It provides a record of the meeting and allows the interviewer to confirm and verify statements.

On the negative side, using a tape recorder in an interview session can cause problems. First, because of the number of times tape
recordings have been highlighted in national news, taping a session,
which is supposed to be an informal get-together, can cause anxiety
on the part of the interviewee. Also, if taping of meetings is not an
accepted practice in your organization, there may be resistance to
such a practice. Even in a firm where sessions and meetings are
taped, perhaps only the very important formal gatherings are handled
that way.

If you decide that your notes will be inadequate and that a tape
of the session will contribute to your success, then permission should
be secured ahead of time. When the permission is granted, it may
be good to watch the person's initial reaction when you actually start
taping. If there is a noncommital or negative reaction, you may be
creating a barrier to open communication. This must be recognized
and dealt with before the interview goes too far.

The alternative to taping the session, of course, is taking good
notes. These can serve well if there is preparation. For example,
if lead-in questions are prepared on paper, leave space between them
for notes about that part of the conversation. A phrase or a couple of
words can serve as a reminder. Use the copies of the format to jot
down recommendations and ideas. Notes can be used to reinforce or
restate an idea, comment, or recommendation to help clarify part of
the conversation. After the session, and before your memory of the
session fades, these notes should be expanded and restated as clearly
as possible. If the outline of the session was generally followed, it
should be no problem to replay the session in your mind directly after
the session. Immediate recording of notes is important, because after
several sessions it can become difficult to remember who said what
and to know the exact point of each comment or recommendation. It
will seem as if all the comments have run together in your memory.
Avoid procrastination, and immediately expand the notes after each
interview. Remember that you are seeking as many separate opinions
as possible. This is the only way to arrive at a correct consensus.

Use of Advance Material

Like the decision whether or not to use a tape recorder, the decision
to send material to the interviewee in advance has pluses and minuses.
The material can be any kind of informational literature to help the
interviewee prepare for the discussion. This material can be an explanation of what will be discussed, formats of the reports, and further

information about the CSDS. The idea is to prepare the interviewee
for the meeting, thereby increasing the depth of the discussion.
Prior preparation and study of the material could provide a better
level of comments and ideas during the meeting. In contrast to using
a tape recorder, sending advance material before a meeting is an accepted practice. The interviewee probably will not feel one way or
the other about receiving such material.

Using advance material, however, creates a problem in the preparation for the interview. The interviewer has to prepare for two types
of interviews instead of one: preparation with the assumption that
the interviewee has read the advance material; preparation for the
manager who has not studied the material. To carry the second idea
a bit further, what motivation does the manager have to read the advance material? Is it more compelling than a production report or a
memo from the boss about a new process? The odds are that unless
the CSDS is already recognized by the manager as something that will
contribute to a better operation, the material will not be read.

The manner in which the advance material is presented can influence the interviewee to read or not read. If the information is brief
(two pages or less), well presented, and fairly specific about the manager's operation, it will have a better chance. If it uses something
like a picture or a question to catch the reader's eye, the material
will at least be glanced at, if not read thoroughly.

The decision factors affecting the use or lack of use of advance
materials are as follows: (1) unless you can persuade each interviewee
to read the material, double preparation will have to be considered;
(2) to increase the odds that the manager will read the material, brev-
ity, reader appeal, and specifics should be used; and (3) a call ahead
to confirm the appointment and to check whether or not the material
has been read may influence the manager to review the material.

The Interview Session: Listen and Understand

Probably the most dangerous potential error during this session is
to be so absorbed in what you are going to ask next that you forget
to listen to what is being said. Remember, listening to comments about
the report formats and about the manager's understanding of the CSDS
is the purpose of the meeting. The reason for the role-playing preparation and the prepared questions is to enable you, the interviewer,
to listen completely.

Careful listening puts you in a position to ask another exploratory
question on the same subject. Additional questions can help clear
away any assumptions made during the conversation. For example,
suppose you ask a question about the CSDS. It is answered correctly
but briefly. It might be good to confirm the information with another
question about that part of the system. In this manner you can learn

the extent of the interviewee's knowledge. This is part of the objective of the interview — to educate and inspire. Fully exploring each answer and making brief notes will keep things flowing.

It is not necessary, when listening, to be concerned with the next question. These are written down on the interview note pad, with spaces between the questions to record your notes. The framework is established. This allows you to listen and to concentrate on what is being said. Try to fully understand this person's ideas and thoughts. One way to get in tune is by reflecting back those thoughts. Reflecting is done by rephrasing the comment just made to double check what you believe you just heard. This confirms and sharpens your understanding of what this person has said on a particular point. It may be that there is a gap in this person's knowledge or perception of the system. It may be that this person is totally unaware of what the system will do or holds a negative opinion about computer data systems in general. The more that is known about why the person holds any given opinion, the better will be your understanding. Through a sharing of ideas can come an attempt by each to see the other's point of view. In some situations, further discussion and education about the system and its benefits will help change a negative attitude to a positive attitude. In the case of misconceptions, it may be necessary to listen, to probe, and to find the basis and background of this feeling or belief. Only when underlying feelings are uncovered can you discuss and share what is true and real about the CSD system.[3]

Obviously, what we have just covered are basic listening techniques. They are points that all of us use in our dealings with people. I believe it is necessary here to stress the importance of these methods because the beginning stage of system development is critical. It is an in-between time. Approval for the CSDS has been granted, but development has only barely started. The system can still be slowed, shelved, or cancelled; so every bit of positive support is vital to continued development progress.

The Interview Result: Useful Data

By the end of the series of informal but structured interviews, you should have a lot of interesting data. The information and comments should break down into several categories: One, the output document format preferences. Two, the interest level concerning the special report capabilities. Three, the knowledge level of management about the system. Four, the specific feelings and opinions about the CSDS. Five, the amount of support for the system as expressed in the interviews.

These five categories will provide an abundance of information to contribute to the actual development of the system. Also, there

are data to use in the development of CSDS information and training
sessions. In addition, the final formats for the regular reports and
the special reports can be developed for another review.

There will be other uses for the information obtained through
these sessions. For example, from the group of interviews you will
know those managers who have the highest interest and who contributed most to shaping the formats of the reports. These managers
are a good panel to use as a sounding board for other ideas and comments. They may be a good support group if you do not already have
one.

There were those from the original interview group who held negative feelings about the system. From these managers there is a sob-
ering point to be learned. Possibly you can find among them one or
two with whom you can discuss their views and learn about their reasons for opposition. In some situations their views may have a stabil-
izing influence on a project or idea that is too optimistic or too radical.

INTERVIEW RESULTS

Development of Conclusions

Once the data have been gathered and put into categories, conclusions
and decisions can be made. A set of report formats can be designed.
Conclusions can be reached concerning managers' knowledge levels
and desires about the CSD system.

Two actions must be taken once the data have been culled through,
an analysis has been made, and conclusions have been reached. One
is to review these new formats with the information systems people.
This is done to ensure that the designs can be run on the system.
If minor adjustments are needed, they can be made easily at this point
without upsetting development. Close coordination with the systems
people is a must. Communications must be open and direct, because
it is they who will be getting the system up and running. They must
know your desires for the CSDS.

Interviewee: Feedback Management Involvement

The second action to take following the analysis of the interview data
is to review it with those same interviewees who contributed their time
to the interviews. This recheck does several good things. First,
it tells the interviewees that you value their opinions and that you
wish to confirm the consensus by showing them the CSDS report format. Second, a short recheck will elicit any other suggestions individual managers may have. Third, for those who were neutral or neg-

ative toward the system, it serves as another opportunity to show off the system and its benefits.[4]

All of these reasons for performing a recheck are directed toward the single purpose of involving management as much as possible in the business of CSD systems development. This is an important concept. Through involvement comes a feeling of shared possession, of contribution, and of authorship. When a person makes a contribution and that contribution is acknowledged, it creates a sense of pride. Pride and involvement reinforce each other and feed the fire of enthusiasm for the project.

Shaping the Final Output Documents for the System

Throughout this chapter we have discussed development of the output documents for the CSDS. The final touches and adjustments to the regular report documents happen at the recheck meetings and through coordination with the information systems people. Another source of feedback is a summary report to management relating the interview process, the report formats, and the conclusions.

As an example of how comments and suggestions from management can shape the report documents, Exhibit 5-2 shows the set of documents adopted from the four trial designs at the beginning of the chapter (Exhibit 5-1). As you can see, changes were made from the earlier output formats, but the information presented remains essentially the same.

SUMMARY

The starting point for development of a CSD system is the design of output documents. Because this is the product of the system, it sets the requirements for the supporting information. The output reports should be easily understood, have a format similar to that of other company reports, and provide comparisons and analyses between the new data and traditional data. To get management involved in the design of the report formats, several sample designs should be developed and presented in a structured one-to-one interview with various managers. Using a structured interview provides standardization of the quality of the session, ensures a high level of information on the specific subjects, and allows the interviewer the freedom to concentrate on the responses. In preparation for the sessions, some of the considerations are whether or not to use a tape recorder and whether or not to send advance material. During the sessions, it is important to listen carefully, reflect the ideas received back to the speaker, and summarize conclusions and comments for clarification. When each session is completed, notes should be expanded and made understand-

```
05/10/XX                        INCIDENT INJURY REPORT           Page XX
Corporation Name                                                 Report No. XXXX
Division Name                   04/01/XX Through 04/30/XX

Expense Ctr. Code XXXX Name
```

Case #	Type of Case	Inc. Date	Time	Days Lost	Case Cost	Freq./ Sev. Potential	Human Factors	Cond. Factors	Root Cause Factors
XXXXX	Temp. Extr.	01/07/XX	1115	05E	$XXXXE	Occas./ Serious	D/N Recog.	Hi Heat/ Cold	Man. Cntl. LTA
XXXXX	Slip	01/13/XX	1455	NWE	$XXXXE	Occas./ Minor	Body Pos. LTA	None	Instr. LTA
XXXXX	Fall DL	01/19/XX	0940	01E	$XXXXA	Rare/ Serious	D/N Use D/N Do. Co. Pd.	Hsekpg LTA	Resp. Unclr. Instr. LTA

Exhibit 5-2A

```
05/12/XX                                    Page XX
Corporate Name                              Report No. XXXX
Division Name

Exp. Ctr. Code XXXX Name
```

<u>INCIDENT/INJURY SUMMARY</u>
01/01/XX Through 04/30/XX

<u>Total Cases by Type</u>		<u>Case Totals</u>		Cost (E/A)
Fall DL		First Aid	(XX)	$XXXXXA
Slip		Medical Treatment	(XX)	$XXXXXE
Struck Against	(XX)	Lost Time	(XX)	$XXXXXE
Overexertion	(XX)	Occasional Illness	(XX)	$XXXXXE
		Property Damage	(XX)	$XXXXXA
		Near-Miss	(XX)	None
		Total Injury Cases	(XX)	$XXXXXE
		Total Reported Cases	(XX)	

Casual Factors

<u>Human Factors</u>		<u>Condition Factors</u>		<u>Root Cause Factors</u>	
Expr. LTA	(XX)			Instr. LTA	(XX)
D/A Recog.	(XX)	Tools LTA	(XX)	OJT LTA	(XX)
Body Pos. LTA	(XX)	Placement LTA	(XX)	D/N Recog. Haz.	(XX)
Equip. Use LTA	(XX)			Tng. LTA	(XX)

<u>Frequency/Severity Case No.'s</u>

Freq./Maj.	XXXXX, XXXXX
Freq./Serious	ϕ
Occ./Maj.	XXXXX, XXXXX, XXXXX
Occ./Serious	XXXXX, XXXXX
Rare/Major	XXXXX
Rare/Serious	XXXXX, XXXXX

Exhibit 5-2B

able. When all the sessions are finished, all of the notes should be analyzed, and the formats of the output reports should be redesigned. Because this is a consensus effort, a recheck should be made with the interviewees. The systems people should also be consulted about the final report formats. When all of the rechecking is completed, a progress report can be made to management discussing the conclusions of the interviews and showing the final output report design.

NOTES

1. John D. Murphy, editor, *Secrets of Successful Selling*, Dell Publishing, New York, 1965, p. 105.
2. Robert R. Blake and Jane Srygley Mouton, *Guideposts for Effective Salesmanship*, Playboy Press, Chicago, 1970.
3. Ralph G. Nichols and Leonard A. Stevens, *Are You Listening?* McGraw-Hill, New York, 1957.
4. Elmer G. Leterman, *The Sale Begins When the Customer Says "No,"* MacFadden-Bartell Corp., New York, 1963.

6

THE USER'S MANUAL: KEY TO THE SYSTEM

Management by objectives works if you know the objectives. Ninety percent of the time you don't.

Peter Drucker

For any complex machine there must be a set of instructions by which a human being can learn to use the equipment. The CSDS is no different. There must be a set of instructions on how to communicate information to the system. For our discussion, this set of instructions is called a user's manual.

A user's manual is arranged to instruct and provide the user with the necessary information to input a case. The manual contains words and phrases that are used to describe facts about an accident, a vehicle collision, or another case, such as a fire, so that it can be entered into the system. The words and phrases are set up in an orderly fashion and correspond to the requirements of the output reports.

DEVELOPMENT OF THE MANUAL IS THE KEY

A computer works with numbers that are translated into electrical impulses. Thus, in communicating with the computer, the words and phrases are assigned numbers. It is these code numbers that are written on the input document to process each case into the computer data base. The user's manual is a code book that lists each word and phrase and the special code number for each.

The manner in which the user's manual is designed and prepared for use will determine whether or not the system will be successful. It is a key function, because the user's manual translates the raw accident data from reports into the code words and phrases for input.

Should the user's manual prove inadequate or incapable of change
or expansion, then the system will be of limited use. If this proves
to be the case, then such a situation will be recognized as a failure.
Such a condition will have a very negative effect on the credibility
of the safety department.

The user's manual must be designed carefully, with several properties built into it: (1) it must be capable of being changed, (2)
it must be designed to be expanded and added to without expensive
reprogramming, and (3) it must meet the needs of the output documents that are designed for the system. When the user's manual
meets these three criteria, the system will be effective and efficient
and will live up to the expectations of management.

USER'S MANUAL DEVELOPMENT: THE NEVER-ENDING TASK

An Outline of Two User's Manuals

The words and phrases in a user's manual are divided into categories
or files. These files are the descriptive subjects under which the
CSD system will maintain information on each case. The titles of
these files and the number of files maintained will vary with the desires of the system designer. Also, the files will differ in accordance
with the industry in which the system is used. Here are two examples
of user's manual subject files. One is from the pharmaceutical industry,
(Schering- Plough Corp., Kenilworth, N.J.) and the other is from the
construction industry, (Construction Safety Association of Ontario).
First is an outline of the file subjects from the pharmaceutical industry
system:

I. Identity

    1. Case Number
    2. Division/Corporate
    3. Organizational Unit (from the table of Organization Codes)
    4. Expense Center Code
    5. Location of Incident
    6. Date of Incident
    7. Time of Incident (24-hour clock)
    8. Property Cost
    9. Medical Cost
  10. Compensation Cost
  11. Other Contributing Costs
  12. Type of Case (three entries available)
  13. Type of Property
  14. Property Ownership

15.   Phase of Operation at Time of Incident
16.   Months Since Last Inspection
17.   Weather Conditions
18.   Specific Operation Involved (pharmaceutical operations)
19.   Specific Component (pharmaceutical machinery)
20.   Year (model or construction)
21.   Contaminant (chemical/physical agent; two entries available)
22.   Recommendations Outstanding
23.   Batch Number
24.   Work Order Number/Appropriation Request Number

II. People

25.   Fatality Date
26—29. (held for expansion)
30.   Social Security Number
31.   Employee Number
32.   Age in Years
33.   Employment Status
34.   Injury/Illness Type
35.   Parts of the Body
36.   Severity of Injury/Illness
37.   Days Off Work
38—39. (held for expansion)
40.   Days Restricted Activity
41.   Relation to Work
41a. Occupation/Job Codes
42.   Employee's Name/Initials
43.   Supervisor's Name/Initials
44.   Supervisor's Employee Number
45—49. (held for expansion)

III. Analysis

50.   Frequency Potential
51.   Severity Potential
52.   Human Factors (two entries available)
53.   Condition Factors (two entries available)
54.   Root-Cause Factors (three entries available)

Next is an outline of the file subjects in a system used in the construction industry. This system was developed by a construction association for its members. This group uses a different approach to portray their data and set data up for analysis. Note the similarities and differences between these two examples:

1.   Firm Number (Workers Compensation Board).
2.   Rate Number (Workers Compensation Board).

3. Subrate Number (Major type of Construction).
4. Worker's Compensation Claim Number.
5. Town (listed by county) Where Injury Occurred.
6. Office of Firm (head office by county).
7. Date of Injury.
8. Time of Injury.
9. Age.
10. Occupation of Injured.
11. Part of the Body Affected.
12. Project Type.
13. Construction Type.
14. Employee Activity.
15. Equipment or Material Acted on (two entries).
16. Type of Accident.
17. Other Involvement.
18. Distance of Fall.
19. Pounds.
20. Working Surface.
21. Condition of Working Surface.
22. Nature of Injury.
23. Causes of Injury.
    a. Environmental.
    b. Equipment.
    c. Human.
    d. Material.

Both of these user's manuals are printed in their entirety in Appendixes B and C.

Files and What They Mean

The user's manual for the pharmaceutical industry divides its files into three subject groups: identity, people, and analysis. Within these three major categories are the files that contain the codes for specifying the exact descriptive words and phrases that can be selected for a specific case. Let us review the subjects that are actually contained in the files of the pharmaceutical industry example.

I. Identity

1. *Case Number.* In a system such as this, each entry is assigned a case number. This serves several purposes: One, it identifies a single case or entry into the system. Two, the number aids in making changes and additions to that case. Three, the case is identified on the output reports. Four, the number can be used to identify all the supporting documents of a particular case.

2—4. Division/Corporate, Organizational Unit, Expense Center
These are company structure designations that are used to specify
exact locations of the people and/or the event within the case. Also,
the designations serve to place the case within the jurisdiction of a
particular corporate unit. This number of breakouts can be modified
to meet the needs of the organization. Note that the corporate des-
ignation is used to specify those cases that occur in the corporate
headquarters or to headquarters people.

5. Location of Incident. This is a six-digit code that is broken
down to specify plant or facility for the first two digits, the building
in the plant for the next two, and the floor for the last digit. The
basement is noted as a zero. Variations on the location are possible:
sidewalk, street, parking lot, grass, and stairs. In these cases the
plant numbers are followed by the number of the variant.

6. Date of Incident. The year is entered first, then the month,
and then the day.

7. Time of Incident. The 24-hour clock is used to avoid a.m./p.m.
confusion. For those who are not familiar with this method of recording
time, it is as follows: The 24-hour time begins 1 minute after midnight, at 0001, and continues to the following midnight, or 2400 hours.
For example, the time of 5:15 P.M. in the afternoon is written as 1715
in 24-hour clock time. The time of 11:13 P.M. is 2313 hours.

8—11. Costs: Property, Medical, Compensation, Other. Deri-
vations of costs are discussed in detail in a later chapter. The costs
associated with the first three categories are direct costs. These
are the costs that are easily identified by medical payments, invoices,
and other fees. The category for other costs includes the indirect
costs. This refers to time spent in connection with the case and other
costs not elsewhere categorized. After the costs are entered on the
computer system, the machine will run totals for each category and
a grand total. These totals will appear on the regular reports.

12. Type of Case. This identifies the case by subject. For this
system, the cases the computer will accept are injuries, illnesses,
fires, inspections, surveys, near-miss incidents, property damage,
outside agency inspections, and security matters. This file provides
a great amount of flexibility for the system. It allows the listing of
more cases than just injuries.

13. Type of Property. This identifies the property by general
or generic name, such as maintenance building or guard facility.

14. Property Ownership. The various ways a property can be
owned are listed here, such as company-owned, contractor, rented,
leased, etc.

15. Phase of Operation at Time of Incident. This file includes
the names of the various shifts, overtime, shutdown, maintenance,
etc.

16. Months Since Last Inspection. This file refers to equipment
and is used to fix the last inspection date.

17. Weather Conditions. If it has a bearing on the case, there is a file to record the type of weather involved.

18. Specific Operation Involved. This file contains the listing of those operations carried on in that company. For example, under the main category of packaging are listed all of the different kinds of packaging performed in that company.

19. Specific Component. Listed here are the items or things that were involved directly in the case. These can also be termed the agent of injury in some circumstances. For example, under ladders, there are fixed, single, step, extension, and step stool.

20. Year. This pertains to the model year for equipment and the year of construction for buildings.

21. Contaminant. This file contains all of the chemicals, physical agents, and energy forms that may be used or may accumulate in the operations.

22. Recommendations Outstanding. This pertains to two points: the original number of recommendations; the number of recommendations that are outstanding and not yet completed.

23—24. Batch Number, Work Order Number/Appropriation Request Number. Batch pertains to a chemical or food process where an amount of material is processed in a single vessel. Each such vesselful is a batch and is given a number. The work order is used by maintenance for tracking the performance of repair work. The appropriation request (there are many names for this) is either an expense or a capital budget item at a high enough dollar amount to require upper-level approval.

II. People

25. Fatality Data. Self-explanatory.

30. Social Security Number. Self-explanatory.

31. Employee Number. Self-explanatory.

32. Age in Years. Self-explanatory.

33. Employment Status. This describes the individual in the case, such as a full-time employee, temporary employee, contractor, vendor, or guest.

34. Injury/Illness Type. This file contains a list of the various types of classes of injuries and illnesses, such as hernias, fractures, burns, and lacerations.

35. Part of the Body. Self-explanatory.

36. Severity of Injury/Illness. This describes where the injury falls amoung categories such as first aid, medical treatment (doctor case), lost time, and occupational illness.

37. Days Off Work. The number of workdays lost due to an injury or illness.

40. Days Restricted Activity. The number of days an employee is restricted from performing all parts of the regularly assigned job.

41. Relation to Work. A determination of whether or not the case is work-related. This file is used for reportability in accordance with government (OSHA) requirements. In all situations, a case can be entered on the CSDS. This file serves as a filter in the search for work-related cases and non-work-related cases, or both.

41a. Occupation/Job Codes. These are job titles for all employees, and usually they are coded automatically if there is an existing personnel data system. If not, the item can be recorded on the CSDS when the case is entered.

42. Employee's Name/Initials. Self-explanatory.

43. Supervisor's Name/Initials. Self-explanatory.

44. Supervisor's Employee Number. Self-explanatory.

III. Analysis

50. Frequency Potential. This is a prediction of the number of times this type of event is *expected* to occur in the future.

51. Severity Potential. This is subjective judgment given the same set of circumstances, concerning the estimated severity of the *next* event of this kind.

52. Human Factors. This lists the types of human errors that can be made by workers performing the job. It could also be called unsafe acts.

53. Condition Factors. This pertains to the hazardous conditions that are factors in the case. It could also be called unsafe conditions.

54. Root-Cause Factors. There are two main categories in this file: Unwanted Energy Flow and Management-System Factors. Unwanted Energy Flow refers to energy that becomes uncontrolled. This can be an electrical short, an exploding grinding wheel, or a pintle hook that breaks under load. The other category, Management-System Factors, describes those omissions, oversights, and errors that are committted by the management system and make a contribution to the accident.

The second example is the construction industry safety data system that is operated by a construction association for its members (see Table 6-1). Because of its association-type approach and the organization's relationship with its members it must differ from the subject files of the pharmaceutical industry.

There are unique parts in each of these two example systems: files that are special to that particular system. Their use, however, need not be restricted to a particular industry. These special features could be used in other systems and, as such, need recognition. They are as follows:

Pharmaceutical Company
   Costs: Property, Medical, Compensation, Other, and Total.

Table 6-1　Construction Industry Safety Data System

Data	Examples	Comments
1. Firm Number	244248F	Taken from Workers Compensation Board of Ontario (W.C.B.O.)
2. Rate Number	753	Taken from the W.C.B.O.
3. Subrate Number	1 (sewer & water-main installation)	General type of construction normally handled by this firm
4. Worker's Compensation Claim Number	1200122	Number unique to each injury; this number is used to find a specific injury
5. County	25 (Lincoln County)	Location of accident
6. Office	25 (Lincoln County)	Head office of firm
7. Date of Injury	08/05/84	
8. Time of Injury	1535	
9. Age	35	
10. Occupation	PIPE (pipelayer)	
11. Part of Body	SKLL (skull)	
12. Project Type	ROAD (road job)	
13. Construction Type	SEWR (sewer installation)	Specific type of construction on this job
14. Employee Activity	CUT (cutting)	Activity at time of injury
15. Equipment or Material Acted on	SEWR (sewer pipe)	Refers back to activity
16. Type of Accident	STRK (struck by)	
17. Other Involvement	BCKH (backhoe)	Refers to type of accident (i.e., tripped over pipe, struck by wrench, etc.)
18. Distance of Fall	NONE	Only applicable for fall from different level

Table 6-1 (continued)

Data	Examples	Comments
19. Pounds	NONE	Only applicable if object was lifted or fell
20. Working Surface	TREN (trench)	
21. Condition of Working Surface	MUD (muddy)	
22. Nature of Injury	FRAC (fracture)	
23. Cause of Injury	ENVR*VISI (enviornmental—poor visibility) EQUP-SLIP (equipment slipped or jammed) HUMN-SUPR (human—poor supervision)	

Type of Case. This file allows the system to include different types or classes of cases beyond the categories of injuries and illnesses. The system can accept a variety of case types, such as fires, inspections surveys, near-misses, and security problems.

Property Ownership. All of the cases in the system may not occur on company property. This can be a check where various types of cases happen.

Phase of Operation. This can be used instead of, or as an addition to, the time of the incident; by identifying the specific type of operation, such as a breakdown, maintenance operation, normal day shift, or holiday shutdown, there is a more precise idea of what was going on when the incident occurred.

Months Since Last Inspection. Cases can be identified in which the status of the inspection can be a significant factor.

Weather Conditions. In summarizing cases, weather can be considered to determine its degree of significance. This can be true in various outside occupations.

Year (model or construction). Useful in fleet operations and to note age of equipment.

Contaminant. Chemicals and physical agents such as noise and radiation are present in many industries. Misuse, poor handling techniques, and incompatible storage can cause and contribute to a wide variety of cases. This file also can be used as a lead into a material safety data file.

Recommendations Outstanding. Follow-up on recommendations is one of the hardest things to control in a safety operation. This file can track all of the cases where recommendations

were made and are still outstanding. A report concerning
recommendations can also include the type of case to identify
the source of the recommendations.
*Batch Number, Work Order Number/Appropriation Request
Number.* These are useful to track recommendations and,
with regard to batch number, to identify similar operations.
Fatality Date. Required for OSHA reporting.
Social Security Number; Employee Number. The Social Secu-
rity Number is an entry on OSHA from 200. The Employee
Number can be used to summarize cases by employee.
Days Off Work. Required for the OSHA log. Also, it is useful
to compare total days lost and days lost per case as an average.
Days Restricted Activity. Required for the OSHA log. Also
identifies a category for comparison with overtime costs and
scrap rate.
Relation to Work. Used to separate those cases that are OSHA
reportable from those that are not.
Supervisor's Name/Initials; Supervisor's Employee Number.
Can be used to summarize cases by supervisor.
Frequency Potential; Severity Potential. Two good parameters
to use to put a priority on cases by class.
Root-Cause Factors. This section is aimed directly at management oversight factors and at uncontrolled energy sources.

Construction Association

Employee Activity. An important consideration is to learn
just what types of tasks are contributing to injuries.
Distance of Fall. Can be used with other variables such as
activity and weather to determine subtle causal situations.
Pounds. The pounds being lifted or moved. Can be used
to compare employee activity versus the incidence of injury.
This can provide clues to causal areas when compared with
other variables.

These unique features, which were built into these example systems,
are useful when used together in other fields. By themselves, they may
not be significant. These systems are included with others in Appendixes B and C. Having a wide choice of different features, or fields,
provides for selection of strong fields for use in a particular application.

Numbering Systems and Identifiers for Code Words

One of the critical points to attend to in developing a CSD system
is to allow for user flexibility, expansion, and change. A system
that does not make allowances for user-induced changes will become
rigid and will require extensive reprogramming any time changes needed

to be made. The users in this case are those in the safety department
responsible for the overall design of the system.

The pharmaceutical company system is designed to be expanded
and changed: it is flexible. These terms have the following meanings:

Expand: To allow users to add descriptive words or subfiles to
the files without resorting to reprogramming the system. There
is also the capability of adding several files to the system.
This would require minor reprogramming.

Change: To allow users the ability to add additional major units
to the system without extensive reprogramming. (see Chapter
12).

Flexibility: To allow the system to accept input of incomplete
cases and to make changes and additions to cases already
entered in the system.

The way a system is designed for user change and expansion
is through the numbering system of the files, subfiles, and descriptive code words. You may have noticed in the listing of the main
files for the pharmaceutical company that there were 11 files left for
expansion. Also, within several files the numbering techniques allow
for expansion.

The flexibility of entering incomplete cases into the system is
another very useful capability. For example, suppose that the close-
out date for case inputs to run computer-generated monthly safety
reports is the third working day of the next month. An injury occurs
at 11:00 a.m. on the last day of the month. The investigation is progressing but is not completed by the second working day of the next
month. The case can be entered in the system without complete data
by using code identifiers. These identifiers tell the computer it is
okay to process the case. This is done by using a "more data" code.
Using this code where there is incomplete information allows the case
to be entered and counted for that month. So that the users will
not forget to complete the report, a More Data Report is automatically
generated each month as part of the regular report system.

Another essential set of instructions to the computer includes
those that tell the system to accept the case even though the data
are estimated. In other words, costs are often delayed; therefore,
the system is instructed to accept the following conditions with regard
to compensation and other costs as well as days lost and restricted
duty. In this example, compensation cost is used.

1. There is no compensation cost involved: A000000 [actual
(A); the 0's are the identifier for the No Factor code, which
is the same as saying no data required].

2. There is compensation cost, but it is not developed yet:
    E000000 [estimated (E), and the 0's in this case indicate more
    data required].
3. There is compensation cost, and it is estimated at $10,645.
    The code would be E0010645. This would appear on the More
    Data Report, because it is an estimate.

The estimated (E) and actual (A) instructions are useful for any
file for entering days lost and days of restricted activity. The No
Factor code is the instruction that tells the computer that for this
case certain items are not relevant. There are, however, files that
must have an entry to be judged as an acceptable case by the computer. This means that to be an acceptable case, something must
be entered in certain fields. For example, for the pharmaceutical
industry system, the entry can be a no factor, more data, an estimate,
or an actual code, but for the system to accept the case, the computer
must receive an entry in the following fields:

Location of Incident.
Time of Incident.
Costs (Property, Medical, Compensation, and Other).
Type of Case.
Type of Property.
Property Ownership.
Phase of Operation.
Last Inspection.
Weather Conditions.
Specific Operation.
Specific Component.
Model Year.
Contaminant.
Recommendations Outstanding.
Batch Number.
Work Order Number.
Fatality Date.
Employment Status.
Nature of Injury/Illness.
Part of the Body.
Injury Severity.
Days Off Work.
Days Restricted Activity.
Relation to Work.
Frequency Potential.
Severity Potential.

Human Factors.
Condition Factors.
Root-Cause Factors.

There is complete flexibility in this system. For example, the
Type of Case file can be coded in three ways to thoroughly identify
the case. The subfiles include not only injury/illness situations but
also security, survey/inspections, fires and their sources, near-
miss incidents, and property-damage cases. There is a potential for
99 subfiles in this field. Note that the employee's name is not a required field for entry into the system. This was intentional because
of the variety of other case types that could be entered.

Development of Unique Industry Files

These are files that identify unique organization, equipment, and processes to an industry in a system. The system from the pharmaceutical
company, for example, reflects that company's organization, equipment,
and processes unique to the industry. The system from the construction
industry association is completely dedicated to the construction trades.
Each of these systems can be adapted to another industry. For example, the pharmaceutical company's system can be used by another
industry by constructing codes in the industry-specific files. The
files that may require change are the following:

   2.   Division/Corporate.
   3.   Organizational Unit.
   4.   Expense Center Code.
  18.   Specific Operation Involved.
  19.   Specific Component.
  21.   Contaminant.

The rest of the files from the pharmaceutical company system can
be used basically as they are in Appendix B. The organizational files,
such as Organizational Unit, Expense Center Code, and Division/
Corporate, can be different in each company. Because these determine the grouping of reports, it is important that the files be set
up to reflect the company organizational structure.
Developing unique industry files turns out to be a rather straightforward exercise. The Division/Corporate, Organizational Unit, and
Expense Center Code are easily derived. Support from operations
is needed to get the Specific Operation Involved, Specific Component,
and Contaminant files organized. Face-to-face meetings are quite
helpful in this regard to answer questions and to speed the process

along. This exercise also helps each operations person to feel authorship for contributing to the development of the system, and it may
encourage further support and usage of the system.

Allowing for Further Expansion and Subfield Development

In a previous section of this chapter there was mention of the use
of a numbering technique for expansion within the files themselves.
This is the use of subfiles to increase the capabilities of a field to
hold a lot of data. This is one way that allows for future expansion.
Within a code system where there are files that will require many code
words and these words can be grouped into categories, then the subfile method of setting up files is used to expand the capacity.

The method used in the pharmaceutical company user's manual
to set up a subfile for expansion is a three-step process. First, the
code words to be used in the file are listed in alphabetical order and
counted. Second, if it appears that the maximum number of code
words will be 999, then three digits are applied to the descriptive
words. However, if it appears that the file may grow and expand
above this figure, then subfiles are formed. This is the third step.
The file of words is separated into natural groups and given numbers.
Within each subfile, the words are listed alphabetically. For example,
in the Type of Case file there are actually seven digits involved:

> Main file code: 12 Type of Case.
> Subfile: 2 digits (99 subfiles potentially).
> Code word: 3 digits (potentially 999 words in each subfile).
> Total file capacity: 98,901 code words.

When assigning numbers to the code words, they should be assigned
considering the descriptive word's place in the alphabet. This is
especially true when assigning a number to a file that will expand
in the future. Those words beginning with letters at the beginning
of the alphabet should be given lower numbers, and those nearer
the end of the alphabet should be given higher numbers. The Weather
file is a good example (Table 6-2). The way the file was set up was
to write down the file in alphabetical order, then count the number
of code words, and divide by the number of code numbers for that
file. To illustrate, the Weather file is divided into the categories
of normal and severe weather. The normal-weather section has 21
code words for 800 code numbers. This allows for approximately 30
spaces between each word. The severe-weather section has 200
spaces for 6 words. This allows for a spacing of 30 also.
Another way to account for expansion is by the use of letters
and numbers, as in the Contaminant file. It has two alphabetic and

Table 6-2 Weather Conditions File[a]

Normal weather

030	Cold, Dry
060	Cold, Rain
090	Cold, Rain, Fog
120	Cold, Snow
150	Cool, Dry
180	Cool, Humid
210	Cool, Rain
240	Cool, Rain, Fog
270	Freeze, Dry
300	Freeze, Snow
330	Freeze, Rain
360	Hot, Dry
390	Hot, Humid
420	Hot, Humid, Fog
450	Hot, Rain
480	Hot, Rain, Fog
510	Warm, Dry
540	Warm, Humid
570	Warm, Humid, Fog
600	Warm, Rain
780	Warm, Rain, Fog

Severe weather

810	Earthquake
840	Electrical Storm
880	Flood
900	Hurricane
930	Tornado
960	Tropical Storm

---

[a]Code: 001, more data; 999, no factor.

five numeric identifiers for each contaminant. This allows for expansion of this file to 67,599,324 contaminants—more capacity than this file will ever require for a single industry.

There are thus three ways to allow for expansion with the files:

1. Leave main-file space for the addition of other main files.
2. Use of subfiles with the main file.
3. Use alphabetic and numberic identifiers together for maximum expansion.

If these rules are carefully followed, the user's manual will be
capable of being expanded and changed as necessary. This is one
of the basic tenets of a good user's manual: that it can be added
to and changed by the user without resorting to expensive reprogramming.

Connection with Existing Data Systems

Most corporations today have computer systems to provide management
with information. Many have added other systems to the accounting
and inventory programs. These other systems further enable management to exercise better and more prompt control over their business
operations. Some of these programs may include a personnel data
system. If this is the case, there is no requirement to add certain
sets of files to the CSDS. All that is necessary is to provide instructions in the program to access the personnel data system for those
pieces of information. This includes

> Social Security Number.
> Age in Years.
> Occupational Code.

All these can be supplied through the personnel data system by using
an instruction plus the employee number.

Conversely, if the company does not have a personnel data system,
the foregoing is not a lot of information. It can be easily obtained
during the investigation and entered on the report form. Thus, the
CSDS can be developed and used whether or not a personnel data
system exists.

THE USER'S MANUAL: A LIVING DOCUMENT

Publishing the User's Manual

The well-designed user's manual is truly a living document. It is
flexible and capable of changing and expanding as new code words
are found. It is never really complete, although after coding about
3000 cases into the system, most of the additional code words that
were overlooked originally will have been entered. Even then, there
will be an occasional addition to be made.

The user's manual is usually typed with the code words in columns
on the page. To avoid retyping such a document over and over, word-
processing equipment should be used. This equipment enables the
user to change and edit the document without the necessity of complete

retyping and reediting. It is a more efficient method to use for a multipage document that may be changed and updated 20 or 30 times in the first 2 years.

As can be seen in the pharmaceutical company's system, each page is identified with the section, the item number, and the page number. The date each page was last changed is also noted below the page number. This method allows for changing the pages and checking manuals to ensure that the changed pages have been inserted. Putting the item number and the section on the top of the page allows for easy scanning of the manual for the correct section and item.

Instructions, Guidance, and Training

In the front of the user's manual, as well as throughout the book, there are instructions for the coder on how to code certain case situations. This allows for better quality in the coding of cases, because it is probable that more than one person will do the coding. It provides a level of guidance.

The aim of the user's manual is to provide standardization in coding and quality. This combination will reduce the cases returned for errors and will give reliable data in the system. The instructions and the notes in the manual should help, but a quality check with the coders themselves will keep the standardization and quality high.

SUMMARY

A user's manual contains information, instructions, and code words and phrases by which the user can input cases into the CSDS. This manual is the key to the system. If it is set up and designed poorly, the system will not be responsive to change and expansion without expensive reprogramming. The inadequacies of the system will reflect directly on the safety department. It is imperative that the user's manual be designed to be flexible and responsive to change and have an expansion capability.

These three properties can be built into a user's manual through the different files set up. Expansion is designed into the manual by providing for extra files. Within files, a subfile system can provide room for expansion. Flexibility is provided by using instructions for the computer that will allow it to accept cases that are incomplete and cost figures that are estimates. Change capability can be built into the system by providing a method by which major units, such as medical information, industrial hygiene, and a material safety data sheet system, can be added to the CSDS. All of these properties can be designed into the user's manual and programmed into the system as it is designed. The result will be a system that is cost-effective and responsive to future needs of the company.

7

A NEW INVESTIGATION FORM MEANS CHANGE

*People may forget how fast you did a job, but they will remember
how well you did it.*

Anonymous

CONNECTIONS: THE TRADITIONAL AND THE NEW

From our research and analysis of the current investigation system,
as discussed in Chapter 2, there is already a file of information to
draw on in our planning for the CSDS. Research on the current investigation system is the basis for preparation of a two-step transition. This is a move from what is currently being done, a revision
and improvement of that system culminating in the new system that
will support the CSDS.

As our analysis has revealed, the present investigation system
will have some excellent points and probably some that are not so
excellent. For the first step of the changeover, the excellent points
will be retained, and the techniques that are not so good will be improved, refined, or eliminated to improve overall system quality.

The analysis of the present system will also indicate what areas
those in management believe should be improved. This is an important
consideration. In using their recommendations, or parts of their
ideas, credit can be passed back to the management contributors.
This will provide the continuing sense of participation, authorship,
and involvement necessary for success of the project. In fact, every
point where a contribution was made should be recognized. Keep a
list of those who contributed, the nature of their ideas, and whether
or not their ideas were used. Give credit to everyone who provided
input, whether it was used or not. If you can give specific praise,
that is even better.

TIMING: CHANGE SLOWLY AND EASILY

As with anything that must be adjusted or changed, revisions to
investigation forms and to the system must be timed properly and
carried out slowly. Complete or partial changes that are set in motion
and pushed through rapidly can meet strong resistance. Rapid changes
can lead to a lot of complaints, and if a change does not produce the
desired result (as envisioned by its originators), this can stall the
entire project. Thus, the slower the transition, with the least amount
of disruption to the operation, the fewer will be the complaints and
the greater will be the acceptance.

For many, the reasons for opposition are difficult to understand
when they believe that the change is so right and logical and in the
best interests of all concerned. In that assumption lies the problem
or misconception. Change, any change, is disruptive and can cause
stress in proportion to the degree of change. Thus, there may be
a certain amount of resentment if those who must operate with a new
system are not part of the change process. This means that those
who will operate the new system should have input into the process
as early in the development as possible; otherwise there will be resistance.

In introducing a change to the investigation system, many people
are affected, and thus the introduction must be accomplished slowly
and with as much participation as possible. The heart of the change
is the new or revised investigation form. The new form must be introduced carefully and over as long a break-in period as practical.

Fortunately, the investigation system will have been analyzed
with input from those who operate the present system. This information, concerning possible changes, as well as the training needs of
investigators, will be well documented in the proposal to management.
It is logical, then to use this to analyze part of the proposal to review
anticipated changes with the operators of the present system. The
ways and means the recommendations can be implemented can be discussed. A time schedule for introduction of the changes can be developed and agreed on as well.

Most of the form design and the revision of the investigative system will be done by safety. That is understood. The fact that the
operators of the investigative system are part of the decision and implementation sessions will provide for better acceptance. Once the
training programs are designed, the investigation form developed,
and the other suggestions prepared for implementation, it is the operators who must contribute their time and use the new system. Therefore, their cooperation is paramount.

DESIGNING INVESTIGATION FORMS

Meeting the New System's Needs for Information

We are still working from the output report back in time to the investigation itself. The user's manual lists the information required for the output reports. The next step is to translate this list of information onto an investigation report form. Because there probably will be an investigation form already in existence, it must be modified to meet the needs of the CSDS.

To accomplish a smooth changeover from the present investigation form to a new form, it is good to take the list of items required for the CSDS and beside this write down the items captured by the present investigation form. In this manner the similarities and differences can be determined. This accomplishes two things: First, it pinpoints exactly the additions and changes that will occur. Second, it identifies those bits of information that are constant in both systems and will pass through from the present system into the new system.

The exercise of listing the similarities and differences can be used in several ways. First, the additions and changes are aids in organizing a design for the new investigation form. Second, the two lists show the constants between the two systems. The number of constants can be converted to a percentage of the whole to show how much of the present system will remain unchanged. Third, the lists of the current system and the CSDS are excellent training examples to show the changes. They will help to explain why these changes are made and what they will do in the way of benefit.

Types of Investigation Forms to Consider

The sole purpose of using a form for anything is to prompt a user to provide essential information. It serves as a checklist. If each person were "super-dedicated" in all phases of work, many forms could be eliminated. A blank piece of paper would be okay. This is not the case, however, in modern industry or other parts of our life where a standardized amount of information is needed. Safety is no exception, and the CSDS requires certain information to satisfy the needs of the output reports. The investigation report form serves as an economical way to gather the needed information.

Exhibit 7-1 shows the form used by the pharmaceutical company as a source document for its data system (as described in Chapter 6). For this form, the originator will in most cases be the plant nurse, who will forward one copy to safety and the rest to the supervisor, who will complete the document's second section and send it on to the department manager. The manager will review the investigation and the recommended actions. At this level, other ideas to eliminate or

SAFETY & HEALTH INVESTIGATION REPORT
(MUST BE COMPLETED WITHIN 2 WORK DAYS OF OCCURRENCE)

REPORT DATE

SECTION ONE

CASE LOCATION: BLDG. FLOOR. PARKING LOT. ETC.

TYPE OF CASE:
☐ FIRST AID ☐ MEDICAL TREATMENT ☐ LOST TIME ☐ FATALITY
☐ INCIDENT (NEAR MISS) ☐ PROPERTY DAMAGE ☐ OCCUPATIONAL ILLNESS

CASE DATE	TIME	EMPLOYEE NAME	EMPLOYEE NO	WORK LOCATION	SUPERVISOR'S NAME

Description by Medical Professional: Include nature of injury, occupational illness, part of body, treatment and recommended restrictions of the patient.

SIGNATURE	TITLE	APPROX. TREATMENT TIME

TO BE COMPLETED BY IMMEDIATE SUPERVISOR

SECTION TWO

Describe what happened and list several probable causes. (See reverse of goldenrod copy for instructions)

In your opinion, based on the probable causes, what actions do you recommend?

SUPERVISOR'S SIGNATURE	EMPLOYEE NO.	NAME (PRINT)	TITLE	APPROX. INVEST. TIME

TO BE COMPLETED BY THE NEXT LEVEL OF SUPERVISOR

SECTION THREE

What changes in methods, procedures or equipment will minimize or eliminate the probable causes identified?
Are there other causes? If so, what should be done about these?

DATE	COMPLETED BY	TITLE	SIGNATURE	APPROX. REVIEW TIME

TO BE COMPLETED BY SAFETY

SAFETY REVIEW

DATE	REVIEWED BY	CASE NUMBER

FURTHER REVIEW

DATE	REVIEWED BY

DISTRIBUTION

WHITE	— Originator ⟶ Location Safety
CANARY	— Originator ⟶ Employee's Supervisor ⟶ Next Level Supervision ⟶ Files
PINK	— Originator ⟶ Employee's Supervisor ⟶ Next Level Supervision ⟶ Location Safety
GOLDENROD	— Originator ⟶ Employee's Supervisor ⟶ Next Level Supervision ⟶ Location Safety ⟶ Division Safety

Exhibit 7-1 Investigation form: pharmaceutical company. (From Schering-Plough Corp., Kenilworth, N.J.)

Form Completion Instructions

Medical Professional ━━━━━━━━━━━━━━━━━━━━━━━━━━━━━━━━━

Complete top third of form. Send white copy to Division Safety Services; send remaining copies to Expense Center Manager.

Injury — Describe fully. If chemicals were involved, list kind of chemicals. If occupational illness, mention possible source, e.g., animal dander, dust, fumes, etc.

Treatment — Describe. Note whether X-rays were taken; if splints, ace bandages, etc. were used. Describe any additional treatment required.

Disposition — Note whether patient was returned to work; sent home; transferred to hospital; given light duty, etc.

Supervisor ━━━━━━━━━━━━━━━━━━━━━━━━━━━━━━━━━

Be complete. List any substandard acts or conditions which may have contributed to incident. Look for basic causes which may have "set up" situation. To assist you, here are some areas to consider in your investigation.

Equipment/Operation — Describe type, placement; correct type; list batch no. if appropriate; current inspection.

Weather — Describe if a factor.

Environment — Adequate lighting; ventilation; work area.

Procedures — Written, understood, followed.

Management — Direction, guidance, support.

Employee — Physical condition, mental condition; job experience, task knowledge; use of personal protective equipment; time spent away from job.

Recommendations

List the items in the methods, operations, equipment or management system which you feel may reduce or eliminate the possibility of the next incident. Note whether or not this situation is economical, controllable.

Manager ━━━━━━━━━━━━━━━━━━━━━━━━━━━━━━━━━

Review incident, investigation and recommendations. Describe what methods, equipment, operations, or management system changes you are making or recommending. If none, so state. If more assistance is required from Safety, request such assistance.

Exhibit 7-1 (continued)

minimize this situation or condition in other areas of the plant or company can be considered and recommended. Following that, the manager will send the report to the safety department, where they will review the document for quality and completeness. At that point, the need for further upper-management review will be considered and acted on if necessary.

The construction association form in Exhibit 7-2 (used by the construction association described in Chapter 6) shows some interesting contrasts in the way the form is constructed. Note that it addresses some of the same things used in the pharmaceutical form,

Construction Safety Association of Ontario

74 Victoria Street, Toronto, Ontario M5C 2A5 • (416) 366-1501

ACCIDENT CAUSAL DATA

CLAIM NO. _______________________________________

EMPLOYEE'S NAME____

DATE & TIME OF ACCIDENT____

BACKGROUND INFORMATION

1.   Type of project where accident occurred (Examples: bridge, industrial, high-rise apartment) _________

2.   What type of construction was your company doing on this project? (Examples: carpentry, painting, home building) _______

ACCIDENT DETAILS

1.   Employee activity at time of accident (Examples: grinding, walking, climbing, drilling)_________

2.   Tools, equipment, machinery involved (Examples: cement truck, hammer, rolling scaffold) _________

3.   What personal protective equipment was being worn? (Examples: safety glasses, goggles) _________

4.   (a) If worker or object fell, how far?________________ft.

      (b) If object was being lifted or fell, how heavy was it?________________lb.

5.   Working surface at time of accident (Examples: ladder, ground, staging, trench) _________

6.   Condition of working surface (Examples: littered; wet, confined and muddy; icy and unprotected) _________

7.   Severity of injury or occupational disease (Examples: cut, bruise, sprain, hearing loss) _________

ACCIDENT ANALYSIS

What were the apparent causes of the accident?

1.   Environmental (Examples: wind, rain, mud, ice) _________

2.   Equipment (Examples: broken shovel, truck gears jammed) _________

3.   Human (Examples: fatigue, lack of training) _________

4.   Material (Examples: lumber piled loosely, cement spilled) _________

5.   Other

RF2

Exhibit 7-2 Investigation form: construction association. (From Construction Safety Association of Ontario, Toronto, Ontario.)

but adds in 4(a) distance of fall and 4(b) weight of the lifted object.
They continue their analysis on the back for prevention points and
approximate costs. Note that in the Accident Details section the form
asks the investigator for very specific information. For example, it
mentions tools, use of personal protective equipment, and the type
and condition of the working surface. These are all important items
for their fine system. In their industry, the person completing the
form may not be a trained investigator. Thus, the form is more ex-
plicit and prompts the person to put in good data. The form becomes
a training tool itself in that it guides the respondent to give the
right answers.

Other investigation forms from other companies are included in
the Appendixes.

Coordination with the System Group

At the same time the investigation form is developed, the input form
must also be created. This requires close coordination with the
systems group. The input form pictured in Exhibit 7-3 is used with
the pharmaceutical company investigation form. Note that this form
is divided into several parts based on the input requirements of
several keypunch cards, because all of it will not fit on one card. A
more sophisticated method of input would be through a video display
terminal (VDT). The reason this method was not selected by the
pharmaceutical company was that their particular system used keypunch cards instead of a terminal system. A keypunch card can be
run again and again to verify the information. The VDT can access
the system and enter the case information directly onto the data base.
In either case, keypunch or VDT, there must be a printout of each
case to verify the information input.

Testing the New Form

It is always nice to think that we do things well and seldom make
oversights. However, mistakes can happen. I am talking about mistakes or oversights that are left and not corrected. Nothing can ruin
credibility and trust in a project quicker than an oversight that is
left uncorrected. Avoiding this opportunity for error is easy, and
it can be done effectively through quality testing and checking.

A new investigation form is like any other development. It is
still a product. Products, especially new ones, should be tested on
a limited basis before being offered for general use. Thorough testing
will detect any flaws and oversights in the system.

The test of the new form should run 2 to 4 months. This period
should be long enough to generate sufficient cases to confirm the

SAFETY MANAGEMENT INFORMATION SYSTEM
SAMIS REPORT

(INJURY, DAMAGE, LOSS)

CASE NUMBER

02651 A

NOTE: ALL SHADED AREAS MUST BE CODED FOR NEW CASES

IDENTITY

Card 1 — CORP. DIV. | ORG. UNIT | EXPENSE CENTER | LOCATION (PLANT, BLDG., FL.) | INCIDENT DATE (YEAR, MONTH, DAY) | TIME (24 HR. CLOCK)

A/R — PROPERTY COST (E/A, AMOUNT) | MEDICAL COST (E/A, AMOUNT) | COMPENSATION COST (E/A, AMOUNT) | OTHER COST (E/A, AMOUNT)

TYPE OF CASE — PRIMARY | SECONDARY | OTHER | TYPE PROPERTY | PROPERTY OWNERSHIP

Card 2 — PHASE OPER. | LAST INSPEC. | WEATHER | SPECIFIC OPERATION | SPECIFIC COMPONENT | MODEL YEAR | CONTAMINANT #1 (ALPHA, NUMERIC)

CONTAMINANT #2 (ALPHA, NUMERIC) | RECOMMENDATION OUTSTANDING (ORIGINAL, REMAIN) | BATCH NUMBER | WORK ORDER OR A/R NUMBER | FATALITY DATE (YEAR, MONTH, DAY)

PEOPLE

Card 3 — SOCIAL SECURITY NUMBER | EMPLOYEE NUMBER | AGE | EMP. STATUS | NATURE INJURY ILLNESS | PART BODY | INJURY SEVERITY

A/R — DAYS OFF (E/A, NUMBER) | RESTRICTED ACTIVITY (E/A, NUMBER) | WORK RELATION | JOB CODE

Card 4 — EMPLOYEE NAME (LAST, INITIALS)

SUPERVISOR NAME (LAST, INITIALS) | SUPERVISOR EMPLOYEE NUMBER

ANALYSIS

Card 5 — FREQ. POTENTIAL | SEVERITY POTENTIAL | HUMAN FACTORS (FIRST, SECOND) | CONDITION FACTORS (FIRST, SECOND) | ROOT CAUSE FACTORS (FIRST, SECOND, THIRD)

(ANALYSIS SECTION IS NOT REQUIRED FOR SURVEYS OR INSPECTIONS)

PREPARED BY | DATE

COPY DISTRIBUTION:
WHITE ——▶ DATA CENTER ——▶ DIVISION SAFETY
CANARY ——▶ DIVISION SAFETY

0507

04–PER400–K01

Exhibit 7-3 Input form: pharmaceutical company. (From Schering-Plough Corp., Kenilworth, N.J.)

validity of the new form and to make any adjustments that are necessary.
The test group selected should be large enough to generate at least
75 to 100 cases. This number of cases should provide enough variety
of accident information. If the group selected is not expected to produce the number of cases by including only medical treatment and lost
time, then first-aid cases can be included during the test period.

To ensure that the test of the new form progresses smoothly, an
orientation should be scheduled for those who will participate in the
test. The points to discuss during the orientation are these:

1. The benefits of the new form and of the CSD system; what
    the test is designed to accomplish.
2. What cases are to be investigated and recorded.
3. The length of the test period.
4. How the new form is to be used.
5. How to make criticisms, suggestions, and corrections to im-
    prove the form.
6. How the final report will be handled.

During the session, it is a good idea to present several example
cases for the participants to complete. This allows "hands-on" experience with the new form. It also may generate good questions
that can be easily answered but that might have caused problems
later. Following the orientation, continued support should be provided through assistance with investigations and completion of the
form. During the test period, every effort should be made to encourage feedback from the users. When ideas and suggestions are received, they can be discussed with other users for their recommendations. This kind of involvement will create a continuing reevaluation
system among members of the test group. The result will be a better
document, one that has a broad base of support.

INVESTIGATION: THE SEARCH FOR CAUSES

Techniques that Reveal Causes

In the beginning of this book, we discussed the difference between
finding just one unsafe act and one unsafe condition versus discovering
as many direct causes and underlying causes of the accident as possible. The second method, searching for multiple causative factors,
offers more opportunity to find sources of harm that have yet to cause
injury. The techniques that are successful in revealing extra causal
factors are rather straightforward and easy to use. One of the best
ways to approach this subject is to divide the investigation into different parts: (1) the scene; (2) the people; (3) machines, equipment,
and supplies; (4) procedures; and (5) the management system.

In all cases, the investigator must maintain an open mind and avoid developing any conclusion or final determination of cause until all the facts have been examined. By doing so, the investigator avoids a serious trap—the trap of finding causes that will fit with the already known outcome.

The investigator must be systematic and attentive to detail, as well as inquisitive. Each of the foregoing five topics should be addressed thoroughly if the investigator is to find the significant causes of the accident as well as those subtle underlying causes that also contributed. Investigative work is tedious, involving systematic review of detail to answer the difficult question: How did the accident happen? The following is a review of the five topics listed.

The Scene. The investigator should get to the scene as quickly as possible. This will afford an overall snapshot impression of the accident scene, a general perspective of the layout. Each impression and idea should be noted down. Careful use of photography and/or video tape can be invaluable in investigations. This is especially true for cases in which the area cannot be sealed off for any length of time but must be cleaned up so that production can be started again. A photographic record will preserve the scene and can provide clues to causes that may not be detected using only pencil notes. Drawings from engineering, obtained later, and sketches of the layout may indicate other points of inquiry.

If possible, it is good to preserve the scene intact for a period of time. Usually this is feasible for only a few hours, except following a major catastrophe. The reason for this is to be able to make several visits to the scene to recheck and reaffirm points of inquiry.

The People. It is vital to gain as much information as possible from all the people who were witnesses to the accident or any events immediately preceding or following. An effective way to speed up the process of interviewing a group of people is to ask each to write a description of what happened. Ask not only for the details of the accident but also for a description of the events in their lives over the same day. Request that they describe events from the previous day to a point 1 hour after the accident. This is done to search for clues to the event that can be found in totally unrelated actions of others.

When a witness or participant completes a handwritten statement, read it and review it before talking with the person. Once the contents of the statement are understood, you should interview the person. The interview can consist of a line-by-line or event-by-event review of the statement. Each point in the description must be clarified and understood. This process is repeated for each person interviewed. The investigator is interested in the facts of the event and how each witness interprets them. This includes what was seen, heard, and felt. The conclusions of witnesses can be misleading, but they are useful as ideas and clues for further investigation.

Machines, Equipment, and Supplies. A machine malfunction or failure due to improper design or operator error poses a question: Why? For example, why should a truck suddenly swerve across the road and collide with an oncoming car? Why should a properly designed forklift mast fail while carrying a proper load? Malfunctions and failures can occur for a variety of reasons. Many failures can be traced to omitted maintenance checks, improper machine assembly, overloading, wrong fluid in hydraulic lines, overloaded circuits, and so on. Cases of failure and malfunction can raise complex technical questions. In some cases, experts in metallurgy, stress analysis, and other disciplines must examine the evidence and offer their opinions on the situation.

Another part of an investigation involving machines and equipment is an analysis of the maintenance records. The timing of the maintenance checks, history of malfunctions, and other such information can contribute valuable clues. In the case of supplies, it is important to know how they are received and handled. Any change can be disruptive. For instance, a change in scheduling, a change in the size of a shipping container, or a change in the composition of a cleaning material or an ingredient in a process all can be contributing factors in an accident and should be checked. In fact, any change that is identified can be a factor and should be thoroughly researched by the investigator.

Another area for inquiry concerns the biomechanical and ergonomic factors involved in the case. For example, a change in the height of a work platform on a production line from $6\frac{1}{2}$ inches to $9\frac{1}{2}$ inches can cause a tripping hazard. The new platform height is higher than a normal step. In one company, this very change resulted in a severe head injury to an operator. Another case involved a control panel that was changed. The dials on the new panel indicated a pressure rise in a manner different from what the operator normally expected to see. On the old panel, a rise in pressure caused the needle to move right, whereas on the new dial the needle moved to the left. To decrease the pressure, the control had to be moved to the left, the opposite of the situation for the old panel. The result was that the operator saw the needle move right and thought the pressure was rising; he turned the control to the right to reduce the pressure. The vessel ruptured under the pressure, and a fire consumed most of the building. During an investigation, if possible, have an experienced operator operate the equipment to demonstrate how it works. If it has malfunctioned, try to repeat the situation.

Procedures. There are many questions to ask about procedures and rules that are supposed to be followed by operators. Operations, maintenance, and general safety rules should be collected for review. In looking over such documents, these are some of the questions to ask:

1. Were the procedures up to date?
2. Were they followed, or had they been modified by operator practices?
3. Has the process changed or have the practices changed since this procedure was written?
4. In what way has the equipment or the material been changed?
5. Was this operation a special condition not covered by the procedures?
6. What other changes have taken place? Is the process still being carried out as it was when the procedure was written? Is the work still being done by the same number of people?
7. Was there a malfunction that required the use of other equipment not covered by the procedure?

All of these questions, and others that may occur during a specific investigation, can be used to generate information and uncover clues that may contribute to the development of data on causes.

The Management System. The inquiry reaches a sensitive stage when questions must be asked about the functioning of the management system in regard to the operation where the accident occurred. People tend to become quite defensive, and rightly so, when they think that their competence may be in question. To avoid antagonizing people, the investigator should avoid considerations of personalities as much as possible and look at the system itself and how it operates. A review of reporting relationships and lines of communication for management decisions can reveal problems. The layers of supervision and the ways people supervise and motivate employees should be part of the considerations. The information developed from the four preceding areas of inquiry will provide indications of where in the management system to inquire.

These five topics of inquiry will generate facts and data to reveal the many types of contributing causes in a single accident. As clues and facts are discovered, they should be traced and verified as carefully as possible. In some cases, learning exactly what happened may be difficult or impossible. But one must get as close to reality as possible; this will, in most instances, be sufficient to determine probable causes for events and to propose remedies that can prevent may future accidents. This is the real goal of injury investigation: prevention of future injuries and loss events.

Knowledge of the Investigator

The investigators in a company usually are first- or second-line supervisors. Many of these people have been making accident investigations for years. In some situations the safety professional will assist

in those cases that are most serious, but usually it is the first- or
second-line supervisor who does most of the investigative work.

Some of these supervisors may have received training in investigation and may have a rudimentary idea of what to do. Others may
have had no training. The level of interest will vary: A few will do
an excellent job in determining the circumstances of an event; others
will simply fill in the boxes on the report. The analysis of the current
investigation system, as described in Chapter 2, will reveal the level
of knowledge among the supervisors. The analysis can be used to
assess the direction that future training in investigation should take.

What Should Be Investigated?

What types of cases should be investigated? This is a question that
all of us in the safety field have had posed to us at one time or another.
Several differing viewpoints can be taken on this question. There
are those who believe that it is important to investigate and report
only those cases that the law requires. The Occupational Safety and
Health Act requires that both lost-time and medical-treatment cases
be investigated and recorded. Certainly, complying with the law and
recording those cases is a prudent decision. However, this approach,
will accumulate investigative data on only about 35 percent of the
cases that occur in a plant. To more closely approach 100 percent,
a much wider range of cases must be investigated. The reason for
investigating a greater range of accidents is to increase the amount
of causal information available for analysis. The number of cases can
be increased by investigating first-aid cases. Really minor cases,
such as paper cuts, can be excluded from investigation, because they
may not provide significant data. Other minor cases can be checked
out and investigated at the discretion of the supervisor. By adding
most of the first-aid cases, this approach will compile investigative
reports on over 90 percent of injury cases. Because we are preparing
to use a computerized data system to analyze the case information,
increasing the number of cases investigated will pay dividends. If
we were still discussing the use of a manual system, more personnel
would have to be hired to analyze the increased number of cases.

Beyond injury investigation, there is the subject of property damage. Some investigate all cases over a certain dollar figure (e.g.,
$500 or $1000). This category is interesting because there may be
no injury involved; however, there can be considerable dollar losses
connected with these cases. Certainly this loss area is worthy of
serious consideration. Most companies have no idea of the extent of
their losses from property damage. The CSDS can handle this classification of cases with ease. Their causes may contribute to revealing
the causes of other mishaps.

Occupational illness is a category that is difficult to work with, because the source of an illness often is obscure. Nevertheless, this category can provide clues to practices and operations that can cause considerable harm or even death over a long period of time. Thus, efforts to track this set of cases can provide excellent data for the prediction of sources of harm.

Prediction of areas where injuries and property damage can occur is the real goal of any safety effort. If sources of harm can be found and corrected before an injury, illness, or property damage occurs, then this goal will be achieved. One way to find potential injury sources is through the reporting of near-miss incidents. This type of investigation and reporting can be an important factor in developing a program to predict illness, injury, and property-damage sources.

The near-miss category is an area that does not receive the attention it deserves. In many instances, in a manual system, it is too much work to investigate and report this category in addition to what is regularly reported. Conversely, the more categories that are investigated, the larger will be the data base, and this is desirable when we are considering a CSD system. Thus, it follows that the larger the data base, the more efficient it is to analyze the collected information to find obscure sources of harm that will not be evident in examining a single case or small group of cases. With the advent of the CSDS, a project can be launched to gradually investigate more and more categories of events. If a plant is investigating only lost time and medical treatment, the next category to add will be significant first-aid injuries. Following that, other categories, such as property damage, can be added. What will occur is that as more cases are investigated, the more sources of harm will be identified. As a result of concentrated computer analysis, fewer injuries will occur in succeeding periods of time. The lower level of injury cases will free investigators to perform investigations on broader groups of cases.

Procedures: How Much Direction?

A person generally will resent being told exactly how to do a job but will appreciate knowing what needs to be done. Company safety procedures and safety rules are generally regarded as telling workers how to do something. In the development of the CSDS, this book has stressed participation and involvement. For some people this may seem a departure from the norm, but it is a concept that seems to work, because it recognizes that each person in the company can contribute to the safety effort.

The procedures for investigation and reporting can be and should be developed through group meetings. A consensus on what the procedures should contain can be worked out. The technique of contribution and participation is directed toward getting a higher-quality investigation and better reports. Only when those who are doing the investigating agree on what should be involved in a procedure will a really high level of quality and contribution be achieved.

INTRODUCTION AND CHANGEOVER

Presentation to Management

The information phase of the CSDS development is important in keeping supporters of the system up to date on the system. During development, several presentations can be made at different times. For example, at this point there is a parallel development process going on: One leg of development is the creation of the user's manual; the other leg is designing and testing a revised investigation form. In the first presentation, both parts of the process should be discussed. The presentation does not need to be extremely detailed, but it should show completed work. For example, a slide showing an outline of the user's manual contents and a slide depicting one or two files to show progress will be sufficient. An example of the proposed investigation form and a description of the form's testing process will also show careful management of the project's development.

Another presentation can be made on the final investigation form and the results of the form's test period. At this juncture, the CSDS development will have progressed to a completed user's manual. This presentation can describe the next steps in the CSDS development process and show that the new investigation form and procedure have been tested and are complete. At this session a brief written report can be distributed to show the development so far. Just prior to this presentation, top management should receive a briefing and an executive summary of the report.

Overall, these presentations provide a continuing information flow concerning the progress of the CSDS project. It is essential during this developmental stage that management, especially top management, know how the project is advancing toward completion. More important, this is another opportunity to reinforce their confidence in the project and what it will do for the company.

Training and the Use of the New Form

Once the test period is completed and the new investigation form is ready, it is time to introduce it to the operations groups that will use the CSDS. The introduction of a new form is like bringing forth any change in a traditional way of doing something: It is disruptive. Even though this form has been through a test period and has been accepted, its introduction to a new group will be disruptive because it is unfamiliar.

To gain the acceptance of the users and attain a high level of quality
from the beginning, two things can be done: One, provide a thorough
orientation on the new investigation form and a refresher session on
investigation techniques. Two, stretch out the introduction period
to allow the users to become accustomed to the change. The orientation program should cover several points:

1. A comparison between the current form and the new form
    showing the advantages and benefits of using the CSDS form.
2. The methods and techniques developed by the user during
    the test period.
3. Examples of the different types of cases that were investigated
    during the test.
4. The improvements in the quality of the reports that were
    attained during the test.
5. How the investigation form will contribute to the success of
    the CSDS.

All of these points in the orientation will prepare the way for
introduction of the new form. The new form should be in use before
the CSDS is put on line. The reason for this is that the new form will
develop the data needed for the system. Reviewing the form in full
use will accomplish another step in the development process.

The methods and techniques of investigation used with the new
system may differ only slightly form those of the old system. However,
many of the current investigators, usually supervisors, may not have
received investigatory training in several years, if ever. On the
other hand, these supervisors have been performing the task and
consider themselves competent. To offer them training would offend
them. With this in mind, the best approach may be to have a program
termed an orientation course. This course will introduce the new investigation form and talk about how it is to be used. Table 7-1 shows
an example of a course outline that can be used for the orientation.
The sessions should be conducted in a conference-style arrangement,
with everyone, including the course administrator, seated around the
table. Note that the leader is referred to as an administrator, not an
instructor. The distinction is important for this style of presentation.
An administrator facilitates the learning process by helping the participants help themselves. The ideal group size is about 8 to 10. This
group size will allow each person to participate.

Follow-up and Feedback for Quality

Once the orientation course is completed and the new form is in use,
there should be follow-up and coaching to continue the learning process

Table 7-1 Orientation Course Outline: Introduction of the New Investigation Form

---

Behavioral Objectives of the Session

Each participant shall understand (1), how to complete the investigation form, (2), the importance of thoroughly investigating each accident to find multiple causes, and (3), the need for promptness in reporting.

Orientation Session: Recommended group size: 8 to 10
Length of Session: 2 hours

   I. Introduction (10 minutes)
      A. Why we are here.
      B. What the session is supposed to accomplish (behavioral objectives).
      C. The organization of the session.

  II. The New Investigation Form (40 minutes)
      A. Review of the new form.
      B. Discussion of the use of each box and section of the form.
      C. Examples of completed forms.
      D. Discussion of methods to derive multiple causes for each case.

  Break - - 10 minutes

 III. Practice Cases (45 minutes)
      A. Continued discussion about how to analyze the cases for multiple causes.
      B. Completion of sample cases. Groups of two's and three's.
      C. Completion of sample cases. Individual work.

  IV. Conclusion
      A. The importance of complete investigations.
      B. Summary - Discussions: multiple causes, importance of form completion and the CSDS, time limits on reporting.

Handouts should include description of the new form, example forms, hints on investigation techniques to find multiple causes, time and reporting requirements and the benefits of the CSDS.

---

and to improve quality. As they are turned in, the safety group should check all forms for completeness and attention to thorough investigative information, especially multiple causes. If a form is incomplete or does not meet other quality standards, safety should get in touch with the originator of the form and discuss the case. This can be time-consuming. It would be simpler to send a note back with the form and ask for the information, but a note is rather impersonal. A telephone call or a visit can provide immediate results and a chance to discuss the difficulties and problems of that case.

Feedback in a one-to-one setting, either over the telephone or in person, usually pays dividends. First, it adds a contact between safety and the investigator. Second, it may reveal other safety problems. Third, it will get the form completed that day. Fourth, it can provide the extra coaching in investigation that will produce better reports in the future. Therefore, the preferred way to achieve quality, timeliness, and effectiveness in investigations is through personal contact, either a telephone call or a personal visit.

SUMMARY

The changes in the current investigation form and procedures must be made slowly and timed properly. The investigation system has been analyzed with the input of the current investigators; therefore, adopting some of their recommended changes will be welcomed. The revision to the old investigation form will not be a total changeover; therefore, listing the information required by both forms side by side will highlight for everyone the changes and the similarities between the two systems. Along with the investigation form, the other form to be developed is the input form for use with a keypunch system. This form is designed through coordination with the systems group. The video display terminal is coming in to replace keypunch cards and is a more sophisticated way to input the cases. Either way, a print-out of each case should be reviewed to verify the input.

The new investigation form, once it is designed, should be tested by a small group to ensure that it works. After 75 to 100 cases have been processed, the form should be declared final and preparations made for general introduction. This type of introduction includes an orientation about the form revision for both management and the investigators. However, the orientation for investigators should be more extensive and should include information about how to investigate for multiple causes. Following the orientation program, a long period of coaching and follow-up support will be necessary to achieve the quality of investigations needed for a successful CSDS.

8

BRINGING THE SYSTEM ON LINE

Experience is a hard teacher because she gives the test first, the lesson afterward.

Tom Dodds

INITIAL CONSIDERATIONS IN CODING AND ENTERING DATA

It is at this point in the development of the system that major problems can be created. In bringing the CSDS on line, there is a great amount of checking, rechecking, and just plain clerical work. It can be boring. Worst of all, the importance of this phase compared with the previous development phases can be underestimated. This happens because all the excitement of development is over, and all that is left is to enter the cases into the system.

Coding of old accident cases, one after another, for example, can be tedious and dull. This kind of repetitive work can cause frustration and can result in errors. Errors in coding cause what is commonly referred to in computer jargon as a GIGO problem; this stands for "garbage in garbage out." This occurs when codes are entered for a case and they are acceptable to the computer, but they are the wrong codes. Then, a lot of misinformation can be entered onto the data base because the codes were acceptable. The dominating goal for this part of the project is effective control for zero error and vigilance to ensure that every case is entered accurately.

Keeping Management Informed of Progress

Sometimes it seems that keeping management informed of what is happening in this phase of the operation is a waste of time. The last briefing brought them up to date about the revised investigation report

form and the user's manual. All that is going to happen here is the
clerical job of coding and entering the cases. This work is not exciting, nor is it stimulating when described. However, there can
be concerns on the part of management about how well this part of
the job is progressing.

A timely presentation to management about the mechanics of coding
as well as the controls that are being used to assure a high level of
quality in the data base can be a confidence builder. Therefore, a
short briefing about this phase of the project will again resell management on the CSDS project and keep them up to date on progress.

CREATING A DATA BASE

Input of Cases

There are several decisions to be made when it is time to enter cases
into the system. First, are cases from previous months or years to
be used and entered? If so, how far back in time will cases be used?
Second, on the initial input, what categories of cases will be used?
Will only lost-time cases be used? Will medical-treatment cases also
be included? Will all cases in the file for the time period selected
(assuming you have decided to input previous cases) be entered?
Third, so that management will know, what is the estimate of how
long it will take to input the past records? Fourth, what methods will
be considered to expedite the case entry process? Fifth, will cost
data be used or developed for the older records? If so, how? These
are tough problems that if considered carefully can be set up to
provide the best of compromises.

Obviously, it is advantageous to initially code and enter as many
cases from previous months and years as possible. This provides an
instant file of data and a good beginning for the new data base. As
the base builds, the statistical validity will increase because of the
larger number of cases on file. The disadvantage is that the old
cases may not be complete and accurate, and those who investigated
them may no longer be with the company. For example, in the pharmaceutical company, about 650 injury cases per year were generated.
They included lost-time, medical-treatment, and first-aid cases. Up
until the advent of the CSDS, only a few property-damage cases had
been investigated and reported. From a review of the case files it
became quite clear that records going back 3 years might be useful.
That meant that about 1950 case records had the potential to be coded
and entered into the system, providing an excellent start on the CSDS
data base. However, there were the disadvantages just mentioned.
Cost data were not available for the most part. About 7 percent of
the cases were incomplete. Overall, however, most of the cases could
be used, even without their cost data.

Coding past cases appears on the surface to be a straightforward clerical job of transferring information from the investigation form to the computer input form. For some cases this is true. However, there are incomplete cases and cases, though complete in information, that are questionable for use. Incomplete information can sometimes be filled in if the case is not too old, but if the supervisor and the injured party have left the company, the retrieval job may be impossible. Using an incomplete case is usually acceptable if the minimum information is there for the computer. Table 8-1 shows a partial list of rules that were used in the pharmaceutical company system for declaring a case acceptable to the system. The entire set of rules accompanies the pharmaceutical company system presented in Appendix B.

The next consideration is to estimate how long it will take to code the past cases. To do this, a test input of several cases can be timed to learn how long it takes to code a single case. There is another variable to account for: rechecking cases for errors prior to approving them for input. These two factors will give an approximation of the extent of the coding job. Another factor is that the safety office still has to function. During the input period, safety will still be expected to carry out its regular duties. Thus, the time estimate must include doing the regular job and coding cases for entry.

This estimating problem was solved in one company in the following manner. Special 1-hour coding sessions were scheduled each week. Each safety professional was assigned a file of past cases to code. Each case file contained a calendar month. New cases from the departments that were the regular responsibility of a certain safety staff member were also referred to that person. After about 100 cases had been coded, rechecked, and approved for entry, a coding time per case was determined: approximately 10 minutes per case. Once the coding time was determined, estimating how long the coding would take was simple. Another part of the problem was the number of coders, the average number of coding sessions per week, and the number of past cases. In this company there were about 1800 past cases to be coded. This number multiplied by 10 minutes was 18,000 minutes or 300 hours straight time. There was an assumption made here that each person on the staff could code about five to six cases in a 1-hour coding session. There were three 1-hour coding sessions scheduled each week. This gave a total capability of about 60 to 72 cases that could be coded each week by a staff of four. Thus, it would have required from 25 to 30 weeks to code the past cases. Another factor to consider was that there were, on the average, 13 new cases occurring each week that also had to be coded. Therefore, in the 25 to 30 weeks estimated for coding past cases, there would be 325 to 390 additional cases to be coded. Therefore, it would

Table 8-1 Pharmaceutical Company Input Error Rules

I. *General*

   A. Case Number
      1. Must be numeric
      2. If transaction code 'A' must be unique
      3. If transaction code 'C' or 'D' must be on file

   B. Transaction Code: Must be 'A,' 'C,' or 'D'
   C. Card: Must be 1–5

II. *Card 1*

   A. Division/Corporate
      1. Must be numeric or blank
      2. Must be on Table #2 (personal DIV)

   B. Organizational Unit
      1. Must be numeric or blank
      2. Must be on table ORG (personnel)

   C. Expense Center Code
      1. Must be numeric or blank
      2. Must be on table EXP (personnel)

   D. Location of Incident
      1. Must be numeric
      2. Must be on Table #5 (personnel)

   E. Date of Incident—will validate
      1. Must be numeric
         a. Day: Cannot exceed for month
         b. Month: Must be 1–12
         c. Year: If leap year can have 0229
      2. Cannot be GT current date

   F. Time of Incident
      1. Must be numeric
      2. Must not be greater than 2400

   G. A/R
      1. Must be 'A,' 'R,' or blank

   H. Property Cost
      1. First position must be 'E' or 'A'
      2. Following positions must be numeric

   I. Medical Cost: Same as II, H
   J. Compensation Cost: Same as II, H

Source: Schering-Plough Corp., Kenilworth, New Jersey.

have required an additional 4 to 5 weeks to complete the coding. The
raw estimate of the time required to code the past cases and keep up
with the new cases was 29 to 35 weeks. There was still another factor
that had to be added for vacations, holidays, and sickness. A 15
percent factor probably was close for estimating. The realistic estimate of the time it would take to code the new and old cases was
about 33 to 40 weeks. At the end of that time, the 1800 past cases
plus the 325 to 390 new cases would have been coded and entered for
a total of 2125 to 2190 cases on file in the data base. Note that this
was in addition to the 100 past cases that were used for the test. The
final total was then between 2225 and 2290 cases.

Over half a year to complete the coding and bring the system
on line is a long time for management to wait for a system they have
heard so much about for so long. It would be better to complete
this part of the project in a shorter period of time. One option is
to assign more than three coding sessions per week. Another method
is to assign one or two staff professionals to do the coding job full
time. Both of these methods can cause disruptions in the running of
the safety department. In turn, these disruptions can indicate to
management that the safety department cannot handle such a large
project. They expect the safety department to provide their normal
service *and* complete the project efficiently.

The example company safety department tried both of these methods.
They found that each method hampered the normal work quite severely.
After a brainstorming session, an alternative was found. A temporary
clerk was hired to code just those parts of each case that were easy
to identify and did not require specific knowledge: name, date of
occurrence, location (if obvious), incident date, time, employee name,
corporation/division, supervisor's name (if known), and employee
numbers for both. This method proved to be advantageous for several reasons: One, the clerk would work with the secretary, who
was keeping track of the case files as they were removed for coding.
Two, the clerk could do the initial coding and checking quite rapidly
and thus was able to find the problem cases ahead of time. It reduced the safety professional's coding time to 3 to 4 minutes per case
This meant that twice as many cases could be coded in a session by
the safety professional. Three, the clerk worked on the cases full
time and coded 3 years of past cases in 6 weeks. The overall coding
operation was cut to 13 weeks. This allowed the system to be brought
on line in half the original estimated time.

For the example company, one knotty problem remained. This was
the decision whether or not to develop cost data for past cases.
There are two choices in this regard: (1) to input the past cases with
cost data contained on some of the cases, or (2) to make an estimate
of the costs and include it for as many cases as possible. In either
situation, there would be some degree of loss either in cost data for
the base or in accuracy.

The safety department decided on the second option: to include cost data for as many of the past cases as possible. Input of the cost data was done after the case was initially entered into the system. The way this was accomplished was by dividing the cases into categories. First-aid cases were estimated using the formula described in Figure 8-1.

The cost analyses for the medical-treatment and lost-time cases were addressed on an individual-case basis. There were about 26 lost-time cases and some 75 medical-treatment cases each year for the 3 years. This was a total of about 225 medical-treatment cases and 78 lost-time cases. A cost-analysis worksheet (such as the one shown in Exhibit 8-1) was used to estimate the costs for these cases. In most instances the medical costs could be retrieved from the medical records, which were kept in the nurse's office. The compensation costs were provided by the insurance department, and the other costs were obtained from those in management who had participated in the cases. It was a slow process, but the work paid off. Fairly accurate costs were generated for the previous 3 years. This gave the new CSD system a good cost base on which to enter even more accurate costs from current cases.

By this pair of methods, almost all cases in the data base contained cost information. Accuracy for the cases of the previous 3 years was estimated at 85 to 90 percent. Thus, it was believed that this was an excellent level, considering that much of the information was estimated. The accuracy goal for the new cases that were to be entered into the system was 90 to 95 percent. This would preserve and continually improve the overall quality of the data base.

Training the Coders

Coding cases for input into the system is a technique that can be taught in a formal way; however, learning by doing seems to accomplish the task more expeditiously. The group coding sessions are excellent settings for a new person to learn how to code cases. Many staff members are present during the session; thus, there are many sources of help and advice.

To provide continuity, one person usually acts as the instructor for the new coder. The first step in the orientation is a review of the user's manual. The manual contains the input codes, an explanation of terms and definitions of words, a list of rules for coding, and a code form with various points to remember. Once the manual has been reviewed, the instructor codes two or three cases. As each case is coded, the instructor explains the process, discusses the use of certain codes, and answers questions. Following this, the new coder begins coding cases while being coached by the instructor. As the new coder learns and gains more confidence, the instructor coaches less and less. They continue to work side by side for the entire session.

The first step is to estimate approximately how many cases it will take to achieve an average first-aid case cost within one-half of a standard deviation. This is done as follows:

Deviation from standard $= \dfrac{\bar{\sigma}}{\sqrt{n}}$

$$\bar{\sigma} = \sqrt{\dfrac{\Sigma \underline{x}^2 - \Sigma \underline{x}\bar{x}}{n - 1}}$$

$\bar{\sigma}$ = sample population
$\underline{n}$ = number of first aid cases in sample
$\underline{x}$ = cost for an individual first-aid case
$\bar{x}$ = average cost (estimate) for first-aid cases
Σ = sum of

As an example:

$\underline{x}$ = \$75.00 (randomly picked case)

$\bar{x}$ = \$65.00

$\underline{n}$ = 50 cases are going to be used for this test

$$\bar{\sigma} = \sqrt{\dfrac{\Sigma 75^2 - \Sigma 75 \times 65}{50 - 1}}$$

$$\bar{\sigma} = \sqrt{\dfrac{5625 - 4875}{49}}$$

$$\bar{\sigma} = \sqrt{\dfrac{750}{49}}$$

$$\bar{\sigma} = \sqrt{15.31}$$

$$\bar{\sigma} = 3.91$$

Dev. from Std. $= \dfrac{\bar{\sigma}}{\sqrt{n}} = \dfrac{3.91}{\sqrt{50}} = \dfrac{3.91}{7.07} = 0.55$

Based on these calculations and estimated first-aid case costs, using 50 first aid cases and averaging their costs will produce an average case cost that will be accurate within about one-half (0.55) of a standard deviation from the actual mean for first-aid cases. Thus, a single group of cases can be investigated to arrive at an average cost and this cost could then be used for the next twelve months.

The average cost can also be used as a set number to enter all past first-aid cases. This could also be adjusted for inflation if it was felt it was a significant factor during that time period.

Figure 8-1 Estimating first-aid costs. (Adapted from John V. Grimaldi and Rollin H. Simonds, *Safety Management,* 2nd ed. Homewood, Ill.: Richard D. Irwin, 1963.)

COST-ANALYSIS WORKSHEET

Complete those parts of the analysis that apply to this case.
Use the worksheet as a tool to develop a good cost analysis.

A. Property Damage Cost

Parts replacement $______   Finished goods damaged $_______

New equipment $______

Outside labor $______   Other property damaged _______

                                                          $______

                            Total Property Cost $______

B. Medical Cost

Nurse treatment time _____hrs x hr rate $_____=$_____

Physician treatment time_____hrs x hr rate $_____=$_____

Supplies, medicines used $_____(est) Other costs not listed____

                                                          $______

                            Total Medical Cost $______

C. Compensation Cost

                            Total Compensation Cost $______

D. Other Costs

Employee time ____hrs x hr rate $____=$____

  Other employees x time____hrs=____hrs x hr rate $____=$____

Supervisor investigation ____hrs x hr rate $____=$____

Manager review time____hrs x hr rate $____=$____

Safety committee review time____hrs x____members =____hrs x hr

  rate $____=$_____

Upper mgmt (Director/V Pres and above) review time ____hrs x hr

  rate $____=$____

Other costs not listed____

                                                          $________

                            Total Other Costs $________

Exhibit 8-1

If a particular problem in coding a case is found by anyone during a session, it is discussed by the group, and a consensus is reached concerning how to handle a like case in the future. Each person notes the pertinent information about this decision in the user's manual. In this way, there is continuity and a good level of standardization maintained in the input of cases. The new coder can also feel free to discuss a problem case and receive a consensus on how to code that case.

Errors and Rechecking: Avoiding GIGO

Mistakes in coding can cause problems in two ways. First, if the error is in a code that is not acceptable to the computer, the case will be rejected and returned for proper coding. This causes a delay in getting the case recoded and reentered on the system. The second problem, which can be more serious, is an error in coding that is acceptable to the computer by the input rules. This mistake is actually an inaccuracy in coding. An example: coding a case as a back strain when it should have been coded as a fall, or coding a fracture as a bruise. Another would be using the code for slippery surface when the correct code should be shielding less than adequate. It is this type of mistake that leads to GIGO or garbage in garbage out. The error is an inaccuracy in coding, but it gets into the system because it is a valid code. This kind of error can have a cumulative effect. The codes stay in the system because they are valid codes. However, improperly coded cases will provide misleading information in the reports processed and sent to management. If enough errors are produced in this fashion and this is discovered by management, trust and confidence in the information the system produces can be eroded.

The efficient course of action is to catch these two types of errors before they are entered into the system. If all coding errors are eliminated prior to entry, only accurate information will be in the system. This is the optimum condition. A 90 to 95 percent accuracy factor is the minimum acceptable. The technique to use to catch the mistakes is a checking system. By checking each case before it goes to data processing, most, if not all, of the obvious errors will be picked up and corrected.

For example, we discussed earlier how a clerk was used to code identification information on cases. She passed her cases to a safety professional to complete the coding. The safety professional then completed coding the case by entering the safety analysis information. The completed case was then passed to another safety professional, who rechecked the entire form for accuracy. If there was a question about why a certain code was used, then the two safety professionals got together. They discussed why a certain code was used and reached

an agreement as to how that piece of data should be coded. In some
cases this discussion revealed the need for a new code or for everyone to make a note about a certain set of circumstances.

Checking and rechecking each case will provide the filters to
screen out almost all errors before the cases enter the CSD system.
One last review of each case is made when the case is entered into
the system and a printout is made listing the codes. The printout
of the codes is checked against the original input form: first, to
ensure that data processing did in fact enter all the codes correctly;
second, to ensure that all the codes were correct and describe the
case accurately. These checks and rechecks provide the expected
level of accuracy for the system. Accuracy is critical in maintaining
management's confidence in the output reports. Each case must be
right.

REPORTS

Trials with Regular Reports

After there are sufficient cases entered onto the data base, the regular computer-generated report sequence can be started on a trial
basis. For the regular reports, it is sufficient that the cases for the
previous year and the current year to date be entered on the data
base and be acceptable to the system. At this point, a trial monthly
report can be produced. It is important to run trial reports for a
month or two to determine if the system will consistently produce
good, accurate reports. We might mention here that these trials with
actual data will follow the tests by the data systems people for de-
bugging of the program and other trials to ensure that the system
will operate correctly.

The following examples show the kinds of reports that can be
generated using the system and user's manual in Appendix B. The
incident/injury report in Exhibit 8-2 lists the cases for the month
of June, while the summary in Exhibit 8-3 provides a ranking of the
cases. This summary tells how many cases of each kind there were,
as well as the costs involved, and it indicates the most frequent
causes. In the three cause categories, the system lists those that
appear in the case report more than once. The limit is that the system will list only the top five causes among those; thus the notation
"Up to 5/2 +." The other parameters are frequency and severity.
The cases that show up here indicate where there may be an immediate
problem and this is another way to determine priorities for the work
of correction. The final report (Exhibit 8-4) is the log of occupational injuries and illnesses (OSHA form 200) that is printed for the
year.

07/05/XX

Packaging 8329

INCIDENT/INJURY REPORT

06/01/XX Through 06/30/XX

Page 4

Report No. 5287

Case No.	Type of Case	Inc. Date	Time	Days Lost	Case Cost	Freq./ Sev. Potential	Human Factors	Condition Factors	Root Cause Factors
00023	Temp. Extr.	6/7/XX	0840	0 A	$114 A	Occas./ Minor	Wrong Place Expr. LTA	Hi Heat/Cold Shield LTA	Instr. LTA Man. Cntl. LTA
00026	Slip	6/15/XX	1000	20 A	$1375 E	Occas./ Serious	D/N Recog. F/T Act.	Light LTA Rough Sur.	Resp. Unclear
00029	Fall DL	6/19/XX	2135	35 A	$3780 E	Rare/ Serious	Body Pos. LTA D/N Warn	Guard LTA Miss Equip.	D/N Corr. Rsk. OJT LTA
00030	Fall DL	6/19/XX	2135	14 E	$2560 E	Rare/ Serious	Body Pos. LTA D/N Get Help	Tools LTA Method LTA	E. Tng. LTA Kn. Crit. Inc.
00028	Strk Against	6/20/XX	1330	0 A	$114 A	Occas./ Minor	Expr. LTA Wrong Seq.	Piling LTA Hsekpg. LTA	Instr. LTA
00041	Strk By	6/26/XX	1750	0 A	$114 A	Occas./ Minor	Equip. Use LTA Omitted PPE	Guard R. Placement LTA	Coord. LTA D/N. Rd. Rsk.
00068	Fall SL	6/28/XX	1150	0 A	$485 E	Rare/ Serious	D/N Use D/N Do Co. Pol.	Steep Sur.	ID Rsk. N. Act.
00039	Strk Against	6/28/XX	1345	0 A	$114 A	Rare/ Minor	Equip. Use LTA	Tools LTA	OJT LTA
00040	Over-Exert. Strk Against	6/28/XX	1945	0 E	$275 E	Rare/ Minor	D/N. Recog. Knowledge LTA	Too Bulky Placement LTA	E. Tng. LTA Instr. LTA

Exhibit 8-2

```
Division                      INCIDENT/INJURY SUMMARY        Page 5
Organization:  Packaging      06/01/XX Through 06/30/XX       Report No. 5286
Expense Center:  8329
```

Cases by Type			Case Totals	Days	Costs	Causal Factors	
Strk. Against	(3)	4 - First Aid		0 A	$456 A	Human Factors - Up to 5/2+	
Fall DL	(2)	2 - Medical Tmt.		O A	760 E	Expr. LTA	- 2
Temp. Extr.	(1)	3 - Lost Time		69 E	$7715 E	D/N Recog.	- 2
Slip	(1)	0 - Occ. Illness		O A	0 A	Body Pos. LTA	- 2
Fall SL	(1)	0 - Property Damage		0 A	0 A	Equip. Use LTA	- 2
Over-Exert.	(1)	0 - Near Miss		0 A	0 A		
Strk. By	(1)	9 - Total Injury Cases	69		$8931 E	Condition Factors - Up to 5/2+	
		9 - Total Cases		N/A	$8931 E		
						Tools LTA	- 2
						Placement LTA	- 2

Frequency/Severity Case #'s

```
Freq./Maj.     0
Freq./Severe   0
Occ./Maj.      0
Occ./Severe    00026
Rare/Maj.      0
Rare/Severe    00029,00030, 00068
```

Root Cause Factors - Up to 5/2+

Instr. LTA	- 3
OJT LTA	- 2
Tng. LTA	- 2

Exhibit 8-3

01/10/XX

LOG OF OCCUPATIONAL
INJURIES AND ILLNESS (OSHA NO. 200)

01/01/XX Through 12/01/XX

Page 2
Report No. 5288

Case No.	Inc. Date	Init	Name	Occupation	Department	Injury/ Illness	Injury			Illness			
							Fat. Date	DY AW	DY RS	Type of Illness	Fat. Date	DY AW	DY RS
00534	5/16/XX	P	Freeman	Incom Insptor	Inspctn Control	Avulsion		0	0				
00538	5/18/XX	J	Polonsky	Pipefitter 2	Plnt Maintence	Burns T		0	0				
00204	5/22/XX	M	Dudas	Maint Mach 2	Plnt Maintence	Strain Bk		4	20				
00548	5/22/XX	D	Ross	Ster Lab Oper 1	Sterile Prodcts	Burns T		0	0				
00935	6/19/XX	S	Cooley	Electrician 2	Plnt Maintence	Contusion		2	0				
00134	6/22/XX	C	Jefferson	Pkg Line Svc 2	Packaging Oper	Strain BK		0	0				
00072	7/ 7/XX	MYR	Kreitzburg	Ster Lab Oper 1	Sterile Prodcts	Contusion		0	0				
01021	7/21/XX	W	Burton	Sr Chem Oper	Chem Manuftrng	Sprain		0	0				
00061	7/22/XX	J	Sullivani	Stat Engineer	Plnt Power	Cut		4	0				
00059	7/26/XX	F	Alcarro	Laboratory Tech	Analytical Con	Burns C		0	0				
00075	8/ 3/XX	M	Acosta		Analytical Con	Internal Inj		0	0				
00081	8/ 5/XX	G	Brown		Plnt Maintence	Contusion		0	0				
00182	11/ 6/XX	G	Brown	Ref Ac Mech 2	Plnt Maintence	Sprain		0	0				
00142	11/11/XX	R	Buchanan	Pkg Mech Fore	Packaging Oper	Contusion		0	0				
00567	11/17/XX	A	Moffat	Electrician 2	Plnt Maintence	Contusion		0	0				

Exhibit 8-4

The "More-Data" Report

This is a very handy report, because it serves as a reminder. The
more-data report lists those cases that were entered into the system
with incomplete information. When a case occurs toward the end of
the month, there usually is not enough information from the investigation to make a full report and code the case completely. All that
is necessary for the case to be entered is to make entries in the code
categories as indicated in Table 8-2. When the investigation of one
of these cases is completed, a change form can be initiated and the
case brought up to date. However, suppose that a case is overlooked
and is left incomplete. Each month the case will appear on the more-
data report (Exhibit 8-5). Thus, it is a reminder to check on the
investigation report and enter a change form to complete the case.

Another use for this report is in the cost development area. Often it is several months before final accurate cost figures can be
provided, especially in the case of workers compensation expenses.
The case can be entered into the system with good estimates and left
coded as "E" in the compensation area. The more-data report will
continue to list the case as incomplete until the code is changed to
"A," for actual.

Follow-up Reports

Probably one of the toughest parts of a safety professional's job is
follow-up. It is a never-ending task to coordinate with managers
to see that recommendations have been carried out. When it is done
manually, it usually cannot be done well. One of the features that
can be included in the CSDS is a follow-up system. The pharmaceutical company included a follow-up feature in its system.

The follow-up feature in the pharmaceutical company system (see
Appendix B) works quite well for all kinds of recommendations. The
input form lists two items: total recommendations and recommendations outstanding. During the coding of an injury case, for example,
the coder will enter the number of total recommendations in the appropriate space on the form. Following that, in the next box, the coder
will record the number still outstanding. In the instance of an insurance inspection, it will be coded as an insurance inspection under
the type of case. The date, time, and location parts of the form will
be coded, and the total recommendations and the recommendations
outstanding will be entered. The computer is programmed to pick
up all cases for which the recommendations outstanding are greater
than zero. All of the cases are listed on a monthly follow-up report.
The report lists case number, type of case, total recommendations,
and those still outstanding. The system is designed to produce this
report by department for easy reference. Exhibit 8-6 shows a sample

Table 8-2 Coding Requirements For Entering Incomplete Cases

Files Needing Complete Coding	Files Needing Additional Information Can Be Coded More Data	Files That May Not Apply To A Case Can Be Coded No Factor
Location	Cost	Model Year
Incident Date	Phase Of Operation	Batch Number
Time	Last Inspection	Work Order Number
Type of Case	Weather	Fatality Date
Type of Property	Specific Operation	Nature Injury/Ilness
Property Ownership	Days Off	Part of Body
Recommendations Outstanding*	Days Restricted Activity	Injury Severity
Employment Status	Work Relation	
Frequency Potential	Human Factors	
Severity Potential	Condition Factors	
	Root Cause Factors	

*Could be entered as O then changed

01/10/XX
Corporation
Division

CASES REQUIRING MORE DATA

01/01/XX Through 01/08/XX

Page 9
Report No. 5290

CASE NO.	INC. DATE	LAST NAME	INIT	PRP CST	MED CST	CMP CST	OTH CST	OPR PHS	LST INS	WTH CON	SPC OPR	SPC CHP	CNT	DYS LST	RST ACT	REL WRK	HMN FTR	CON FTR	RT CS
00610	5/03/XX	Moscaritola	A		XX	XX	XX												
00556	5/10/XX	Culver	H		XX	XX	XX												
00558	5/11/XX	Jackson	R		XX	XX	XX												
00605	5/15/XX	Dowd	A		XX	XX	XX												
00606	5/18/XX	Carabello	D		XX	XX	XX												
00552	5/22/XX	Clutter	J		XX	XX	XX												
00607	5/25/XX	Pulice	A		XX	XX	XX												
00609	5/30/XX	Cusack	A		XX	XX	XX												
00608	5/31/XX	Samer	G		XX	XX	XX												
00035	6/07/XX	Whitehead	C		XX		XX												
00037	6/08/XX	Green	H		XX		XX												
00114	6/12/XX	Grimaldi	C		XX		XX												
00115	6/13/XX	Poggioli	M		XX		XX												
00117	5/23/XX	Singelton	R		XX		XX												
00113	6/26/XX	Raimo	F		XX		XX												
00045	7/18/XX	Roth	I		XX	XX	XX												
00043	7/20/XX	CRONIN	B		XX	XX	XX					XX							
00042	7/24/XX	Anderson	W		XX	XX	XX												

Exhibit 8-5

07/05/XX
Organization: Packaging
Expense Center: 8329

FOLLOW-UP REPORT

06/01/XX Through 06/30/XX

Page 6
Report Number 5290

Case No.	Type of Case	Inc. Date	Recommendations	
			Total	Outstanding
00023	Temp. Extr.	6/07/XX	2	1
00026	Slip	6/15/XX	4	4
00024	Fire Carrier	6/16/XX	3	2
00029	Fall DL	6/19/XX	5	3
00030	Fall DL	6/19/XX	5	3
00028	Strk Against	6/20/XX	1	1
00068	Fall SL	6/28/XX	4	1
00072	Govt. Agency	6/29/XX	3	3
		Totals	27	18

Exhibit 8-6

report. Note that the injury report's case numbers appear again,
and the outstanding recommendations are given. Note that case number 00024 has fire carrier as the type of case. Also, case 00072 is
identified as government agency. These two cases concern inspections that occurred during the month in that department. When requested, the system will list the inspections that occurred in the
entire plant for a certain period. This would be a special search.
As a special search, follow-up information can be listed by type of
case, by expense center, and by department. It is a very effective
way of tracking recommendations.

This follow-up system serves as a reminder for the safety staff
and for the department. Both get copies of their recommendations.
This is especially helpful in tracking insurance company recommendations. In that type of inspection, many parts of the plant are usually
involved. Having a computer report and knowing that last month
there were 27 recommendations in packaging and that of that number
18 are still outstanding and need follow-up is a big help in two ways:
One, it indicates what department needs follow-up help. Two, the
report can be discussed with upper management to illustrate the volume of follow-up activity in which the safety department is involved.

Still, even with the report there is the manual work of getting
the file on the inspection, injury, or property-damage case, reviewing
the recommendations, and calling or visiting the department to check
on progress in clearing the recommendation. However, the department
manager is receiving a follow-up report each month, and this can
produce action, because the manager knows that upper management
will also see the report.

Updating the report when recommendations are completed requires
only the initiation of a change form and the entering of the new figures for recommendations outstanding. The next month's report will
reflect the progress. The CSD system serves as an aid in making
follow-up a regular routine. It is a manageable part of the job. It
adds a little more efficiency to the organization by tracking outstanding
recommendations until they are removed from the system.

In the development of the CSDS for the pharmaceutical company,
it was contemplated that the system would generate not only the number of recommendations outstanding but also a list of the subjects of
the recommendations. An attempt was made to develop a file that
would list the various subjects of recommendations. This was an excellent idea, but there were so many different recommendations from
so many different sources that the file became so large that it was
unmanageable. They considered an attempt to summarize or shorten
the subjects to reduce the size of the file. This process would have
worked, but still the department and the safety professional would
have had to review the case file to see exactly what the recommendation
said. Also, considering that a case could have had as many as 100

recommendations, as with a fire carrier report, for example, the number of case inputs would have been enormous. It was decided that the benefit would not be worth the effort. Therefore, for the pharmaceutical company's system, the report on recommendations was kept at straight numbers.

Special Search Reports

One feature that is part of the basic design of the CSDS is the special report capability. This feature turns the regular monthly safety report from just a single report into a springboard document. For example, when the monthly report is received, the user, rather than simply reviewing and filing the report, can request more information and a special report on a particular question. The regular report pertains only to the current month, current year, and previous year. The user may want to search the entire data base for information about a specific point. For example, the user may want to know about the incidence of a particular type of injury on a certain machine or process. The user may need to know the most prevalent human factor involved in injuries in a specific operation. The special report can do these things and more. The following chapters will describe in detail the methods used in setting up special searches.

A CSD system provides design for this special search feature by including it in the original program. The system uses random-access methods for searching the data base. This means that the system will allow access at any point in the file without sorting through the entire file.

SUMMARY

After the excitement of the development of the user's manual, the investigation form, and other details in creating the framework for the CSD system, it is a letdown to perform the clerical job of checking and rechecking cases for entry into the system. This is perhaps the most critical phase of the work. If errors arc madc in coding that are acceptable to the computer, there will be incorrect information on the data base. If enough inaccuracies are identified, management confidence in the system will be seriously eroded. Thus, methods must be devised to check and recheck each case to eliminate as many errors as possible.

For new cases, this procedure of checking is relatively simple. However, in entering past cases, there are problems of incomplete records, investigators no longer with the company, and other inaccuracies that can cause headaches when trying to code such a case. It is an advantage to enter past cases, because this increases the size

of the data base and provides almost instant searching capability.
One method to reduce errors and to check omissions in past cases
is as follows: One, have a clerk code all the identity facts for a case.
Two, a safety professional will check those codes and enter codes
concerning the safety analysis of the case. Three, a second safety
professional will review and recheck the case. It is at this point
that the case is entered into the CSD system. When a printout of
the case is received from data processing, the codes will again be
checked for accuracy.

To keep management informed of progress and to provide a project completion date, an estimate of the time it will take to code the
cases must be made. An accurate estimate will indicate if the coding
and entry period is too long and must be shortened.

Including cost information from past cases is another consideration. If it is decided to include such information, the next step is
to obtain reliable information based on actual cost analysis of serious
cases and an estimate of the costs of first-aid cases. Training coders
to enter cases is best performed on the job during regular coding
sessions; the new coder can ask questions, observe others, and receive
coaching in coding cases. Once the cases have been entered in sufficient quantity, trial reports can be generated. To allow incomplete
cases to be entered, as well as estimated costs, a more-data report
is printed monthly to remind safety that there are cases in the system
that are incomplete. Another reminder feature of the system is the
follow-up report. This report lists the numbers of recommendations
made and those still outstanding in each department.

The critical feature in any CSD system is quality control. Without this feature, management confidence in the information will dissi-
pate through the gradual discovery of system inaccuracies. The key
to the system's longevity is a high level of accuracy and quality.

PART III

SYSTEM INTRODUCTION AND APPLICATIONS

9

INTRODUCING THE SYSTEM TO MANAGEMENT

Men are not against you; they are merely for themselves.

Gene Fowler

CHANGE: THE CONTINUING BEGINNING

Everything changes; nothing remains as it was. Change is a constant
in our lives, and it is unsettling. Change means beginning something
new. It breeds fear of the unknown, the unfamiliar. Past habits
and familiar practices must be adjusted and altered to accommodate the
new way of doing things. Change requires extra energy and extra
thought to do something in a different way. A change is a beginning
of something different and an end of what is familiar.

The introduction of a CSDS to company operations, whether it
is for the entire company or for just one facility, means a change.
Even though the CSDS project has been discussed, reviewed, and
talked about for months, its impact will not really be experienced
until it happens. When the CSDS is finally introduced to the users,
it is new and real, and it means changes in the ways they do their
jobs. The introduction of the change must be performed carefully;
otherwise it will not produce the desired results and may be less
successful than expected.

Opposition to Change

When anything that is new necessitates changes in the traditional way
of doing something, there will always be some opposition. The strength
of the opposition often is directly proportional to the level of knowledge about the change. This comes from the reality that people use
their imaginations to fill the gaps between the facts. The outcome of

these imaginings is heightened fear and anxiety about how the change
will affect their work groups, themselves individually, and their
specific jobs.

In a business operation there can be many forms of opposition.
Many of these appear to be variations on a theme that has two main
forms. One form derives from inadequate knowledge combined with
imagined adverse consequences of the change. The other kind of
opposition also derives from fear, but fear of loss of power or control. Both of these conditions feed on rumor and hearsay, which
become converted to statements that sound factual. This kind of
misinformation can cause a ripple effect among the group who will be
involved in the change. The net effect is a higher level of anxiety
and resistance.[1]

Why New Ideas Are Rejected

When an idea, such as the CSDS, is rejected, this produces an uncomfortable situation. The idea may be innovative and potentially
quite useful to the organization; nevertheless, such good ideas sometimes never get farther than the proposal or discussion stage. They
simply die. Why?

There are several reasons. No case or situation is clear-cut or
simple. Some ideas die or are rejected because of other competing
ideas. Others are not successfully implemented because of business
conditions. Others are not accepted because of the personal preju-
dice of someone in power. Others are stopped for a combination of
these conditions.

There is another reason why new ideas can be rejected. This
reason originates with the person who is proposing the idea. It, too,
comes in several forms: First, overconfidence that the idea is so
good that once it is heard it will be immediately accepted; this can
lead to underpreparation of the proposal and poor delivery. The
second variant is underestimation of the opposition. The third is
poor timing for introduction of the proposal. One of these, or a combination, can bring about rejection just as surely as can a determined
opposition.[2]

A Bridge to Acceptance

At the beginning of this chapter we discussed change, opposition to
the new, and reasons for rejection of ideas. This, may seem to be
a digression from the main theme: the selling of a completed computer
system to management. In earlier chapters we also discussed opposition to change and new ideas. Why this renewed emphasis? Is it not
too late at this stage for strong opposition to stop the introduction of

the CSDS? Yes, it is. But it is not too late for failure because of
poor results and reduced usage.

As mentioned at the beginning of this chapter, the full impact
of a change is not realized until the change actually occurs. This
means that all the careful preparation to sell the proposal and keep
the project alive through the development stages will have been a worth-
less exercise if the users fail to accept and use the system when it
is introduced.

The way to promote interest and confidence in the CSDS is through
detailed preparation for its introduction. This means acknowledging
that the two types of opposition still exist and still have to be addressed. Opposition arising from fear of the unknown, from anxiety
and uneasiness about the change to the CSDS, can be dealt with
by providing convincing data to illustrate its advantages. A good
information program has been suggested for use throughout the project. At this point, the information flow should continue and intensify.
A strong presentation of the benefits to be derived will help each
work group toward understanding and acceptance. Repeating again
how the system will serve them, how it works and what will be expected
of each user, can close the gap between fact and fiction.

The second type of opposition also derives from fear, but a different fear. It does not come from lack of knowledge but from fear of
the consequences of loss of power. This fear of loss of power can
take many forms: apprehension among top management because another department is gaining access to the computer system; fear that
the CSDS will show up past mistakes and management errors; concern
that the system will lead subordinate managers to focus on safety
problems to the neglect of production goals.

To individuals who feel this way, a new user of the company's
computer system is a real threat. It is important not to ignore the
potential danger in these situations. Some of these fears and concerns
can be overcome, and others can be diminished, but some will remain.
A way to reduce the fear of loss of power is to show how the system
can contribute to a manager's effectiveness and efficiency. This can
best be accomplished by one-to-one meetings with key managers who
have the capacity to disrupt CSDS implementation in their operations.

A well-prepared list of personalized benefits to be gained from
the system can contribute to frank discussions with uneasy managers.
Because an analysis of the power and authority structures will give
good information about each manager and about the functioning of
each department, the discussion of the CSDS can be personalized to
emphasize specific features. For example, the discussion can emphasize the special search capability to identify costs and causes, from
which specific improvement approaches can be chosen. In another
context there can be discussion about how the CSDS reports can be
used with other reports for better overall decision making. Thus, a

well-prepared, personalized approach to those who may oppose the
system can turn a negative position into one of cooperation during
the implementation phase.

If this sounds like a repeat of the same type of information process
discussed several chapters earlier, it is because we must emphasize
the ongoing nature of the process of keeping everyone informed. The
reason for this is that people tend to forget. When we talk about
what is going to happen or what new project is coming on line in 3
months, they listen and understand. Then the time passes, and the
project is no longer something in the vague future; it is here and it
is real. When these same people actually see the working system,
all their objections and questions start popping up all over again,
as if they had never heard of the CSDS. Therefore, we mention the
need for a continuing information process again, because it is vital
to the success of the project, especially during its introduction.

CHANGING OVER

Do Not Scrap the Old System Yet

Many times in our enthusiasm to get a new system running, we are
tempted simply to discard the former method and switch to the new
one. But in many cases a new system will develop small problems
and will not run well. If the former method has been completely aban-
doned, there will be nothing to fall back on until the new system has
its "bugs" taken out and is running well. In the case of the CSDS
it is prudent to keep the old reporting system functioning for a
period of time as a backup system. The users will feel more comfortable seeing a familiar report along with the new for a period of time,
and any temporary inaccuracies in the new system will not disrupt
the normal reporting sequence. Also, with the old system still
functioning, it will be less embarrassing should something go wrong
with the new system. This will show management that you are prudent in your management style.

Comparing the Results of Both Systems

Operating the old reporting system for 3 to 4 months after the CSDS
is on line is useful in another way. The users can compare the reports of both systems and see how much better the CSDS handles
the same accident data. During this transition period, comparisons
will help demonstrate the benefits of the CSD system. For example,
ask managers to place the regular reports side by side and compare
them. Review the types of special searches available. Show how
the follow-up report can be used as is and in addition can be tailored
for more specific information through a special search.

During the transition period there should be careful coordination with management. They should be informed of the length of the transition period and when the old system will be phased out.

Training Sessions and Seminars for Users

When introducing the system, the initial way to encourage its use is by training users in ways of reviewing the regular reports and then requesting special searches. Training sessions should include as many managers and supervisors as possible. The more management people who know how to use the system, the more use the system will see. These training sessions can be conducted for each department. An ideal group is six to eight participants. A larger group does not give each participant sufficient time for speaking and participating during the session.

The initial training session, where each new user learns how to use the system, should be a round-table discussion meeting in which there is no real leader, or teacher, but one person who acts as an administrator. This person, from the safety department, is to act as a facilitator to encourage the exchange of ideas. Once the system fundamentals have been described, the idea exchange can begin in earnest. It is good to stress that the CSD system truly puts control of the safety program in the hands of line management. This is because it allows the generation of adequate decision-making information.

Presentation, Meetings, and Conferences to Build Support

The criterion of success of a CSD system is the amount of use it gets. Just announcing the fact that the CSD system is on line and ready for use may not do much. Seeing the new report format may remind management that something has happened to the regular reports. This is not really active use; it is passive acceptance of the new regular safety report. The fact that the report has more information than the former system may be recognized as a novelty, but that may be all. For the system to be used to its design potential, the users must understand how it can be used and what can be gained by using it. Again, this may seem to be a repetition of what was done to present the project proposal in the first place, and another review of the types of meetings held prior to the system going on line. And it is a continuation of the information program begun at the initiation of the project. Why? Because there are a lot of demands on a manager's time. The benefits of using the CSD system, now that it is on line, must be repeated. Why? Because the system is in competition with other priorities for a manager's time. Thus, the presentation must be compelling. Its goal is to encourage use of the system, and use means requesting special searches, a capability beyond mere passive reading of the regular reports each month. There are ways to increase use. Here are a few ideas:

1. Organize weekly sessions to brief users on ways to make special searches, on ways to read their monthly reports, and on methods to employ the special searches in connection with other reports and information.
2. Schedule special short briefing sessions so that users can describe how they have used the special search technique.
3. Have round-table meetings between various departments to share techniques for using both regular and special search reports.
4. Develop a series of one-to-one meetings between a manager and safety to set up a special set of searches with a new emphasis.
5. Prepare a loose-leaf notebook with special search ideas. Keep sending out new pages with ideas, or pass them out at meetings and briefings.
6. As time passes, obtain reports from individual managers discussing their results in cost savings and injury reduction from using the system. This should be repeated with as many managers as possible. In addition, the reports should be put in the loose-leaf notebook for future reference.
7. Regularly inform top management about the use levels of the system and the successes that can be attributed to the system's use. Encourage top management to use the system and to encourage its use.

These are just a few ideas to increase the use of the system. As you develop your own system, many more ideas can be added to this list. It is important to devise as many ways as possible to encourage use of the system. To keep its use at a high level will require constant encouragement and enthusiasm about results.[3]

Testimony of a User

As in selling any tangible product, the testimony of a user who is a respected person can have an influence on other potential users. In the case of the CSDS, testimony should be handled in a dignified and diplomatic way, with an eye to encouraging cooperation and mutual gain. For example, testimony can be handled by setting up an advanced seminar in which two or three users are asked to present their problems and the solutions they found through use of the system.

The advanced seminar should be arranged around a table, conference
style. This configuration puts everyone on an equal basis and facilitates discussion. Another way to use testimony is to collect special
search techniques and give credit to each contributor in a memo that
is three-hole punched for entry into the notebook for users. These
success reports can be used to advantage to emphasize the effectiveness of the system. Such success can provide encouragement for
other users to develop uses in their own work situations.

Use is the key word for the whole program once the system is
on line. Upper management has authorized the expense of developing
this system, and only through its use can the investment pay off.
Thus, encouraging and maintaining a high level of use is the main
task for safety. This will require constant tending and administration to ensure that the system is effectively used. Many of the techniques of information flow, testimony, and training may have to be
employed over and over again to keep the users motivated to use
the system.

COMPLETION OF THE CHANGE

Status Report to Management

The project development phase is at an end when the completion report is submitted to management. This report discusses the introduction phase and reviews use of the CSD system. It will cover
progress in training users, methods employed to encourage use of
the system, and examples of both the regular and the special reports.
These reports can be compared with the former reporting system
format. It is a gratifying experience to complete a development project and begin to use it after so many months, and the report should
reflect your excitement and enthusiasm.

The status report can give a scenario of the plans for meetings,
seminars, and conferences to encourage use of the system. This
review of the short-range planning for the CSDS can include estimates
of how the use levels are expected to increase. These figures should
be conservative. This will allow a good margin to attain the target.
More important, if the work is well done, the use can exceed the
mark. Potential problems and possible solutions can be discussed,
if they are appropriate. This status report should be basically one
of good news, positive signs, and expectations. Management likes
a winner and a contributor to profits. The CSDS can fill that need.
The report can also give credit to those in management who contributed ideas and are actively supporting the system. The wider the
circle of recognition, the more support the system will have.

Personal Consultation with Users

Everyone enjoys personal attention. A good way to encourage greater
use of the CSDS is through private meetings, or consultations, with
users. These informal sessions can be used to discuss special search
methods and analysis of current projects where the system can provide input. This idea exchange can lead to new uses for the system
as well as encourage more traditional uses.

As in the previous sessions to develop the report formats, careful planning of these sessions is vital for success. Thorough preparation in advance will make the session relaxed and easy. This pre-
planning should include the following:

1. Thoroughly review the several monthly reports of the particu-
    lar department. Make an analysis of the reports for ideas
    for special searches.
2. If possible, preplan several special searches. At the proper
    time in the session, these can be suggested.
3. Check the follow-up report for the number of outstanding
    recommendations.
4. Determine the extent to which the department has employed
    special searches of the system.
5. Try to learn the special problems associated with the particu-
    lar manager's operation.
6. From these analyses, develop other ideas for using the sys-
    tem, ideas that will enhance the meeting and encourage use.

Once preparation is complete, the meeting should be scheduled.
Although there has been careful preparation, the meeting is not a
formal session, but an informal get-together to encourage use of the
system. There is a mutual interest. Both safety and the manager
are interested in the success and effectiveness of this particular
operation: specifically, to improve the operation through better management information based on increased use of the CSDS. This will
occur only if the manager believes that the time spent will be of benefit. The session, therefore, should center around two themes:

1. The benefits of the system. How others in management are
    using the CSDS to their advantage. Provide examples of the
    searches and the results.
2. How the manager can use the system to best advantage. This
    part is based on the preplanning analysis of the manager's

operation and of his CSDS reports, as well as carefully listening to the manager's problems and concerns during the meeting.

The discussion of these two topics can be kept in a positive and
supportive mode quite easily, because the conversation is directed
toward developing ways to improve the manager's operational efficiency
through use of the CSDS. Given the goal of helping to improve management efficiency for the particular manager, it would be unusual
for that manager not to be interested. Each session can emphasize
increase use of the system through special searches.

Periodic Reminders about Special Uses

As with any system whose use requires effort, the CSD system's
special search uses will decline unless the users are periodically reminded of the benefits. In many cases, operating managers are well
aware that the results of special searches are effective, but those
who do not share this opinion will not take the time to use the system. This means that safety must be the catalyst to maintain and
increase use of the system. One of the means available is to circu-
late new search ideas among managers. This was mentioned earlier
in this chapter when we discussed the system introduction phase.
This technique should certainly be continued and used to best advantage. But as time passes and the CSD system becomes more and more
a part of the overall safety program, there will come a time when the
number of special search requests may start to decrease. It is advisable to keep an eye on these figures and know what is happening.
Safety will have the job of monitoring the level of requests for
searches encouraging use. This means reviewing the regular reports,
spotting conditions where a special search may be indicated, con-
tacting the manager, and discussing the conditions with him. It is to
be hoped that this little assistance with the review will be appreciated
and will generate renewed interest. If necessary, the safety professional can design several special searches and ask if the manager
wants them run. When the safety professional obtains the reports,
they can be reviewed with the manager. The goal is to get managers
sufficiently interested in such results that they will begin requesting
such searches on their own. By providing special searches, the safety
professional may stimulate managers to discover other areas in which
searches will be useful.
These reminders and close support for managers in performing
special searches will serve to maintain or increase the level of use
of the system. They will also serve as tools for keeping in touch
with the managers and their operations.

IMPROVING THE SYSTEM

Modifications and Their Effect

In Chapter 12 we shall discuss the major modifications that are possible to increase use of the system to provide a wider range of information to management. These modifications, such as an industrial hygiene module, can be introduced in much the same manner as was the CSDS originally. In this case, the justification and presentations will be easier, because the system will already be operating and will have proved its cost effectiveness.

In addition to major modifications and additions, minor changes can be made in the system: the addition of a new file or addition of a subfile. Both of these modifications will have to be considered carefully and approached with caution. By adding a new file or a subfile, another source of good information is added to the system's searching capabilities. On the other hand, if there are, say, 4200 cases in the data base, how does this additional information get added to those cases? If the change will not affect every case, a special search can be initiated to determine how many cases will be affected and what kind of work load will be involved in entering changes to the affected cases.

Adding a single code word will not be as consequential and therefore will not be a major consideration unless it will affect a great many cases. Here are some discussion questions to review prior to committing any resources to make an addition to the system:

1. What is the proposed change or addition?
2. How many codes, files, and subfiles are to be added to the system?
3. How many cases are affected by this change?
4. Will any reprogramming of the system be involved in this modification? If so, how will it affect other parts of the system?
5. What will the proposed change or addition do for those who use the system? How will it make the system more effective? In what way will it improve decision making?
6. What will be the costs in time and actual cash outlay to make this addition?

By running through this exercise, the merits of a proposed modification can be evaluated on balance. The proposed change must be so compelling and useful that it will make the effort justifiable. Any such change or modification must stand on its own merits. The CSD system is a successful system; a poorly calculated change or modification could lead to mistrust and loss of confidence. The key is careful decision making.

Follow-up on Quality of Input

Probably the most important consideration in maintaining an effective
computer system is the quality of the data. If it is suspected that
the information contained within the system is not completely accurate,
users will hesitate to spend their time doing special searches. They
will reason that if the data are not accurate, then it will be better
not to use the data than to risk censure by management for using
poor information. This situation can be fatal to the system and to the
reputation of the safety department. Thus, after the system is on
line, effective quality-control checks are necessary to keep the system accurate. These quality checks can be performed in several
ways:

1. To keep investigation techniques sharp, safety can, on a random basis, join in on investigations of accidents.
2. If the investigation report appears to be less than adequate
   for effective coding, safety can contact the investigator,
   clarify the situation, and improve the report.
3. When a case is coded, it should be rechecked by another coder
   before initial processing.
4. When the case is printed out from information systems showing
   the actual input codes, these should be checked against the
   original input form and the original investigation form.

Quality must be stressed at each point from initial investigation
to data input. If this kind of precision checking is performed routinely,
90 to 95 percent of inaccuracies and oversights will be discovered and
corrected before they contaminate the system. The quality-control
accuracy goal should be 95 to 98 percent. Not all oversights will be
identified and changed, but with faithful rechecking, more than enough
errors will be discovered and eliminated to maintain the desired accuracy levels. That high level is quite sufficient to keep users confident
that the information they use is accurate.

Stressing quality is a confidence builder. Management should be
continually apprised of the efforts that safety is making to assure
quality in the CSD system. Users in management should be encouraged
to maintain quality control during investigations. If the investigator
provides thorough reports on every case, this will improve the quality
of the rest of the system. With a thorough, well-written investigation
report, a safety professional can easily code the case into the system
with a high level of accuracy.

A quality-control system within the CSD system will pay dividends
by assuring the user and upper management that the system is a
reliable tool. The information that is retrieved in reports can be re-
lied on for making good decisions. This type of effort will add another

layer of confidence for the users and will remove another reason for
not using the system. Good quality in any product promotes confidence and pays more dividends in good will than the effort required.

SUMMARY

Introducing the CSDS entails changes. Even though the system has
been announced and discussed, only when it is experienced as a
fact, on line, will its impact be fully realized. Understanding what
the change means and how the new system works can reduce the
fears of those who might oppose the project because of lack of understanding. Keeping the former reporting system operating during
the start-up period of the CSDS will reduce anxiety somewhat and
provide a system backup, as well as facilitate comparison between
the old and the new. During the transition, a series of training sessions should be held to introduce the system and explain how to use
the special search capability for the most benefit. Other meetings
will be orientation sessions to discuss specific problems, and there
will be advanced sessions to encourage use. The important factor
is to see that the system is used as much as possible. Reports of
users, publishing of special search methods, and sessions to discuss
special search ideas will encourage use of the system. Also, to keep
usage high, one-to-one sessions can be used to discuss problems
that might not be aired in a group meeting. Always the goal is increasing use of the system. A user's notebook can be provided, including a section for special search methods to encourage new ideas.
The user's manual is a living document. There will be minor code
additions from time to time. Should a later change be proposed, it
should be examined carefully before being implemented. As the system is used, the quality of the input must be continually stressed to
maintain confidence in the information that is produced. With good-
quality input and a high level of use, the system will provide a valuable service to the company.

NOTES

1. Peter F. Drucker, *Management.* Harper & Row, New York, 1974.
2. J. D. Batten, *Developing a Tough Minded Climate for Results.*
    American Management Association, New York, 1964.
3. Uris Auren, *Mastery of Management.* Dow Jones—Irwin, New
    York, 1968.

10

APPLICATIONS: IMAGINATIVE SPECIAL SEARCHES

*Every time a man puts a new idea across, he finds ten men who
thought of it before he did . . . but they only thought of it.*

Henry G. Weaver

USING THE SYSTEM

Special Searches Encourage Use

To be an effective tool in conserving company assets, the CSDS must
be used. To achieve a level of use that will make the system effective, safety must employ many methods, such as training, seminars,
conferences, and one-to-one sessions. All of these gatherings are
aimed at instructing and encouraging management to use the system.

When the system is widely used, two things are accomplished.
First, through effective use, each manager obtains better information about injury and property-damage cases. Thus, with better information, better accident control is achieved, and company assets
such as property and skilled workers are conserved. Second (and
this accomplishes the first), through using the CSD system, each
person will come to realize that it is designed as a personal research
system. It allows each user to tailor special searches for the specific information that will be most useful to an operation or function.
Therefore, by encouraging special searches, a climate of self-motiva-
tion is created. The user makes a special search, is encouraged by
the results, and requests another search. The cycle will repeat,
providing better and better information that will create enthusiasm
and support for the CSDS.

In this chapter we shall discuss special search techniques and
ways that special searches can be used in conjunction with other company reports.

Giving Management the Tools to Use the System

Discussing the benefits of the CSDS introduces potential users to
the system. However, to use the system, managers and supervisors
must be equipped with certain tools. In the last chapter we discussed
a notebook for managers. This is not to be confused with the user's
manual, which contains the code files. Rather, it is a notebook con-
taining instructions, directions, and forms necessary to prepare
investigation reports, review the regular CSDS reports, and request
special searches. The notebook for managers can contain the following:

1. Copies and descriptions of all the forms used in the system.
2. A list of data needed from an investigation to maintain the
    high quality of accident reports.
3. A discussion of the quality-control system used to maintain
    accuracy in the system.
4. A discussion of the regular reports and how they can be
    analyzed.
5. A description of the special search form, if one is used, and
    complete instructions for its use.
6. Special search routines and examples of results.
7. Completed special searches performed by other users. These
    are to be used as guides and idea sources for special searches.

In Chapter 9 we reviewed the techniques of continually providing
managers with examples of special search reports from other departments. Many of the searches performed by each department will provide excellent idea material. All of these can be distributed to be
inserted in the notebook.

Regular Reports: Only a Carrot

The regular report system is a means for the user to determine that
there may be a need for more information. The regular report shown
in Exhibit 10-1 shows brief information about each case. The information is certainly superior to reviewing pure incident rates or raw
case numbers. This is especially true when the case-by-case report is combined with the summary report, as shown in Exhibit 10-2.
The neat part is that these two reports are not the last word. They
provide a starting point.

For a moment, let us take the part of the manager of the maintenance department and review these reports. Let us see what further
information, if any, will be helpful. Here are some observations; take
a moment and see if you can find others:

Case No.	Type of Case	Inc. Date	Time	Days Lost	Case Cost	Freq./ Sev. Potential	Human Factors	Condition Factors	Root Cause Factors
0113	Fall SL	8/4	1015	ØA	114A	Freq./ Minor	Opr. Too Fast	Light LTA Piling LTA	No Haz. Rev. Hsekeep. LTA
0126	Lifting	8/5	1510	18E	1265E	Occas./ Serious	Body Pos. LTA Wrong Place	Too Bulky Vis. LTA	Instr. LTA Proc. LTA
0128	Fly Part.	8/9	1145	ØA	295E	Occas./ Serious	Omitted PPE Knowledge LTA	Guard R.	D/N Corr. Rsk. Kn. Prec.
0137	Slip	8/16	1335	ØA	114A	Rare/ Minor	D/N Use. Opr. Too Fast	Miss. Equip. Surf. LTA	Sched. LTA D/N Det. Haz.
0139	Lifting	8/16	1615	ØA	680E	Freq./ Serious	D/N Check. Omitted Step	Method LTA Too Heavy	Arrange. LTA Instr. LTA
0142	Lifting	8/22	1535	7E	950E	Freq./ Serious	Knowledge LTA Wrong Place	Method LTA Miss. Equip.	Instr. LTA E. Tng. LTA
0143	Pres. Extreme	8/23	0910	ØA	475A	Occas./ Serious	D/N Get Help Omitted PPE	Hi Heat/Cold No Energy Bar.	Instr. LTA Kn. Prec.
0151	Struck by	8/24	1040	ØA	114A	Rare/ Minor	D/N Check D/N Use	Guard R. No Energy Bar.	Arrange. LTA
0164	Material Move.	8/30	1610	ØE	275A	Freq./ Serious	Equip. Use LTA Oper. Too Fast	Equip. LTA Vis. LTA	D/N Corr. Rsk. Kn. Crit. Inc.

Exhibit 10-1

Division
Organization: Maintenance
Expense Center: 8347

Cases by Type		Case Totals	Days	Costs		Causal Factors
Lifting	(3)	3 - First Aid	ØA	342A		Human Factors up to 5/2 +
						Opr. Too Fast (3)
Fall SL	(1)	4 - Medical Tmt.	ØA	1225E		Wrong Place (2)
						Omitted PPE (2)
Fly. Part.	(1)	2 - Lost Time	25E	2215E		Knowledge LTA (2)
						D/N Use (2)
Slip	(1)	Ø - Occ. Illness	Ø	ØA		
						Condition Factors up to 5/2 +
Pres. Extreme	(1)	Ø - Property Damage		ØA		Vis. LTA (2)
						Guard R. (2)
Struck by	(1)	Ø - Near Miss		ØA		Miss. Equip. (2)
						Method LTA (2)
Material Move.	(1)	9 - Total				No Energy Bar. (2)
		Injury Cases	25	$3782E		
						Root Cause Factors up to 5/2 +
		9 - Total Cases	25	$3782E		Instr. LTA (4)
						Kn. Prec. (2)
						D/N Corr. Rsk. (2)
						Arrange. LTA (2)

Frequency/Servity Case Nos.

Freq./Maj.	0
Freq./Ser.	0139, 0142, 0164
Occ./Maj.	0
Occ./Ser.	0126, 0128, 0143
Rare/Maj.	0
Rare/Ser.	0

Exhibit 10-2

(a) The summary (Exhibit 10-2) shows three lifting cases and one case involving material movement. What does the first half year of experience show for this type of case? Concerning location, department, and supervision, could there be a common connection? What were the three most frequently listed causes of these material-handling cases? How many recommendations were made, and how many are still outstanding? What were the average cost and the total cost of these type of cases over the past half year? What were the total days lost?

(b) Among the cases listed in Exhibit 10-1, there are three cases in which Operating Too Fast is identified as a human factor. How many cases list this as a cause this year? Can you rank them by type of case? Do location, department, and supervisor have a connection with this cause? What are the connections between the condition and root-cause factors and this cause? Do time of day and day of the week show any interesting parallels?

(c) Visibility and Lighting LTA is shown as a condition factor in three cases in Figure 10-1. How many cases list this as a cause for this year? Is this condition common to a location, department, or operation? How many recommendations were made in these cases? How many of them are still outstanding? Does the time of day have a connection in these cases? How many of these cases have the potential to be serious? What are their case numbers?

From these questions, special searches can be designed as follows:

(a) *Case-by-Case Report*. Time period: 01/01/XX to 08/31/XX. Type of case: material handling and lifting. Report headings: Case Number, Date/Time, Location, Department, Supervisor, Frequency/Severity Potential, Human Factors, Condition Factors, Root-Cause Factors.

Summary Report. Time period: same. Report headings: Total Number of Cases, Total in Each Type of Case Category, Average Cost/Case, Total Cost, Total Recommendations and Total Outstanding, the five cause factors that appeared two or more times in cases during the period.

(b) *Case-by-Case Report*. Time period: 1/1/XX to 8/31/XX. List all cases with the human cause factor Operating Too Fast. Report headings: Case Number, Date/Time, Location, Department, Supervisor, Human Factors, Condition Factors, Root-Cause Factors.

Summary Report. Time period: same. Report headings: Total Number of Cases, Total Cost, five cause factors that appeared two or more times in cases during the period.

(c) *Case-by-Case Report*. Time period: 1/1/XX to 8/31/XX. List all cases that have Visibility and Lighting LTA as a condition factor. Report headings: Case Number, Date/Time, Location, Department, Frequency/Severity Potential, Recommendations Outstanding.

Summary Report. Time period: same. Report headings: Total
Number of Cases, Location (two or more times), Department (2 or
more times), Frequency/Severity by Case Number, Total Recommendations, Total Outstanding.

These are examples of some of the varieties of searches that are
possible. Obviously, these searches can be used in many different
departments and are not confined to maintenance. For example, the
special search described in (a) could just as easily be requested in
the warehousing department as in maintenance. The results might
be altogether different, but the beginnings would be similar.

Special searches provide a new source of information for the
manager. This new information can be used by itself or compared
and contrasted with other information about the group that is being
managed. A single search can be used as is, or other searches can
be developed for more precise information.

Cost Data Have an Impact

Cost development is a difficult task, because safety is interested not
only in direct out-of-pocket costs, such as medical and compensation,
but also in indirect costs. Indirect costs include such items as how
much time supervisors spend investigating cases and how much time
safety committees and higher management spend reviewing past cases.
We discussed suggestions on the generation and management of cost
data in Chapter 8. Through this technique, costs, both direct and
indirect, can be developed with good accuracy. The way these costs
can be used effectively is to design special searches that include
cost data. This will enable the manager to analyze cost—benefit situations. The cost data analysis can then be used as part of a persuasive
presentation for the introduction of courses of action to reduce or
minimize injuries and property losses through engineering control,
employee training, and safety procedures.

For example, a manager in a production department had reports
for the month that showed only one medical-treatment injury. The
cost was $277. The injury prompted him to wonder if there had been
similar cases. He searched the computer's file and found that in his
department there had been 11 first-aid and medical-treatment cases
with the cause of either Omitted Step or Did Not Shut Down, all on
the same piece of equipment, and ususally with the same supervisor.
The cases had occurred over the previous 18 months. The total costs
during the period were $1894. The Frequency/Severity Potential data
listed for each case showed that Occasional/Serious was indicated in
about 50 percent of the cases.

In his presentation to management, the manager stated that in
spite of efforts to curb these injuries, they continued to recur. The
recommendation was to perform an engineering study of the machine

and its controls, as well as a review of the injury records of similar
operations in other companies in their industry. The total cost of
the study project was calculated to be about $5500. The cost of any
engineering changes would be additional. The manager cited the
fact that over the next 5 years the injury cost alone, if the cases
continued at the first-aid and medical-treatment level, would be
over $6000. In addition, each case involved an average of 35 minutes
lost from production time. Lost production time cost an average of
$850 per case or $9350 for the 11 cases. For the coming 5-year period,
these cases would be expected to cost over $31,000. If a lost-time
case should occur during the 5 years, the cost could rise one-third
higher. Thus, over the next 5 years, the overall costs could range
from $37,600 to $55,700. The proposal was accepted.

The result of the study was a change in the machine's controls,
as well as special sequencing for emergency shutdown of the line.
As a result, the incidence of injuries on this machine dropped to zero.
The savings from the study amounted to over $20,000 per year.

In this example, the manager used costs developed routinely by
the system to prove that the study and subsequent changes would
be cost-effective. To find the specific causes and then to develop
the costs manually would have been nearly impossible, for a number
of reasons: The injuries were different; there were several different
case types. The cases were spaced over 18 months. They were relatively minor: only first aid and medical treatment. They were usually
filed by date, not by machine. Thus, a manual search would have had
to cover almost the entire file for the previous 18 months. Without
the CSDS, the job probably would not have been done. In fact, without the computer capability, the manager would not even request such
a search.

After all, they were only first-aid and medical-treatment cases—
nothing serious like an amputation. In other words, without the CSDS,
nothing would have been done until a serious injury had occurred.
But as the example indicates, the costs were continuing to mount.

TECHNIQUES FOR DEVELOPING SEARCHES

Using Logic

Developing special searches requires the user to employ logic or in-
ductive and deductive reasoning. Although this may sound difficult,
in practice it is quite manageable. It involves looking at the regular
reports, like those depicted in Exhibits 10-3 and 10-4, and determining
what additonal information can be found.   In Exhibits 10-3 and 10-4
there are some points that could be explored further.

05/06/XX
Warehouse 8336

INCIDENT/INJURY REPORT
04/01/XX Through 04/30/XX

Case No.	Type of Case	Inc. Date	Time	Days Lost	Case Cost	Freq./Sev. Potential	Human Factors	Condition Factors	Root Cause Factors
0247	Lifting Over-Exertion	04/05/XX	1035	Ø	114A	Rare/Ser.	Body Pos. LTA Equip. Use LTA	Method LTA Placement LTA	Instr. LTA Sched. LTA
0248	Material Mvmt. Reaching	04/09/XX	1850	Ø	114A	Freq./Minor	D/N Get Help D/N Check	Lift Fac. LTA Miss. Equip.	D/N Con. Rsk. E. Tng. LTA
0291	Splash Liquid (contact)	04/11/XX	1620	Ø	277E	Occ./Ser.	Omitted PPE Oper. too Fast	Ground LTA No Energy Bar.	No Factor
0295	Struck by Material Move	04/19/XX	0815	7E	2147E	Occ./Sev.	Expr. LTA Opr. Too Fast	Vis. LTA Placement LTA	D/N Corr. Rsk. Kn. Prec.
0304	Struck Against	04/25/XX	0925	Ø	114A	Occ./Minor	D/N Recog. Wrong Place	None	Instr. LTA Plnd. Chg.
0327	Fall SL Slip	04/27/XX	1430	Ø	114A	Rare/Minor	Distract. D/N Warn	Hsekeep. LTA Rough Sur.	Resp. Unclr. D/N. Det. Haz.

Exhibit 10-3

Cases by Type		Case Totals	Days	Costs	Causal Factors
Material Move	(2)	4 – First Aid	ØA	456A	Human Factors up to 5/2 + Opr. Too Fast (2)
Lifting	(1)	1 – Med. Tmt.	ØA	277E	
Over-Exertion	(1)	1 – Lost Time	7E	2147E	Condition Factors up to 4/2 + Place. LTA (2)
Reaching	(1)	Ø – Occ. Illness	ØA	ØA	
Splash	(1)	Ø – Prop. Damage		ØA	Root Cause Factors up to 5/2 + Instr. LTA (2)
Liquid (Contact)	(1)	Ø – Near Miss		ØA	D/N Corr. Rsk. (2)
Struck by	(1)				
Struck Against	(1)	6 – Total Injury Cases	7	2880E	
Fall SL	(1)	6 – Total Cases	7	2880E	
Slip	(1)				

Frequency/Severity Case Nos.

Freq./Maj. - Ø
Freq./Ser. - Ø
Occ./Maj. - Ø
Occ./Ser. - 0291, 0295
Rare/Maj. - Ø
Rare/Ser. - 0247

Exhibit 10-4

For example, of the six cases reported from the warehouse department, there were only four causes that were indicated two or more times. Because the summary will print only those causes listed two or more times in the case-by-case report, only those four causes were noted in the summary. Also, in the Type of Case column, only Material Movement showed up twice. A careful review of the case-by-case report should be made to determine if parallels and connections exist and to reveal the need for more information. For example, Material Movement, Lifting, Overexertion, and Reaching all apply to the handling of material. This would be an expected injury form in a warehouse activity. The causes associated with the cases, however, could indicate a need for more information about other cases in the data file. A search for that information could be designed as follows:

Case-by-Case Report. Time period: 01/01/XX to 12/31/XX (previous year) and 01/01/XX to 04/30/XX (current year). List all cases of Material Movement, Lifting, Overexertion, and Reaching. Report headings: Case Number, Date/Time, Days Lost, Specific Component, Human Factors, Condition Factors, Root-Cause Factors.

Summary Report. Time period: same. Report headings: Total Number of Cases by Category, Total Cost by Category, Specific Component (up to 5/2 +), Causal Factors (up to 5/2 + for each factor).

Another search could be initiated to determine the extent of occurrences of chemical splashes in the warehouse, as noted in case 0291. The search could be set up as follows:

Case-by-Case Report. Time period: 01/01/XX to 12/31/XX (previous year) and 01/01/XX to 04/30/XX (current year). List all cases of the type Splash, Liquid (contact) and Chemicals. Report headings: Case Number, Date/Time, Frequency/Severity Potential, Specific Operation, Specific Component, Human Factors, Condition Factors, Root-Cause Factors.

Summary Report. Time period: same. Report headings: Total Number of Cases by Category and Total, Frequency/Severity Potential by Case Number, Specific Operation (up to 5/2 +), Specific Component (up to 5/2 +), Causal Factors (up to 5/2 +).

The intent of this search is to learn if there is a connection between the incidents of this type and the specific operation, component, and causal factors. Another avenue of inquiry could be commonality of location, time of day or time period range, such as 1400 to 1800 hours, and the supervisor involved. Such reports can turn up subtle situations, such as causes connected to certain operations and components, plus the job experience levels of the work crews during those periods. Note that the job experience levels can be retrieved from the employee information system data base.

Another thing to remember is that it is not necessary to have a current safety report to decide to search the data file. The only

requirement to perform a special search is to have a concern about
a potentially hazardous situation. The results of an injury investigation, for example, may indicate that there is a deeper problem that
should be searched out. The introduction of a new procedure or
process might suggest the need for a search. A reported near-miss
can also suggest a need for more information on which to base management decisions. All of these are valid reasons to initiate a search.

The Special Search Will Lead Somewhere

The reason for searching the file is to obtain more information about
a specific problem. A search is used to narrow down the set of information parameters to extract precise data. But suppose the searches
that are performed yield no definitive information on which to make
good judgments. What then? There are two choices: One, to conclude that the data are bad and not worth any further effort. Two,
conclude that the data in that specific area are not sufficient and that
perhaps the search should be redefined. This means either broaden
or narrow the next search, depending on the results of the initial
unproductive search. The second choice is obviously the more positive approach. Also, there is another aspect: Suppose sufficient
searches have been made and the net result shows no definitive pattern. What then? It is possible that there is no real problem. This
is a conclusion that must be accepted as an answer at times. It should
be recognized that occasionally there will be no definitive answer to
indicate a problem, or perhaps only a minor one, has surfaced. If
sufficient searches have been made, then that answer is fine too.

From time to time, when the safety data base is new and growing,
there will be searches that will net only sparse information. In the
last example, suppose that the first search showed only one or two
incidents of chemical splashes over the last 18 months. Thus, to
determine if there is any possibility of a real problem, the following
search could be initiated:

Case-by-Case Report. (1) Expand the time period by another
year back in time (30 months total), (2) expand the type of case to
include Vapors and Fumes, and (3) report headings and summary
report will be the same.

From this set of reports, one of three outcomes is possible: One,
in this area of the data base the data are insufficient to be useful.
Two, the information shows no problem. Three, the data give enough
information to be useful or at least to indicate the direction for a
further search. This third option is where the searches should lead
each time. The limitation is the size of the data base. A base of
2000 cases is a good start, because that puts 72,000 to 86,000 bits
of information on the files. However, remember that there may or
may not be 43 bits of information on *each* file, and therefore, when

searching for a specific point, it may be at a low level. Thus, as
was said before, it is correct to begin to search quite broadly to
learn how many cases are in that particular file of interest.

Multiple-Qualifier Search

In the previous search examples our desire was to discover broad
areas of information for analysis. This will net good results and is
an essential part of learning to search the data base. However, suppose that more specific information is desired. Can a search be designed to produce very specific information? Once a broad search
of the data base has determined that there are sufficient cases in
this part of the base, a technique called a multiple-qualifier search
can be made. Your system's design should be capable of performing
this kind of search; however, this technique should be discussed
with the systems people during the initial program development to
assure that this capability is designed into the system.

In reality, the multiple search involves having multiple qualifiers
and making a narrow search for very specific data. Let us refer
back to the reports in Exhibits 10-3 and 10-4. Suppose that there
had been two or three cases in which workers had been struck by
forklift trucks, as in case 0295. A search could be designed to find
the specific exposure: Struck By. The specific component would
be Lift Truck, and the time period would be the entire file. The
hour of the day could be restricted to 0600 to 1300 hours, for example, which could have been the time period in which the last several
cases occurred. This type of search will yield very specific and restricted data.

This type of search accomplishes two things: First, the possibilities of finding useful information are increased. Second, the full
capabilities of the system are better used. The possibilities for this
type of search are endless. The number of cases in the file is the
only restriction. The beginning logic for searching is the same.
Begin with a broad search to determine the size of the interest area
in the base. Following that, narrow the next search as necessary to
obtain more precise information.

Using MORT

The management oversight and risk tree (MORT) was developed by
W. G. Johnson. It has been used by the Energy Research and Development Administration, U.S. Department of Commerce. Part of
the MORT logic tree is depicted in Figure 10-1. The MORT system
was developed to reveal oversights and omissions as well as improperly
accepted risks within a system. In accident investigations, MORT can
be used to define where oversights could have occurred. The investi-

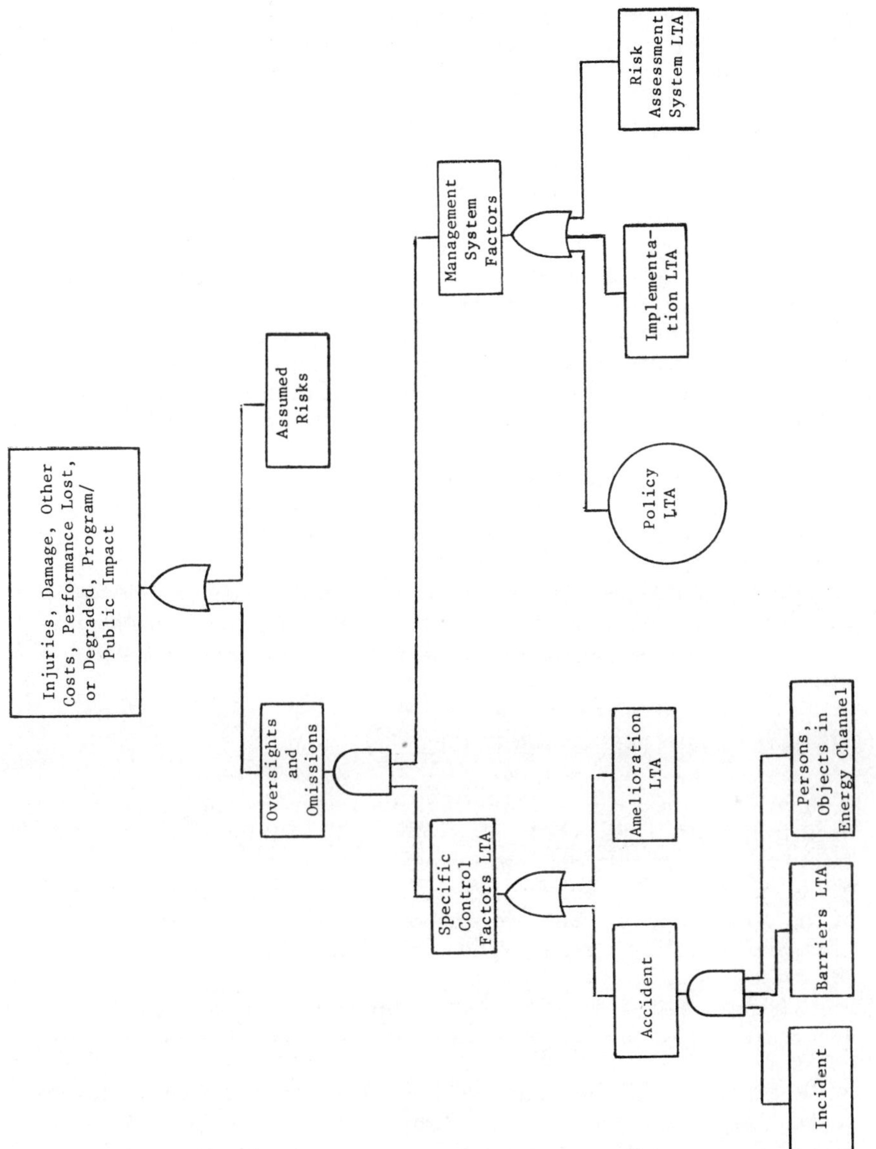

Figure 10-1 MORT. (Adapted from U.S. Department of Commerce Publication SAN 821-2.)

gator then can check out various conditions where indications are that an oversight could have occurred. Table 10-1 is a memory jogger concerning fault tree symbols.

The primary use of the MORT system is to confirm that no detail is overlooked in an accident investigation. The two branches of the top of the tree just below Oversights and Omissions are Specific Control Factors LTA and Management System Factors LTA. These two areas correspond to the causal factors of the CSDS. The Specific Control Factors are the Human Factors and Condition Factors of the CSDS. The Management System Factors correspond to the Root-Cause Factors in the safety data system.

The MORT can be used in searching the system to develop more information on where to investigate. For example, the search for material-handling cases described by the reports in Exhibits 10-3 and 10-4 could have revealed that the majority of the cases showed causes where the supervisor "Did Not Correct the Hazards." Referring to the MORT system, we see that below this cause there is a tree that depicts further sources to investigate as to why these situations occurred. Figure 10-2 shows this segment of the tree.

The MORT system can be used to find and define omissions and oversights in the operating system and to promote better efficiency. This is the reverse of the data system, because the MORT system is an investigatory tool and thus is used to expand our knowledge of a situation.

Table 10-1 Safety Fault Trees and Their Symbols

The diagram of the serious-injury fault tree contains two symbols used to indicate logic relationships for events within a systems safety model. In this manner, events lower in the fault tree can occur and depending on the logic symbol, can affect the execution of the upper events. The two symbols that are used in the diagram are called logic gates. These two gates are called AND and OR gates.

The AND gate, which looks like a tombstone, is the indication that all events immediately below it must occur to allow this gate to operate.

The OR gate, on the other hand, which looks a bit like a Buck Rogers rocket ship, needs only a single event to occur immediately below this gate to allow it to function and cause the event above the gate to occur. This means that in a fault tree, an OR gate signifies a greater degree of risk than an AND gate in the event sequence. Therefore, depending on whether the outcome of the top event in the tree has a good or bad consequence, either the AND gate or the OR gate will be preferred. The AND gate reduces the probability of the event, whereas the OR gate increases the probability.

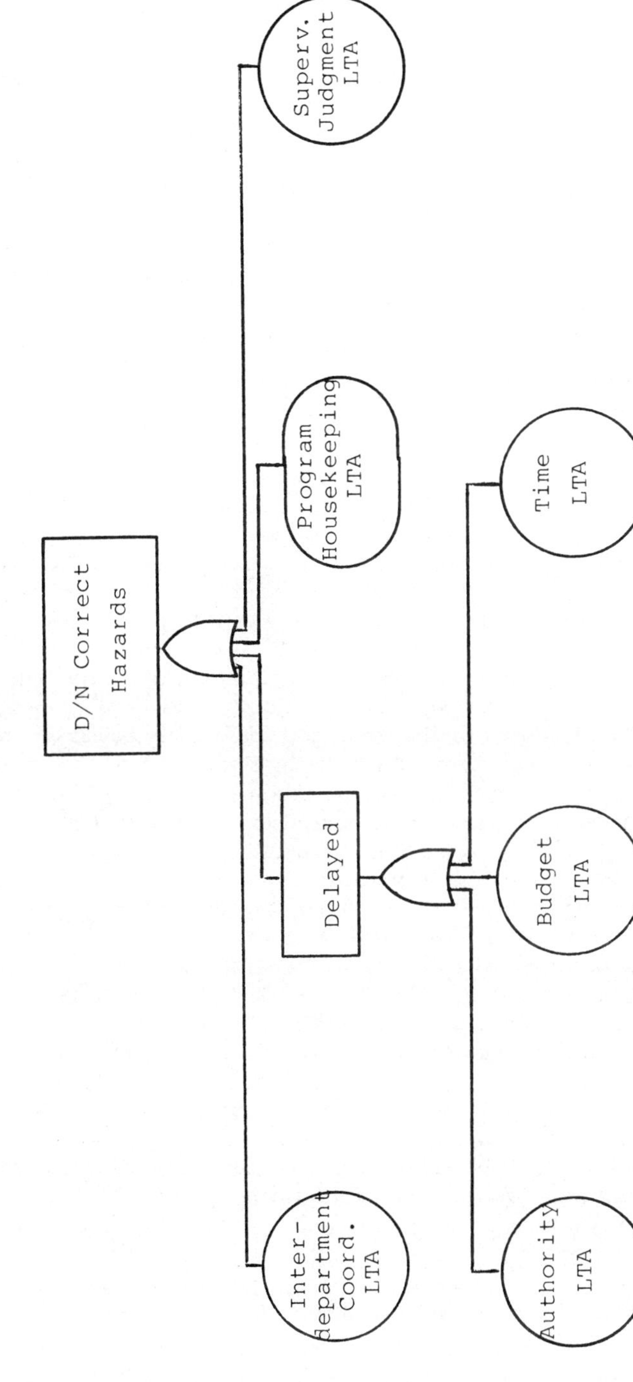

Figure 10-2 Areas to investigate from the cause "did not correct hazards." (Adapted from U.S. Department of Commerce Publication SAN 821-2.)

PERFORMANCE MEASUREMENT

Typical Safety Performance Measures

In many organizations, management will measure safety performance
in terms of results: numbers of accidents. This means that a certain manager or group is measured by the number of reportable accidents, property-damage cases, or collisions that occurred during a
given time period. This sort of measurement, whether it is done in
raw numbers of cases or with calculated rate system, such as the
OSHA incidence rate, causes frustration among managers. This
frustration is evidenced in an apathetic attitude toward safety. This
is not because managers and supervisors do not believe in having a
safe plant or a safe department, but because they feel they are being
measured by a parameter over which they have little or no control.

Over the years, there have been analyses of the causes of injuries
and property damage. These studies have traced the major causes
of these cases to human error. In addition, once the accident sequence
begins, the severity of the outcome of events in an individual case is
the result of random chance. For example, suppose a worker takes
a risk, takes a shortcut in doing a job, such as not stopping a machine
to clear a jam. The shortcut is successful. The worker repeats the
shortcut without being injured. Several days or weeks later the
worker is not as quick as before, and a finger is cut. Had the worker
been slower or distracted longer, the injury could have been an amputation. The severity of the injury, therefore, is based on the
chance circumstances surrounding the individual event that was finally
harmful. Therefore, measuring performance by numbers of accidents
is actually measuring performance based on random events. If severity is added as another appraisal factor, it just adds another chance
sequence to the pile of random accident events.

When you first look at the formulas for accident frequency rates
and severity rates, it would appear that the measure counts events
that occur at a standard rate over time. Instead, it is a measure
that counts random events as standard. It is like being given a job
of counting rejects from a production line for a month, then predicting
how many rejects will occur next month. Frustrating? Yes, it would
be.

Consider the case of the company that measures its managers
solely on the basis of accident frequency. A manager has no idea
how many random events will result in injuries in any month, but he
is held accountable for the number that occur. There is frustration
here. For example, a concerned manager puts effort into a safety
program, with meetings, films, prompt reporting, surveys, and one-
to-one contact with workers. Still there are two reportable injuries
in his department for the month. The manager in the next department,

however, does nothing but put up a few safety posters. There is
not a single injury in her department. The second manager receives
compliments from management for good safety performance, but the
concerned manager is questioned about the accidents in his department. Random events and chance appear to be poor performance
indicators. Yet each year some companies set goals to reduce injuries
by a certain percentage over the previous year. Individual goals for
managers are set the same way. This is an acceptable objective if
it is part of an overall, carefully rounded program that includes attainable and controllable goals as part of the safety performance
measure. Otherwise, if the only safety goal is a percentage reduction in injuries, there will be frustration over the goal. The reason
is that, as was said before, we are dealing with random events. If
injuries are the only measure, this can lead managers to feel that
they cannot control the safety program or their safety performance.
So why bother?

It is a dismal task to try to overcome apathy. A goal based on
chance and random events over which an individual manager has
little control is an apathy producer. After more than 50 years of
safety efforts, it would appear that someone should realize that there
might be alternative performance measures that would reduce injuries
and keep the interest of managers.

Goals: A Good Beginning

Many companies today use goals as measures of performance for management personnel. Management by objectives (MBO) has been the watch-
word for many years. If done effectively, this is a good way to generate positive direction and purpose in any department. Goals for safety
are also a good way to maintain a high standard of safe performance.
However, like other objectives and goals, they must be directed at
improving performance.

It may appear that the last section pointed out that goals in safety
are frustrating, and now this section is saying that goals are good.
Contradictory? Not really.

It might seem that a goals program for safety could be set up in
much the same way as for production or other departments. It can.
There is only one condition: The goals must meet the same criteria
by which other goals are measured or tested for validity.

To be valid, a goal must meet four criteria tests. First, it must
be specific. This means that the goal must be stated in precise terms
to accomplish something. Second, the goal must be measurable. In
other words, there must be a way of quantifying the results to determine if the goal has been accomplished. Third, the goal must be attainable. The person or department who has the goal must have the
power, ability, resources, and control over the operation to accomplish
the goal. Fourth, there must be a completion date.

Let us consider a typical safety goal and test it. Better yet, let us use a goal we discussed in the preceding section: the goal of injury reduction. Try this goal: "The maintenance department will reduce injury cases by 20 percent during the coming year." Certainly this goal is specific, and it is measurable. Knowing what the department did last year and taking 80 percent of that number will provide the goal for this year. It has a completion date of year's end, and that is good. There are three of the four criteria. But is the goal attainable? Can the maintenance department manager directly control the number of injuries? Not likely, unless the manager stops the operation and holds his employees in a classroom for the year. The manager could try to watch every worker all the time. That is also an impossible solution. The maintenance department has to operate and keep the plant running. Maintenance is a high-risk department, but progress can be made. Perhaps this goal is not a goal after all. Perhaps injury reduction is an overall guiding objective. Perhaps it is something to strive for, but the manager should not have his feet held to the fire for not attaining it.

Other goals can be written to push toward an overall objective. However, these goals must meet the four criteria. Perhaps this could be one: "Using a checklist, make a hazards survey of the department once each month. Prepare a written report of recommendations, and complete each within 30 days unless engineering work is required." This goal will stand up to the four criteria of a good goal. It will contribute to reduction of injuries and will improve the overall operation as well. The manager will be confident of accomplishing such a goal, and so a positive feeling toward safety will be created.

Using the four criteria and setting goals is one way to measure performance in production, in sales, and in safety. Basing the performance measures on doable parameters that will lead to a reduction in injuries is a good, positive way to approach the problem.

> Management can measure line supervisors to see whether or not they are utilizing such techniques of accident control as toolbox meetings, job hazard analyses, inspections, accident investigations, incident reports, safety committees, and safety meetings. Management may require line supervisors to submit activity reports. When management measures these activities, it is setting up a system of accountability for activities. It is emphasizing to the supervisor that management wants performance from him in safety. It is, in terms of our model, raising the supervisor's perception of the probability that if he gives effort, reward will follow.[1]

These types of measures are the kinds for which goals can be developed and the kinds that can be tracked and measured. The measures

are acceptable to supervisors because they are attainable. This is the
kind of activity that, when combined with the other safety program
elements, will bring lasting reductions in injury events.

Using the Safety Data System

Many of the bits of data contained within the CSDS can be used in
performance measurement. These can be analyzed and used by a
manager to evaluate supervisors for their performance in safety. For
example:

1. The number of safety surveys/inspections performed for the
    period.
2. The number of recommendations made on accident reports,
    near-miss incidents, and surveys.
3. The number of recommendations accomplished during the
    period versus those remaining.
4. The number of accident causes identified per case (averaged).
5. The number of near-miss incidents reported for the period.
6. The top three causes of injuries in the department.
7. The top three causes of near-miss incidents in the department.
8. The names of the supervisors in the department and the num-
    bers of accidents in their crews.

It should be obvious from previous search procedures how these
searches can be executed. The important thing to remember is that
the emphasis is on performance measurement. Using the CSDS, the
manager can track recommendations, surveys, development of causal
data, reporting of injuries, and listing of near-miss incidents. These
measures can be added to reports of toolbox meetings, safety meetings,
and one-to-one contact sessions. The total effort by a supervisor
can be evaluated against the actual number of injuries, their severity,
property-damage events, and the costs of all cases that occurred in
the department or section. This summary, and comparisons between
the positive efforts and the injury and damage results, will provide a
good picture of sustaining performance.
    During a performance review session, the analysis from the CSDS
can be used to develop a constructive give-and-take discussion on
performance and ways to improve. The data on causes of injuries and
near-miss incidents can provide recommendation sources for courses
of action and other points of discussion. If the positive indicators
are not up to the level of the goal, the manager can discuss improvement plans. both the supervisor and the subordinate have access to
the data in the CSDS, and both can track the progress. This provides
a healthy situation, because each knows what the data say and can
verify the information.

CORRELATIONS WITH OTHER REPORTS

Personnel Actions: Additonal Indicators

The personnel department has the requirement to produce many reports and track many conditions within a company. When some of
these reports are combined with the data from the CSDS, they can
be quite helpful in determining causes of certain situations and sug-
gesting alternative courses of action.

The following are classic cases: Suppose an employee who is a
subperformer suddenly has a back strain and calls in sick 2 days
prior to a 3-day weekend. And suppose a good employee with many
year of service calls in sick with an injury that he says he sustained
the week before. Not unusal. But for this employee it is the third
time the same thing has happened in 2 months. Both are suspicious
cases. In both of these circumstances it is wise for the supervisor
to begin a thorough check involving as many parameters of performance as possible.   Some of these are the past injury record from
the CSDS, the illness record from the medical department, the attendance report and lateness record, performance evaluations, and hospi-
talization benefits records. These reports and records can provide
clues to indicate whether or not this current incident fits a pattern
from previous months or is an isolated occurrence. Either way, the
record can aid in determining the appropriate action.

Production Reports

The CSDS reports can be compared with production reports to provide
information that can be useful to safety efforts, as well as production.
For example, for safety, the time of day, day of the week, and day
of the month may be important in determining if there is a relationship
between injuries, certain time periods, and production levels. The
investigation can include production parameters such as line speed,
type of product, quality of maintenance, reject rate, and downtime.
The results of the inquiry can identify factors such as maintenance
and repair schedules and line speed that have a relationship to injuries,
for example. Injuries can also be related to downtime, reject rate,
and higher per unit costs. the injury causes on production lines can
lead to consideration of operational line changes to minimize injuries
and reduce downtime. Injuries on certain shifts may provide clues
to factors such as quality of maintenance, adequacy of operator training, inappropriate line speed, or improper positioning of operator controls.

All of these types of problems can provide incentive to search the
CSDS for information. The searches can be aimed at several points.
Here are some examples:

1. Does a time of day, a day of the week, or a certain shift have a higher incidence of injury cases?
2. Are injuries happening only to new employees or only to experienced employees or both?
3. Based on an evaluation of causes of injuries and near-miss incidents, what causes are listed most often? Is there a connection between certain frequently occurring causes and certain supervisors? Certain employees?
4. With the information from item 1, do the injuries occur when a certain type of product is run?
5. Are injuries predominantly to one part of the body (hands, arms, feet)?
6. Is the site of these injuries a factor (i.e., going to and from the rest room)?

These searches and the resulting information are not new or revolutionary in safety work. These are the types of factors the safety professional always begins to look for in attempting to pinpoint a problem. The differences lie in the speed of the computer and the many combinations of data parameters the computerized system can provide. With a manual system, most of these searches would not be done because of the amount of time required to find the data. With the computer these searches take microseconds. This tool allows safety professionals to do the things for which they were trained: analyze data and design methods to minimize or eliminate accidents. Thus, by correlating the CSDS data and product information, the safety professional can advise those in production management. Together, they can go about finding solutions to problems in production through injury analysis.

Maintenance Reports

Maintenance is interested in keeping the operation running smoothly without unscheduled breakdowns, and maintenance tasks must be accomplished at the lowest possible cost. The injuries that occur to maintenance workers are of interest to the maintenance manager because they reduce the available work force and increase costs. Why those injuries occur is of greater interest.

The CSDS can be searched to aid maintenance in its quest for both reduction of injuries to maintenance personnel and reduction of costs through better preventive maintenance. Identification of injury causes is accomplished by searching the CSDS in the ways just discussed. The improvement of maintenance efficiency is another matter. Most maintenance operations code their work orders on a maintenance computer system. The way to correlate information to improve efficiency is to compare injury times, types of cases, and

specific components with the work-order system. In addition, a
search by component or equipment can be compared with failure rates
to determine abuse and operator error levels. Other file classifica-
tions can be used, such as parts of the body, crew supervisor identity, time of day, day of week, and causal factors. All of these, when
searched in different combinations and compared with maintenance
failure rates and work-order levels, can help determine points where
preventive maintenance periods need to be adjusted.

Because maintenance is concerned with costs, operating efficiency,
and work-load factors, the management of this function should be
quite interested in the report comparisons. Once they understand
the magnitude of the possible cost reductions, they will form a strong
support element for the CSDS.

Sales and Scrap Reports

The number of sales is a function of the number of customer calls
per day, per week, or per month. The greater the number of calls,
assuming that a quality presentation is made each time, the greater
the number of sales. Therefore, a sales manager is interested in
the call-to-sales ratios for salesmen plus the sales volume that each
territory and each salesman is producing.

The CSD system can generate information that can contribute to
the sales manager's analysis of sales. The system is capable of accepting
vehicle-collision data as well as injury information. Searching for
collisions per sales person, the manager can obtain another bit of
information about his marketing operation. For example, if there is
a sales person who is in a good productive territory but for some
reason is not making the expected sales, the number of collisions plus
other factors may provide clues to the problem.

The CSD system can be used in the overall analysis of other problems. For example, the scrap report identifies the number of rejected
units on each product line. If this rate is compared with the rates
of injuries and near-miss incidents during the same period on the
same lines, this may indicate possible causes. The injury causes can
also provide clues to the scrap rate. For example, a comparison of
the level of employee job experience, the injury-case level, and the
scrap rate on a certain line may indicate a control or line-speed malfunction. A special product run with experienced operators on overtime can detect the reason for an unusually high scrap rate. Again,
the CSDS can be used to examine multiple parameters and compare
these with other reports to provide evidence of the source of a problem.

OTHER MANAGEMENT ACTIVITIES

New Developments, Products, and Ventures

It is always exciting to produce a new product, develop a new process,
or move into a new field of endeavor. Always there are risks and
pitfalls. The astute manager tries to anticipate as many problems
as possible, to make allowances and develop alternative courses of
action.

The CSDS can make a contribution in this area with careful searching of the files. For example, a manager is moved to head a new department that is to run a new process. The CSDS data base can provide the following:

1. The types of injuries, the level of severity, the parts of
    the body, and the causes involving similar machines or com-
    ponents or chemicals.
2. Injury histories for the employees in the new department.
3. The numbers of injury cases listed by supervisor.
4. The types of cases that occurred using similar components
    or the same chemicals.

A similar search can be made for a completely new venture. In
the operation to be acquired or set up, many of the potential injury
sources will involve similar functions, such as lifting, material handling,
production line operations, chemical handling, and so on. Searches
of current operations can alert management to potential problems that
can be checked out with managers of these functions as to specifics.
In this manner, another parameter can be added: the ability to analyze
the profit potential of the new venture by predicting potential injury
and damage sources.

Production Changes

Production changes can involve the switching of products or production from one plant to another, one building to another, or one department to another. This is similar to the development of a new product, but not exactly the same. In many cases the problems are more
insidious. For instance, it is usually assumed that a proven process
is always a safe process, having been performed safely in another
area for so long. The manager of this process or production line may
consider moving a process a routine change and therefore may spend
little time in discussing the special problems, other then the ones involving the moving and installation of equipment.

The CSDS can be queried for a historical case-by-case readout of injuries as an intial overview of the product or production line. Central or major causes can be identified to indicate former problem areas. Other areas of special searches could include the following:

1. Chemicals identified as accident causes.
2. The day, time, or day of the week predominant in a majority of cases.
3. Cases with high severity and frequency potentials identified by type.
4. Causes of cases; predominant causes using both injury and near-miss data.
5. Supervisors on the former line who are staying on in the new line. Their crews' injury case records need to be reviewed, with the causes of each case.
6. Employees with two or more injuries within 24 months. Their entire personnel files should be reviewed to see if there are unknown stresses or problems.

This analysis of search data can reveal potential problem areas that may be transferred to the new operations area along with the equipment and the process. It can raise questions regarding work practices and methods. The data can indicate to the manager various sets of conditions that could cause future problems. The CSD system is a valuable information source for the manager that can aid in sound decision making. Good decisions are made better by the use of more complete information.

NOTES

1. Dan Petersen, *Safety Management: A Human Approach.* Aloray, Huntington, N.Y., 1975, p. 60-61.

11

APPLICATIONS: SERIOUS-INJURY SOURCE PREDICTION *

I think that we safety professionals do a pretty good job of predicting
injury sources right now. Much of this predictive ability has been
acquired through our experience with what has happened in the past.
For example, we know that chain-and-sprocket drive mechanisms without adequate guards can cause serious harm if placed near operator
positions. When we see a new piece of equipment that has such an
ungarded drive mechanism near the operator's station, we know that
it has the potential to cause serious harm. Further, during routine
safety surveys we all see situations that, based on our experience,
can reasonably be expected to be sources of harm. In reviewing a
set of construction plans for a warehouse, we may find that part of
the building will have flammables stored there, and proper precautions
have not been taken for this storage. Our experience again indicates
that under the right circumstances a fire could occur. In each case,
the safety professional uses judgment and evaluates a situation based
on past experience.

Taking this logic a step further, we can say that the safety professional's prime task is finding sources of harm before injury occurs
and then minimizing or eliminating those situations. Finding and predicting are much the same thing in this context. On too many occasions

---

*The majority of this chapter is based on an article I wrote for the
December 1981 issue of *Professional Safety*, the journal of the American Society of Safety Engineers, entitled "Serious Injuries Are Predictable."

we spend our time and effort reacting to an injury case in an attempt
to prevent recurrence. This is a worthy effort, and it must be a
continuing job as long as people are being injured. But we can be
more successful if we develop an organized effort to predict sources
of serious harm. If we had to choose between the two approaches,
it would be a tough decision. These two approaches can best be used
together: one, investigating and controlling situations that have already produced accidents; two, analyzing and predicting future
sources of harm and controlling them. Through the rest of this chapter, we shall explore this second approach, introducing an efficient
way to predict sources of serious harm.

If we study any mass data, we readily see the types of accidents
that result in temporary total disabilities are different from the
types of accidents that result in permanent partial disabilities,
in permanent total disabilities or fatalities. The causes are different. There are different sets of circumstances surrounding
severity. Thus, if we want to control serious injuries, we should
try to predict where they will happen.[1]

From a personal study of fatal injuries in one company over a 3-
year period, I found that these cases had different cause factors than
the more frequent lost-time cases. When I compared the fatalities
and the cases of lost time (temporary total disabilities under the American National Standards Institute Z16.1 standard) in which the
workers recovered and returned to work, there were causes and factors in the fatal cases that either were not in the lost-time cases at
all or were in a different sequence. Fatalities occurred mostly in certain kinds of tasks or job categories and were associated with higher
levels of available energy. This is interesting as a study, but what
is its practical use? Generally, the safety professional tries to remedy
a wider range of injury sources than just the serious ones. Why do
we tend to concentrate on these?
   I studied this question for a while and came up with two reasons:
First, if we concentrate on the causes of serious-injury cases, this is
an easy point to sell and promote. Management can readily identify
with a focused effort to find sources of serious harm. The workers
and the labor unions also can support such an organized program,
because it affects their well-being. Second, the justification and reasons for a program to predict serious-injury sources can be well documented. The support of the local community, from a public-relations
standpoint, is a good selling tool for company publicity. Impressive
facts are readily available to help promote this program: Less than
20 percent of all injuries fall in the serious category, but 80 percent
of medical and workers compensation costs center in this area. Also,
management will be concerned because these serious events usually

cause some degree of business interruption and loss of production
and result in high per unit costs. In addition, OSHA and other reg-
ulatory agencies become concerned and send inspectors to plants
where fatalities and multiple injuries have occurred. Therefore, a
program aimed at prevention of the serious event can generate a lot
of support from many sources and can be used as a rallying point
for other safety programs.

   Just to be sure that we are using the same definition, serious injuries include fatalities, amputations, major second- and third-degree
burns, and injuries to two or more persons during a single event.
In other words, a permanent total disability or fatality, according
to the ANSI Z16.1 standard, will be termed a serious injury.

SEQUENCE OF EVENTS

Sources of Serious Injury

Like most safety professionals, I tried for many years to find sources
of serious harm or injury by analyzing the file of lost-time cases.
When a department would show up as having a high number of cases,
I would work with that area. Always there was a variety of slips,
trips, and falls, hand and back cases, lacerations and contusions,
and a few others. Serious injuries happened so rarely that I never
had a chance to find a pattern. I simply had to use all my cases and
hope I could control some part of the problem that way while other
safety program elements like training and inspections did their part.
Overall, I achieved a modest degree of success. The serious injury,
however, still occurred from time to time. This bothered me. I believed that there had to be some orderly way to discover categories
of sources that would lead to better control.

   In reviewing studies by insurance carriers along with a study I
had done, I thought I saw a trend in two areas: maintenance and
construction activities. Then I read a passage in Dan Petersen's book
Safety Management: A Human Approach that indicated that "studies
in recent years suggest severe injuries are fairly predictable in cer-
tian situations."[2] Sure enough, the list in the book contained these
items: unusual, nonroutine work; nonproduction activities; sources
of high energy; certain construction situations.[3]

   I considered these sources carefully. What could be done with
this information? Could I pass it directly to management as a new
program? Probably not, because it was only four pieces of information with no plan for use. How could it be used? Also, the four categories (nonroutine work, nonproduction activities, high-energy use,
and construction situations) seemed to be missing something. I felt
that they were not in harmony with one another as similar categories.
Then I realized how to make the change.

First, I made sources of high energy a separate category, because this is not a work activity. Second, I added a new fourth work activity, called extra health hazard. This is a situation in which there is more than the normal risk to worker health. The following are some examples within this category: using materials whose characteristics are unknown to the worker; mixtures that occur when cleaning chemicals and production chemicals are mixed during the cleaning of production equipment; entering contaminated confined spaces; substitution of gasket material that is incompatible with the liquid in a pipeline, leading to leakage of toxic material.

The first activity (construction) is one that has a certain degree of risk associated with it normally. However, there are high-risk tasks, such as working over water, working in a trench or pit, and working with flammable liquids or materials, that present risks that are higher than normal and should be highlighted as serious-injury sources.

The second category (nonproduction activities) includes groups such as maintenance and repair, research and development operations, and technical services groups that handle special projects in plants for corporate or division engineering staffs. These people get out into the production areas where they normally do not work. Securtiy is another nonproduction group, as are cleaning crews, if those employees are assigned to these crews as their only job duty.

The third activity (rarely done, unusual, nonroutine work) is the kind of job that is done once a year or once every half year. It can be a maintenance job to perform a one-time installation of a machine or other equipment. A breakdown, especially in production line operations, is a nonroutine event. A special one-time production requirement, where a machine must be adapted to do the work for which it was not intended, is nonroutine. A project in which someone must work alone on an experiment is nonroutine, as is other laboratory work that extends into off-hours. Also in the rare, nonroutine category is the production run of a product that is made only once a year. Another rare event is the removal or demolition of equipment. For example, the removal of asbestos piping insulation in a chemical plant would be considered nonroutine.

Unwanted Energy Flow

The harnessing of energy sources to do work is what began the Industrial Revolution. Energy sources provide the power for everything we do in industry, and without abundant energy resources our economy would collapse. Therefore, it is important to see that energy sources, especially sources of high potential, are channeled for safe uses. When energy is not properly directed, it becomes unwanted energy. The level of potential energy that we shall consider to be high is any level

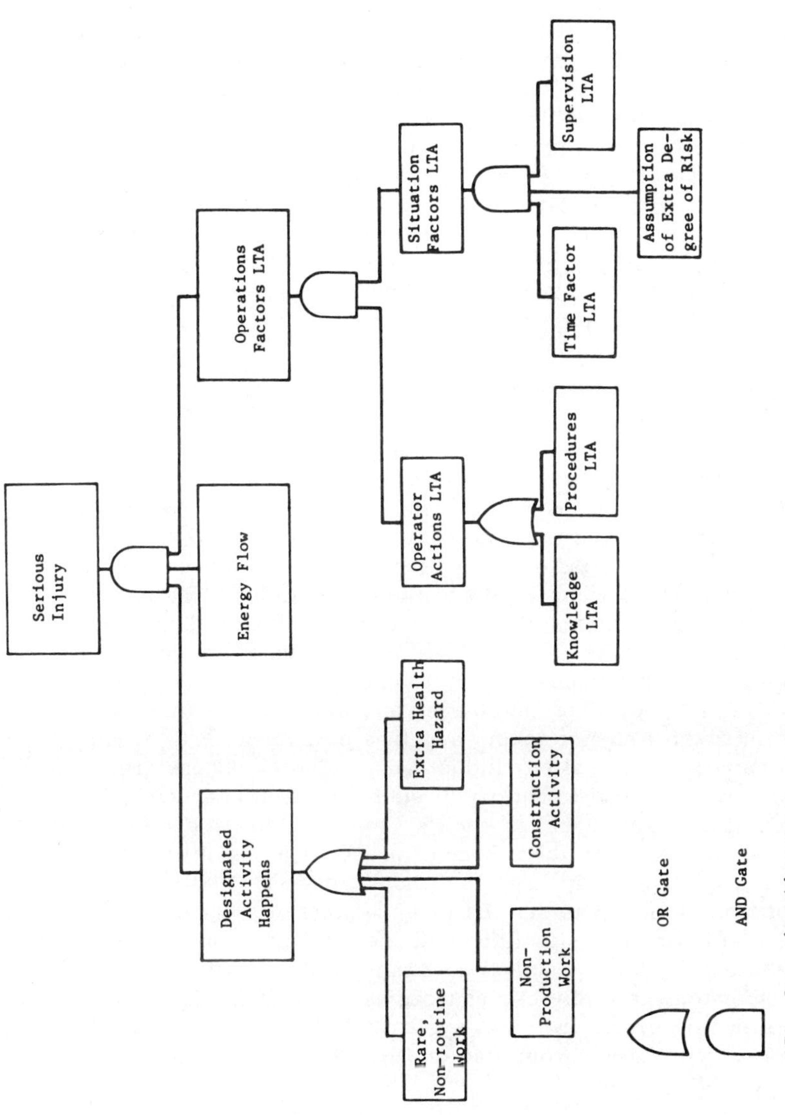

Figure 11-1 Serious-injury fault tree. (Charles Ross, adapted from U.S. Department of Commerce Publication SAN 821-2.)

that, if misused, has the potential to cause serious injury or death. To help organize my thoughts concerning the causes of these serious-injury events, I developed a fault-tree model that I call the serious-injury fault tree. The fault tree depicts three main subevents that must occur together to produce a serious injury: the specific work activities, the use of energy from a high-potential source, and operations factors. Energy sources can be separated into four categories: electrical, chemical, mechanical, and radiative. The serious-injury fault tree (Figure 11-1) explains the relationship among work activities, high energy, and operations factors. All three factors must be involved to produce a serious-injury situation.

Operations Factors Less than Adequate

In our discussion thus far we have talked about the four work activities that are most likely to be sources of serious harm. In conjunction with the work activities we have said that a high-energy source must be involved to cause the main event: a serious injury. The other event, operations factors less than adequate, must contribute equally to cause the serious-injury sequence to occur. The operations-factors event contains elements that are familiar to everyone in safety. These elements are the well-known errors and omissions that both safety and management try to control. Ineffective procedures, inadequate operator knowledge, and inadequate supervision, when combined with chance taking and time pressures, are all contributing causes in most of our industrial accidents. These are also the factors that can lead to control of such events. Therefore, if the source of a serious event is known, control is relatively easy. Predicting the location or the source of an event is the difficult part. Why? Because we normally lump the whole (sources of both serious and nonserious injuries) under the two headings of plant safety inspections and injury investigations and then hope that we find enough physical hazards and malpractices to eliminate all hazards, and in the process catch the really bad sources as well. In conjunction with these two programs, why not systematically concentrate on the sources of serious harm through an organized plan?

FINDING THE SOURCES

Prediction Is Possible

In our model, we have identified four work categories as having the highest potential as sources of serious harm. In other words, to have a serious injury, a work activity must be going on where energy is being used and is inadvertently misrouted and the operational factors

connected with the job are less than adequate. When these factors
exist together, then the potential for serious injury is quite high.
The occasions when all these elements work together are rare. However, when they do, the consequences of the events are quite severe.
All of the theory we have discussed thus far is fine, but it is only
theory until put into practice. For the rest of this chapter, we shall
discuss a plan to use this theory in a practical approach that is designed so that the department manager and the supervisors can make
an effective analysis and develop controls for their serious-injury
sources in their own department area. This is the real payoff.

Effective use of this technique requires a three-phase process.
By following this process, department management can successfully
identify the potential sources of serious harm, analyze the current
control level for each source, and apply additional controls as needed.
The three phases are (1) the introduction, (2) the source list, and
(3) the analysis.

Phase 1, the introduction, is composed of four instructional and
practice sessions on how to effectively carry out phases 2 and 3.
Each session is both an informational presentation and a training
course on how to use the analysis forms and what to look for in developing the source analysis. The informational presentation is designed to stimulate the participants to use the system. Because the
emphasis is on serious injuries, it is relatively easy to develop strong
interest and a desire to use the system. This kind of approach is
different from the usual methods of hazard identification, which stress
reduction of injuries as a generic problem. As a contrast, the purpose
of this program is only to eliminate the serious-injury potential. In
practice, many minor-injury sources may also be uncovered. This is
a bonus to the main purpose and will strengthen the overall safety
and health program. Table 11-1 shows a training outline that is
suggested for use in phase 1, the introduction.

Phase 2 is more difficult and may seem tedious, because it requires an inventory of tasks. Using the first form, the potential
serious-injury source analysis, the department manager and the supervisors list the various jobs or tasks that, when coupled with sufficient
energy use, might cause serious harm. It is here, in actual practice,
where the explanations and discussions of phase 1 will be fully appreciated. To make a list of tasks, the manager and supervisors must
understand what types of tasks they are looking for. In a maintenance activity, for example, those in management might list the types
of activities noted on the form in Exhibit 11-1.

To support the search for significant tasks, the CSDS can be
searched for maintenance cases in which there is the potential for
severe injury. These searches can contribute ideas about the tasks
being listed on the form shown in Exhibit 11-1. The following are several searches that are designed to reveal potential injury sources in
a maintenance operation:

Table 11-1 Introduction to Serious-Injury Analysis (A 4-hour training course)

Session 1. Welcome and introduction of the subject (1 hr)
a. Why analyze only serious injuries?
b. Past cases used as examples of causal relationships
c. A brief overview of the rest of the sessions

Session 2. The source form (1 hr)
a. Brief review of the two forms that are to be used
b. Discussion of the potential serious-injury source analysis form
c. Examples and a practical exercise

Session 3. The analysis form (1 hr)
a. Discussion of how the source form is used with the analysis form
b. The analysis form
c. Examples using the source form from session 2
d. Practical exercise

Session 4. Practical exercise (1 hr)
a. Case exercises using the two forms
b. Review and discussion of the exercise
c. Discussion of plans to use the technique

---

1. List all cases where Frequent/Serious, Occasional/Serious, Frequent/Major, and Occasional/Major are noted (last 24 months). List: Case Number, Type of Case, Weather, Specific Operation, Specific Component, Human Factors, Condition Factors, Root-Cause Factors. (Note that Days Lost is not used, because the intent is to look for potential cases.)
2. List all property-damage cases above $5000 in last 24 months. List for each: Case Number, Type of Property, Location of Incident, Phase of Operation, Weather, Last Inspection, Human Factors, Condition Factors, Root-Cause Factors.
3. List all injury and near-miss cases noted with serious potential involving maintenance. By supervisor, list for each case (last 12 months): Case Number, Type of Case, Phase of Operation, Date/Time, Contaminant, Specific Operation, Specific Component, Root-Cause Factors.
4. For electrical maintenance, list all cases with serious potential: Case Number, Type of Case, Weather, Location, Specific Operation, Specific Component, Nature of Injury/Illness, Human Factors, Condition Factors, Root-Cause Factors. (Could also search for HVAC, plumbing, carpentry, and sheet metal in the same way.)

POTENTIAL SERIOUS INJURY SOURCE ANALYSIS

PAGE 2 OF 4

| LOCATION/DEPARTMENT | DATE *May 17* |

EXAMPLE ACTIVITIES	SPECIFIC ACTIVITIES
<u>Rarely done, non routine work.</u> – Installation of equipment. – Removal and demolition of equipment. – Semi-Annual inspections and maintenance – Break down maintenance <u>Construction Activity</u> – Build support equipment requiring construction. – Install equipment requiring construction of base structures and scaffolding.	7. Install new compressor in Building #4, second floor. 8. Remove the insulation from the steam lines running between Building #6 (Power) and Building #8. 9. Remove the old Nash vacuum pump from the first floor, Building #4 and install the new one which is on the loading dock. 10. Repair the impeller in #8 Process Tank. High priority job. 11. Remove and replace the post indicator in #3 underground fire water line at Building #2 west side. Bad leak. 12. Install new milling machine in the Maintenance shop. Includes building a concrete pedestal and providing electrical power. 13. Remove and replace the work platform which is broken at the 2nd level on #2 power boiler.

Exhibit 11-1

5. For all leg- and arm-injury cases that list serious potential, for maintenance operations over the last 24 months, list for each case: Case Number, Type of Case, Location of Incident, Specific Operation, Specific Component, Human Factors, Condition Factors, Root Cause Factors.

These searches are directed toward discovery and confirmation of those types of activities that have the potential to produce serious injury. The CSDS special searches can provide additional information by requesting summaries of each of these searches. The tasks thus identified can be evaluated, and those that appear significant can be added on the form shown in Exhibit 11-1.

Some of the activities fall into each of the four designated activity categories. The main emphasis in phase 2 is not to analyze but to list as many tasks or jobs as possible that fall into the activity categories. Phase 3 is the point to be concerned about the amount of potential energy that is available.

Phase 3 is the analysis phase of the project. For each task or job listed on the source form, an analysis form is prepared. On this form it is necessary to describe briefly the type of job or task that is to be performed. The example form in Exhibit 11-2 indicates a completed analysis of a welding job to be performed in a chemical tank (listed as item 10 on the source form in Exhibit 11-1). In section 2, the preparer checks the activity category or categories that apply. The total number of the activities that are checked is put in the box on the far right. In this case, three boxes were checked: The job is a nonproduction job; it is rarely done; it presents a potential extra health hazard. Next, section 3 is divided into 3A (Energy Input) and 3B (Energy Sources). For this job, the welder is lowered into the tank, thus entering a confined space, and he is joining two parts together, for a total of three boxes checked. In section 3B, Chemical and Electrical were indicated as energy sources because of the former contents of the tank and the use of the arc welding equipment. These two sources are totaled in box 3B on the right. In section 4, the total of the boxes checked in sections 2, 3A, and 3B is noted and summed up to arrive at the figure for the Serious-Injury Potential for that task. This number is also written in the box for the Serious-Injury Potential in Section 6.

Section 5 is the other half of the analysis exercise. This involves a listing of the amount of operational control that is exerted over this job or task. Within this section are the accepted methods of control with which we are familiar. These include use of procedures, permits, checklists, knowledge, supervision, and safety features. The total of the boxes in this section that are checked as being done during the task or job is entered as the total in the section and in the box labeled Operational Control in section 6. In this case, five boxes were checked.

POTENTIAL SERIOUS INJURY ANALYSIS

DATE _May 17_

1. Describe the task or job to be performed. *10. Repair the impeller in # 8 process tank. Requires welding two gusset plates to each portion of the shaft where it joins the blade. The tank has been washed down but there may be residual process chemicals still in the tank.*

(Definitions on the back of the form)

2. **ACTIVITY CATEGORY** (Check all that apply) ☒ Rarely done, Non-Routine Work ☒ Non-Production Work

   ☒ Extra health hazard ☐ Construction Activity ☐ Other (Specify) _______________

   TOTAL OF ACTIVITIES CHECKED ☐ 3

3. **ENERGY INPUT** (Check all that apply)

   A. Does this job or task require the part, item, machine, material and/or person to be:

   ☐ Elevated ☒ Lowered ☐ Moved on same level ☒ Enter a confined space ☒ Joined/connected

   ☐ Changed in size or shape ☐ Work in pit or trench. TOTAL CHECKED 3A ☐ 3

   B. **ENERGY SOURCES** ☐ Mechanical ☒ Chemical

   (Check all that apply) ☒ Electrical ☐ Radiative TOTAL CHECKED 3B ☐ 2

4. **ACTIVITY/ENERGY ANALYSIS**

   Total boxes checked in: 2. ☐ 3 + 3A ☐ 3 + 3B ☐ 2 = TOTAL ☐ 8

   Serious Injury Potential Enter total in same box in Section 6.

5. **OPERATIONAL CONTROL** (Check all that apply)

   Procedure: ☒ There is a procedure for this task/job ☐ Procedure is current ☐ Operator has/has access to a copy

   Permit: ☒ Welding Permit Required ☐ Lockout Procedure Required ☐ Confined Entry Procedure Required

   Checklist: ☐ There is an operational checklist for this job. ☐ Operator has a copy

   Knowledge: ☒ Operators have been thoroughly trained ☒ Operators have thorough knowledge of Job/Task

   Supervision: ☒ Prior to start, supervisor checks the Job/Task each time ☐ F/u by supervisor is done each time

   Safety Features: ☐ There are Interlocks and Fail Safe mechanisms in operation to minimize harm

   TOTAL BOXES CHECKED ☐ 5 Operational Control — Enter this total in same box in Section 6.

6. **OPERATIONAL CONTROL VS. SERIOUS INJURY POTENTIAL**

   Serious Injury Potential ☐ 8

   ______ = ☐ 1.6 Level of Control

   Operational Control ☐ 5

   If Control Level is:

   < 1.0 it is adequate

   > 1.0 refer to item 7

7. **ADDITIONAL OPERATIONAL CONTROLS TO BE USED.** When the Level of Control is greater than 1.0 more control should be applied. Refer to Section 5 and provide a plan of what other controls will be used and how they will be accomplished. Refigure the Level of Control with the addition of the other controls.

   Procedure • Has been revised and is now current
   • Operator has a copy

   Permit • Confined Entry Permit will be used

   Supervision • Supervisor will F/u.

   Serious Injury Potential ☐ 8

   ______ = ☐ 0.8 Revised Level of Control

   Operational Control (Revised) ☐ 9

   LOCATION *High Point Plant* NAME *Carl A. Smith*

W9-2308 5/80

Exhibit 11-2

Section 6 is an estimation of whether or not the level of Operational Control is adequate, in view of the estimated Serious-Injury Potential. By dividing the Serious-Injury Potential total by the total for Operational Control, a level of control can be estimated. If the result is a number less than 1, then control is considered to be adequate. If the number is equal to or more than 1, then additional controls are necessary to reduce the potential for serious harm. Note that the calculation in section 6 shows a 1.6 quotient; thus, not enough controls are being used. The additional controls are applied and noted in section 7. These include additional procedures, a permit, and additional supervisory follow-up.

The analysis form can be used with an initial study of potential sources. It can also be used as other jobs or tasks are recognized as potential serious-injury sources. This allows two things to happen: One, an initial analysis of the known sources of harm in the department is performed, and this analysis is documented as to the level of control that should be exercised during these jobs to make them safe. Two, additional tasks or jobs can receive the same scrutiny. Thus, the analysis serves as a check for future tasks of the same nature.

Prediction Is the Key Point

Controlling potential sources of harm before they cause injury is what safety professionals have been doing for years. It is an important aspect of our work. In many cases, however, injuries occur before discovery of the potential source. The injury requires the safety professional to react to the situation and work to prevent further injuries from that source. It seems that we never get ahead of this problem. It is a position of reacting to unexpected situations. The better way of working is to act on the basis of planning that leads to control of the situation.

With a technique such as the one described in this chapter, the safety professional can develop a program to plan a control strategy and then act, instead of always reacting. Admittedly, all sources of harm will not be found. However, with a good level of effort, persuasiveness, and a plan, many serious-injury sources can be identified and controlled. This technique is designed to predict those sources of harm that stem largely from human error. Because this is the major source of serious injuries, this is a results-oriented approach. This technique will generate support from both management and workers, because it is a planned, specific approach to a single problem. Serious injuries are feared by all people; therefore, there is a natural desire to eliminate these types of injuries. Of course, all of the injury sources that are identified will not have a high serious-injury potential; therefore, there will be a broader base than just the serious category.

As the program progresses, management should be informed of
progress. Because the program is based on the two forms discussed,
it is statistically easy to track. For example, an analysis program
for serious-injury potential can be effectively developed and reported
to management and to employees, if appropriate, using some of the
following parameters:

1. Raw numbers of tasks listed by department.
2. Tasks listed that have been analyzed.
3. Of those analyzed, the number that are in the serious cate-
    gory (From the work thus far at one company, seven or
    more are likely to be serious.)
4. The number of nonserious sources found.
5. The average level of control required for both serious and
    nonserious sources.
6. The most dangerous tasks (Note: Caution should be exer-
    cised in using this measure, because a union may insist on
    special pay for such work.)

A report such as the one described here can give management an
indication of the potential savings involved in the program and lead
to more enthusiastic support. Potential savings can be estimated by
using the costs of injuries from the National Safety Council's current
Accident Facts booklet. The average cost of a lost-time injury can be
derived from the estimate of total industrial accident costs. For 1980,
for example, this is found on page 24. This number is divided by
the number of industrial lost-time cases to arrive at a per case figure.
This case cost is then multiplied by the number of all the injury-source
tasks that have been identified and are under effective control to
arrive at what the approximate dollar cost would have been if such
sources had produced injuries. Because this is a discussion of serious
injuries, the total cost figures, which are for all lost-time injuries,
will be quite conservative and thus believable. Cost savings can be
a powerful factor to keep up support for the predictive approach. It
is an approach that management will find to be sound, because it begins with careful planning; it is not a sweep up the mess reaction
after the fact. For once the safety professional will be acting, not
reacting.

Prediction of serious-injury sources is a practical and useful approach. This system, which is currently being used at one company,
can be used by other companies to gain the same benefits. The concept is what is important. This is a technique for predicting sources
of serious injury that begins with an action posture, not a defensive
reaction. It is a program that can be planned and carried out. The
results of the cost savings can be conservatively estimated with a

fair degree of accuracy. Along with management, every worker in
the facility can identify with this type of safety program, because
no one wants to be killed or maimed, nor does anyone want such a
fate for a fellow worker. Therefore, overall, a predictive program
to identify and eliminate serious-injury sources can serve as an
action base for a strong, positive, well-managed safety effort that
is action-oriented.

ACKNOWLEDGMENTS

Special recognition for contributions in the development of the analy-
sis form is due to Charles E. Vejvoda, CSP, Division Safety Manager,
Westvaco Corp., New York, New York, and Patrick J. Evers, Mill
Safety Director, Westvaco Corp., Wickliffe, Kentucky.

NOTES

1. Dan Petersen, *Safety Management: A Human Approach.* Aloray,
    Huntington, N.Y., 1975, p. 19.
2. Ibid., p. 19.
3. Ibid., pp. 19–20.

12

FUTURE DEVELOPMENTS

Real joy comes not from ease or riches or from the praise of men, but from doing something worthwhile.

Sir Wilfred Grenfell

This book has been written with one purpose in mind: to provide the safety professional with the ideas, materials, and methods necessary to design, evaluate, and promote a computerized data system to track injury cases and identify potential sources of injuries. For this reason, there has been little mention of other applications of the system and additions that can be made to the system.

This chapter will discuss one application and two additions that can be used with the basic safety data system: One, the application of identifying additional sources of harm through the technique called incident recall. Two, the addition of industrial-hygiene data and a collection-and-storage module to the safety data system's information files. Three, the addition of a module for collection and storage of employee medical and health data. Of these three, the incident-recall technique application will be covered in detail, because it can be applied directly to the normal safety data system operations without additional programming. The latter two are additions to the system and will be reviewed broadly regarding what is necessary to develop such modules for this system. Complete details for the development of these systems are outside the scope of this book.

THE INCIDENT-RECALL TECHNIQUE

Incident Recall Can Reveal Causes

Every day, workers in plants everwhere observe many conditions that could cause injury or property damage. These situations are commonly referred to as near-miss or near-loss incidents. "Unfortunately, the majority of people who witness such incidents as these do not report them to someone who could utilize the information in time to save lives or avert other serious loss."[1] Why is this so? How do we know this? From attitude samplings and employee questionnaires, the following reasons have emerged in the majority of studies:

> The employee did not recognize the condition as a hazard. It had been there before he began work in the department. He was told that it was part of the job and was told to be careful.
> Peer pressure from the work group not to report: "Where will it get you? They won't do anything about it anyway."
> If it does not interrupt work, let it go.
> Reporting may affect the safety record in the department and lose a prize or an award.
> Fear that the company provides bad medical treatment.
> Reporting a hazard may get someone in trouble with management.
> Reporting means paperwork and red tape. It is not regarded by supervision as important. Not much is done when hazards are reported.[2]

Everyone in the safety business recognized that incidents occur every day that under different circumstances could have caused injury, death, serious property damage, work stoppage, or worse. These near-miss situations are brought on by the same symptoms, malpractices, and bad conditions that cause accidents. The trouble is that they occur time and time again without correction until finally there is an accident or property damage. The investigation of the accident usually confirms that this practice or condition had recurred many times in the past without reports.[3]

Thus, it is logical to use some method to try to find out where these practices and conditions are occurring and to correct them. In Chapter 11 we discussed another technique with a similar purpose: concentration on identification and discovery of sources of serious harm. The incident-recall method is an expansion of this method. Its purpose is to detect additional causes and sources of harm on a broader scale, not just to identify the serious ones.

The Interview: Heart of the Technique

As has been said, the purpose of the incident-recall technique is to detect causes of near-miss and near-loss incidents and situations that have yet to cause harm or damage. This means talking to workers about potential accidents. But there can be problems. Workers may not realize there is a danger. They may not understand what the

interviewer wants. They may believe that the operation is safe: "You
have to be careful at work anyway." In addition, workers may feel
that the things they must watch out for are "givens," part of the
job, and that discussion of such conditions will cause fellow workers
to be disciplined.

There are several ways to approach the problem of finding the
information. One of the more successful ways is to create an atmosphere of mutual trust in an interview, where ideas can be exhanged
and good information can be obtained. The level of trust depends
on several factors:

Privacy of the interview session.
Anonymity for the interviewee.
Length of the interview (15—25 minutes).
Absence of any blame fixing.
Degree of importance attributed to the information received by
    the interviewer.
Feedback on corrective actions taken.
Recognition of a contribution to the safety of the plant.

One of the best people to conduct these interviews is the first-
line supervisor. Although busy and often overworked, this supervisor has the most to gain in making the department safe. The first-
line supervisor understands the operation and knows the people and
is the one who will initiate corrective action. The supervisor investi-
gates most injury cases and has been trained in interview and investigative techniques. To develop incident-recall interview techniques
might require a brief orientation and practice session. If the orientation session is scheduled as a follow-on to a successful safety program,
such as detection of serious-injury sources, then acceptance will be
high for the session and for the incident-recall technique itself.

By following the items that have been listed as those that will
develop trust, an interviewer can develop good rapport with a worker.
This is fine, but the hard part is getting the worker to discuss specific hazards, malpractices, and near-miss incidents. To prepare for
the interview, and to overcome this problem, there are three things
to do: One, prepare a list of the general parts of the process, and
list the worker's job tasks. Second, design questions that will prompt
the worker to describe near-miss incidents. Three, develop examples
to help the worker remember other near-miss events.

The first part of the interview preparation process is to develop
a list of operations that go on in the department. This can include
specific job duties in which the worker will be personally involved,
as well as co-worker duties. Also, the list can contain the operations
that this worker can observe during a normal workday. This will

include both maintenance and cleaning operations. All of these can be made into a checklist to review during the session to make sure nothing is missed.

The second part of the preparation is the design of questions to let the worker understand what you are looking for. The worker may not realize that a procedure or practice is unsafe. The question, therefore, should be designed to encourage the worker to discuss the kinds of events that have happened, particularly those that have seemed unusual, strange, or bizarre; For example, incidents in which someone has had a close call or has almost damaged something, but "luckily" nothing happened. Putting things in this context, the interviewer is more likely to generate discussion. The worker must understand clearly that there will be no blame fixing. The events can also be discussed without names, if necessary.

The third step is to develop a list of examples that may jog the memory of the worker to recall additional events. The CSD system is ideal for this kind of project. Searching for potentially serious injuries and property damage can bring up a list of cases to use as examples. Another use of the CSDS is to search for those areas where there have been the highest incidences of minor, but potentially serious, cases. What will this do? This search can identify the areas of high potential in which to conduct interviews. The report will help select the workers to be interviewed by virtue of the fact that they work in the high-potential areas.

We said that the first-line supervisor is the logical person for the interview task. In most cases this is true. However, depending on internal structure, policies, and company philosophy, this supervisor may not be the logical choice. The incident-recall system is based on developing a trust relationship between the interviewer and the worker. The worker must have confidence and trust in the interviewer. The worker must believe that the information given will be used to make the operation safe, not as a means of spying on fellow workers. Therefore, the decision who the interviewer should be will depend on the factors mentioned earlier and others, such as these:

> The union/management relationship (if there is a union).
> The individual department relationship between management and labor.
> The company political climate (Is the supervisor regarded by the workers as one who will support them?)
> Community factors and relationships within the company.

These are some of the factors that should be considered when making the decision who should be an interviewer. For example, if the union/management relationship is strained by many past differences,

there may be a low perceived level of employee trust. However, if
in an individual department the manager has treated the employees
fairly and the supervisors work at keeping good worker relationships,
then, in spite of union/management differences, there may be enough
openness to allow these supervisors to conduct the interviews. In
other departments the decision may be just the opposite. Therefore,
the decision to use the first-line supervisor as the interviewer must
be assessed each time in each department.

The alternative to using the first-line supervisor is to use safety
staff members, professionals from the personnel department, and the
plant nurses. All of these professionals can perform interviews and
are attuned to observing and listening well to responses. For the
plant nurses and the personnel professionals, some orientation to the
techniques will be required, but, using the three points of preparation, the interviews should go smoothly.

So far we have discussed the most likely candidate to perform the
interview, preparations for the session, how to use the CSDS in finding
areas of high injury potential, and some specific employees to begin
the interview process. We have also discussed alternative candidates
to serve as interviewers. Another part of the interview process is
how to record the data that will be received from the employees. Probably one of the better devices to use is the standard injury report
form. This form can be modified if it does not have a space to record
incident-recall events. Therefore, by completing an accident report
form, the information can be entered into the CSDS and a proper
follow-up initiated. This way, the employee can receive feedback
about the information. This also adds another case to the data base,
which is always a plus.

Surfacing More Incidents through Additional Interviews

One aspect of this technique that may have the highest payoff in terms
of information about near-miss incidents is the second and additional
interview sessions. The lead-in for the second session, for example,
can be feedback about the action taken or contemplated concerning
the incidents discussed in the previous interview. In the second and
additional sessions the interviewee may remember some occurrences
not mentioned in the first session. Also, because the data will have
been analyzed from initial sessions conducted with other employees,
more examples can be discussed. These new ideas and examples can
lead to more incidents that had been forgotten. Between the feedback
and the additional cases discussed, the employee may be able to provide several more incidents and ideas. This is where the employee
may begin to discuss the subtle happenings in the department. This
is especially true of obscure health effects that the worker believes
are just part of the job. These may not have been mentioned because
"that happens to everyone here."

The format for the second and additional interviews is basically
the same as for the first, with one exception. The events identified
in the previous session are discussed, and feedback is provided regarding action taken or contemplated. Once the employee sees that
the reported near-miss incidents are seen as important and corrections are being made, the level of trust between interviewer and
employee will increase proportionally.

Using the Safety Data System for Correlation

We discussed how the CSDS is used to prepare for the initial incident-
recall interview. Following the interview, the incidents are entered
into the CSD system in the near-miss file. Thus, the CSDS can be
searched and can provide ideas for the additional sessions. As the
process proceeds, it will generate a data base of near-miss incidents,
This part of the base can then be searched, with reports being generated to correlate these cases with the injury-case data base. Both
files can be searched together to gain insight into the actual levels
of incidents that are occurring. In addition, the technique can be
a major source of data on which action can be taken before injuries,
property damage, and production losses occur. The incident-recall
technique, like the serious-injury source analysis, is an active re-
sponce technique whereby the safety professional acts in advance of
the events. Effective use of both techniques can change the safety
department from a reaction organization to a pro-action group.

As the data base on near-miss incidents grows in numbers of
cases, it can provide an interesting kind of information as a resource.
It can provide causal information about events that have the potential for harm. By concentrating resources on this type of case, the
safety professional moves into the area of accurately predicting and
controlling potential sources of harm and loss. It is exciting to con-
template the time at which 90 percent of the cases reported and analyzed by the CSD system will be near-miss events. The cost savings
in production can be significant, and the reduction in human suffering
will be extremely gratifying. ●

THE INDUSTRIAL HYGIENE MODULE

Health-Hazard Tracking

Worker exposures to hazards that affect health are of concern to all
safety professionals. Chemicals are used in every industry. This
means that workers are exposed to many hazards from toxic chemicals.
Therefore, a natural addition to the CSD system will be an application
to track and measure these exposures of workers. An industrial

hygiene application can provide the capability to record measurements of worker exposures to chemical materials such as chlorine and benzene and to record exposures to physical agents such as noise and radiation.

The collected data can be analyzed and entered on the CSDS to be used to determine the most feasible method of control. Some of the information required for the industrial hygiene data base is already on line in the CSD system:

Organizational Unit.
Expense Center.
Location of Incident (sample).
Date/Time.
Type of Case (three ways).
Phase of Operation.
Specific Operation.
Specific Component
Contaminant (two materials).
Recommendations (completed and remaining).
Batch Number.
Social Security Number.
Employee Number.
Job Code.
Employee Name.
Supervisor Name.
Supervisor Employee Number.

The previously established CSDS data will be included on the industrial hygiene input form when it is established. The additional data files that need to be considered for the industrial hygiene application are as follows:

Sample Number (for reference to the work sheets on that sample).
Duration (in minutes).
Methods: the sample method, the analytical method, and the
    laboratory codes (internal, AIHA-approved, outside, etc.).
Personal Protective Equipment Used (type and approval of the
    equipment).
Level: the measured level of the sample; the file designates
    whether it is an area or personal sample.
Overstandard: designates the standard and the approval level
    (OSHA action level, TLV, STEL, company standard).

In Appendix H are two articles from the *American Industrial Hygiene Association Journal* that discuss computerized systems used in two companies. These systems can be used as idea sources when deciding how to set up an industrial hygiene application. These articles, which were printed in the journal 2 years apart, show the changes in data processing capability from keypunch batch processing to user-oriented systems using intelligent video display terminals (VDT). There are ideas and formats in both systems that can be employed in the CSDS application. The variables are the desires of the system designer and the needs of the users of the particular system.

MEDICAL DATA APPLICATIONS

A companion to the safety and industrial hygiene data files is the medical information on exposed workers. This information is valuable from the standpoint of diagnosis of potential problems in the workplace and review of the overall health of the employee population. Through the use of computer tracking programs, the health and prevention data can be updated and expanded.

In addition to the usual sorting of data, computer routines can be used to detect subtle problems, as well as hypothetical conditions. The caution in this area of medical information, however, is for protecting the worker and avoiding problems of unauthorized access. With the advent of privacy legislation and government requirements for access to medical records, methods of safeguarding employee privacy must be considered when developing this module.

Medical Review: Trigger for Further Industrial Hygiene Studies

A computerized medical file of employee health information is useful in two ways. First, the file can be searched for an overall review of employees working in certain departments where toxic materials are being used. This can be used as a check to see that existing controls are effective. Depending on the outcome of the overview, further industrial hygiene studies may or may not be indicated. Second, searches of the medical file can show abnormalities in employee health status. If these persons work or have worked in the same department area, this may indicate a need to perform industrial hygiene studies in this department area to determine what action may be needed.

In addition to the foregoing, the computerized medical data file is useful in performing epidemiological reviews of the work population. These studies can reveal chronic exposures to chemicals or other subtle problems not readily identified through industrial hygiene reports based on formal TLV or STEL values. In short, the computerized medical data file can provide checks and balances against industrial hygiene work.

The computerized medical data file can also have the normal uses of ensuring that all employees who work with toxic materials, such as lead, or who are exposed to high noise levels receive periodic screening tests to monitor their health status. Thus, the system can act as a scheduler and reduce the administrative burden on medical personnel.

Medical Records: The Risks

To protect themselves from fraudulent claims, many companies develop and maintain certain amounts of medical data and records. Records are also required to be kept by government regulation. For example, federal requirements mandate that records must be maintained for hearing-conservation examinations and for lead exposure screenings. Companies in the chemical and pharmaceutical industries screen their employees who are working with materials that have been found to be or are suspected of being toxic and are not regulated by law. These screenings are not only to protect the worker but also to protect the company from costly future claims because of health impairments.

On the other side of the argument, the law states that if companies maintain medical records, the employee, or his designated representative (with the employee's permission) must be allowed to see and have copies of the medical records. This presents a risk to the company from two directions: One, if there has been an exposure, the employee can obtain copies of the records and use them as evidence to support a claim. Two, if no records are kept and the employee believes that there is a significant health-damaging exposure, then a complaint can be filed with federal or state OSHA. This action can trigger an inspection, and if the exposure is confirmed, there can be a fine and a claim by the employee is there is health impairment. It is plain to see there are risks on both sides of the argument.

Another aspect of the problem is the involvement of the labor union. Their concern is worker protection, and to show their membership that the union is doing something about it. If the medical records show exposures in certain departments, the union may ask for hazardous-duty pay or other benefits for the exposed members. They may call in OSHA to inspect or may ask to have the union's industrial hygienist study the problem. They may do all of those things. Whether or not records are kept, there are risks to the company from these outside forces called in by the union. If records are not kept or exposure measurements are not made, the union can file a complaint with OSHA and have them perform an inspection and take measurements. The results can be citations and fines. This can be followed by the union's action to bargain for more worker benefits because of the added hazards.

Probably the better choice of action is to make exposure measurements and keep medical records of preemployment physicals, medical histories, and periodic screening tests for various exposures. Although these records can be obtained by the employee and probably the union, this can be an example that shows genuine concern for employee welfare and a responsible corporate attitude. It appears that having records and doing what is necessary to protect the employees who are exposed is an easier position to defend than the position in which the company can be accused of doing nothing.

Preemployment Physicals and Job Specifications

The preemployment physical and the specifications for the individual job go hand in hand for hiring employees. There are so many regulations and laws that protect the employee after hiring that the company must perform adequate physical assessments of candidates seriously considered for employment. Many required parameters of the physical can easily be set up in data files within the medical system module based on assessments of job requirements and limiting factors.

The job specifications constitute a factor that sometimes is taken quite lightly and is a source of controversy when marginal or handicapped individuals are, or must be, considered for employment. There is a link between job duties and injuries. Many times, the way the job is designed sets up a booby trap for an injury pattern. Many back injuries, for example, are caused in this manner because inadequate attention is given to the biomechanical aspects of the job.

There are two choices in this kind of situation: One, to continue as before and provide physicals, but make little tie-in with regard to job duties and specifications. The result of this course action is almost predictable. There will be the same or higher compensation-claim costs. In addition, there may be discrimination charges and investigations. To combat this set of results, hiring can be narrowed to only those physically strong enough to do the job without injury. Of course, federal and state requirements that mandate the hiring of women and the handicapped may interfere with this approach. The other option is to establish a process to provide job specifications that accept a wide variety of applicants. This set of specifications can be based on present-worker job-performance studies of the actual physical requirements of the job. If the studies show that only a few candidates can fill the job, then modifications should be made to widen the acceptability level. For example, mechanical aids in manual handling tasks can change the work condition on the job to allow women to be in the acceptable range for candidates to perform the work without the potential for injury. It is a large task, but many companies have already successfully developed job specifications that allow selection of entry-level job candidates who have demonstrated the basic skills

necessary for the work. The selection criteria are sufficiently broad
to accept women and, in some cases, handicapped persons into positions formerly held only by men. Companies such as Georgia Pacific
and Western Electric have ongoing programs to achieve a balanced
selection of candidates.

FINAL THOUGHTS

Not a Conclusion, but a Beginning

This book is an introduction to computerized safety data systems.
Its intent is to show the reader how to go about developing a system.
At the very least, the book is designed to inform the reader of what
is possible with such systems and why it is advantageous to use data-
processing equipment in safety work. The book is written as a basic
how-to text, assuming that the reader wants to learn how a system can
be developed.

The reader's own needs constitute the determining factor. If,
because of this book, more CSD systems are created and new and
better techniques are found, then its purpose will be accomplished.
This book is not intended to be a final document, but a motivator. It
is intended as another contribution toward creating a larger body of
knowledge about the uses of computerized safety data systems. It
is to be hoped that this book will encourage others to contribute their
experiences and knowledge in the form of articles and books on this
important subject. This is the challenge.

The Challenge Ahead

As computer equipment becomes more user-friendly and less costly,
safety professionals will have easier access to develop their own CSD
systems. From the existing body of knowledge will come more creative
systems. For example, to reduce paperwork, more efficient methods
of recording injuries from investigation information can be found. A
computer program can be written to train investigators in the techniques
of investigation through a question-and-answer format displayed on
a user-friendly VDT. A program can be developed to aid investigators
to find clues and causes to accident cases. A link can be made between
a minicomputer in the manager's office and the company's main data
base. This link will allow the manager to use investigative models and
hypothetical programs in which various conditions and situations can
be entered, with the model providing the potential risks and hazards
associated with the various actions. A section of the model can provide options and alternatives to be used to cope with the hypothetical
situation. Another part of the computer system can direct activities

in checking out potentially dangerous processes. Using the CSDS
case files against the hypothetical mode, the system can identify potential failures. Armed with this type of information, the manager
can prepare detailed operational procedures and train operators to
recognize potential failures and be prepared to take recommended
courses of action based on this analysis.

In the natural expansion of computer accessibility and better
data-base management, the personal computer terminal is the next
step. The manager will soon have a personal VDT or minicomputer
in the office. It is estimated that savings of 15 percent or more are
possible in the use of management time through the use of computer
terminals linked to the main data base of a company. In dollars, a
15 percent conservation of management time can show up in increases
in production via thorough planning and better decision making from
more complete and more up-to-date information.[4]

For the safety professional, providing the monthly safety reports
to managers via a VDT allows instant response and information. Following a review of the report, a manager can access the main computer data base and query it directly for additional information on
specific points of interest about the report. The terminal will allow
the manager to take the specific data on injury causes and compare
them, for example, against per unit cost, scrap rate, and manpower
absentee rates to obtain a projection of what effects these factors
are having on the operation. The different variables can be analyzed
to consider ways to increase production without injury or increases
in the scrap rate. The greater the flexibility and availability of this
type of information, the better will be the decision-making ability
of management.

There is no question that there will be productivity paybacks
from the wider accessibility of personal computers and VDTs. To be
a contributor to productivity, the safety professional must be able to
provide effective information about cases and their causes quickly
and efficiently. Being a part of the electronic revolution that is
sweeping across industry will require boldness and persuasive arguments to convince management. Modern safety professionals are such
people. They are able to stand before the leaders of industry and
present compelling reasons for accepting their ideas. These safety
professionals will develop safety data systems for their companies
and write their own papers and books on this exciting subject. They,
too, will be contributors to the continuing beginning.

NOTES

1. Frank E. Bird, Jr., and Robert G. Loftus, *Loss Control Manage-
    ment.* Institute Press, Loganville, Ga., 1976, p. 215-216.

2. Frank E. Bird, Jr. *Management Guide to Loss Control.* Institute
    Press, Atlanta, Ga., 1974.
3. William E. Tarrants, *The Measurement Of Safety Performance.*
    Garland STPM Press, New York, 1980.
4. *The Kiplinger Washington Letter.* Washington, D.C., December
    1980.

APPENDIX A

ANALYSIS OF THE CURRENT REPORTING SYSTEM

When making an analysis of the current reporting system, one can uncover areas of inquiry that were originally not directly addressed. If
these points will contribute to the overall review of the reporting system or to better understanding of the operation of the safety department, they should be included in the final analysis. The following outline of pertinent questions provides us with a set of practical guidelines
for an effective and complete analysis of the reporting system. The
answer to each of the questions should be as thorough as possible
to ensure that no point is overlooked.

A. THE OUTPUT DOCUMENTS

1. Describe by title and content the summary reports that safety
   prepares for management's review. (Include copies of these
   reports with this analysis for reference.)
2. How is the information about numbers of cases presented in
   the report? Are only raw case numbers given, or is there
   a frequency rate computed as well? Is there an analysis of
   the rate or raw numbers?
3. Is there a comparison of frequency rates used? In other words,
   is this month's rate compared with that of last month? A
   year ago? If this is done, what does it show to management?
   What can a manager do with the information?
4. From these statistics, can a prediction be made as to where
   future accidents will happen? Can the sources of serious harm

be identified? Try this: Take a report from 2 years ago
and predict the injury sources and trends for this month;
then confirm your findings with the current report.

5. In the report, what kinds of cases are reported? Does the
    list include lost-time, medical-treatment, first-aid, near-miss,
    property-damage, and other incidents? All or just some? If
    the current system includes just some, find out why.

6. What is the history of these reports? How did they come into
    being? Who developed them? If possible, find out why these
    particular reports are used. If you were the originator of
    this set of reports, describe the process of development, the
    decisions and the compromises made during the beginning
    period. For example, during the development period, were
    any compromises made that drastically changed the original
    concept? If so, these should be explained and included in
    the examination.

7. Are accident costs computed and included in the report? How
    are these costs computed? What is the degree of accuracy
    of the compiled costs? (Are they estimates or actual figures?)
    Are both direct and indirect costs included in the report?
    If not, why?

8. Make a block diagram based on the organizational chart for
    distribution of the output reports. Describe why the routing
    is organized in this manner.

9. What redistribution is made by those on the distribution list?
    What is the effect of the redistribution? That is, what happens?
    For example, feedback to the original senders, calls to safety,
    nothing, what?

10. What kind of feedback comes from these reports?

11. Describe how costs are compiled and how they are handled in
    the report. What is the accuracy of these costs? Are they
    all estimates, or are they actual costs? Are indirect costs in-
    cluded in the report, or are only direct costs used?

B. THE INVESTIGATION SYSTEM

1. What accident events get investigated? List and describe the
    types of events that are investigated and the thoroughness
    of the investigation.

2. What is the history of the current reporting system? Who
    started it? How did it develop? Describe in detail.

3. How are investigators trained? What type of training course
    is used? Is there a course outline and training materials?

4. What kind of information is asked for by the investigation
    form? (Include a copy.)

5. How is the investigation form routed? Include the routing
    from notification of the event to filing of the report. All
    review levels should be listed.
6. What type of insurance or compensation forms are used?
    Describe their use, routing, and contribution to the reporting
    system.
7. How well does the system function? Rate the quality of the
    system.

C. FILING AND USING CASE RECORDS

1. Describe the system into which the cases are put.
2. How are requests for information concerning cases handled?
3. In developing reports on the case files, how long does it take
    to make a simple search? A complex report search?

D. GENERATION OF ACCIDENT COST INFORMATION

1. What types of direct costs are compiled regularly? Where do
    these costs come from? How accurate are they?
2. If indirect costs are developed, how is this done? What are
    the sources of such information? What is the level of accuracy?

APPENDIX B

A PHARMACEUTICAL COMPANY'S SAFETY DATA SYSTEM

On the following pages are the user's manual, the input error rules, the investigation form, and the input forms which are adopted from a system used by Schering-Plough Corp. The system they use is called SAMIS, which stands for Safety Management Information System, and is the system that is used as an example system throughout this book. It is a third-generation system, having evolved from two previous systems, and thus it is flexible and easy to adapt.

The input forms used with this system are designed for card processing. As such, the forms are color-coded. Green ink is used with the prenumbered case input form, and maroon is employed to identify the change form.

In Section A, Identity, there are two versions of file 18. One is the file that was used originally in the SAMIS system. This is the file called Specific Operation Involved, and it lists all the questions within a manufacturing operation. The other file, Physical Movements or Operations, has been used by others in place of this file. Both files are useful. One must decide whether one file is more useful or whether both should be used. There is no right answer: it is a judgment issue.

SAMIS

Safety Management Information System

User's Manual

INSTRUCTIONS FOR CODING CASES

1. <u>Definitions of Terms</u>

   Case - Any input into the system, e.g., Injury, Safety Survey,
         etc.

   D/N - Did not

   Dn - down

   Fite - Fighting

   Gds - Goods

   Gnd - Ground

   Hvy - Heavy

   Loc - Location

   LTA - Less than adequate

   Mtr - Motor

   Pdt - Product

   PPE - Personal Protective Equipment

   WX - Weather

2. <u>Coding Rules</u>

   a. Always code cases in pencil

   b. All alphabetic letters are written in capital letters

   c. A zero is written Ø and an alpha O is as written

   d. The letter Z is written Ƶ

   e. The letter i is written I. The numeric one is as written

SAFETY MANAGEMENT INFORMATION SYSTEM
(SAMIS)

Section A Identity   Items 1 to 8   Page 1

1. Case Number (Preprinted)

   a. Type

2. Division/Corporate

   Code

   04 World Headquarters
   03 Research & Development Division
   01 Pharmaceutical Manufacturing Division
   02 International Division

3. Organizational Unit obtained from table of organizational unit
      codes.

4. Expense Center Code obtained from table of company expense center
      codes.

5. Location of Incident Plant/Facility (2), building (2), floor (1)*
      *Note: Basements are coded Ø.

   Code   Code

      00   Not Co. Prop.   10   Gainesville
      01   Bloomfield   11   Twin Brooks
      02   Union
      03   Kinelworth
      04   Niles
      05   San Leandro   88   Beisa
      06   Dallas   89   Manati
      07   Chamblee   90   Florham Park
      08   Madison   98   Overseas, International
      09   Lafayette   99   Sales Force, U.S.

   Other than building locations

   Code

   XX991   Sidewalk
   XX992   Street
   XX993   Pkg. Lot
   XX994   Grass
   XX995   Stairs

6. Date of Incident Year (2), Month (2), Day (2)

7. Time 24-hour clock

8. Property Cost

   Code for 8, 9, 10, 11

   A000000 No Factor (same as no data)   E - Estimated Cost
   E000000 More Data   A - Actual Cost

<u>Section A Identity</u>   Items 9 to 12   Page 2

9. <u>Medical Cost</u>

10. <u>Compensation Cost</u>

11. <u>Other Contributing Costs</u>

   Total of 7+8+9+10 = Total Cost will be coded and totalled by the
          machine.

12. <u>Type of Case</u> (3 selections available)

   Primary must be completed. Secondary and Other are important and
   should be coded if one is available. Be concerned with the source
   of the incident and not the result; i.e., broken leg is the injury
   but the type should be Fall SL.

Code   Code

  <u>20 Injury/Illness</u>

020   Breakdown (machinery or   640   Qual. Prob. (Quality)
        Equipment)   660   Reaching
040   Caught (In or Between)   700   Slip (or Twist)
060   Chemicals   720   Spill (Splash)
100   Dust   740   Sports (Company sponsored)
120   Elect. Sh. (Electric   760   Strk. Against (Struck)
        Shock)   780   Struck By
140   Explosion (See 80 for   800   Temp. Extr. (Extremes)
        Source)   840   Traf. Col.
160   Fall DL (Different Level)   860   Vapor
180   Fall SL (Same Level)   880   Veh. Damage (Vehicle)
200   Fal. Obj. (Falling Object)   900   Vibration
220   Fire (See 80 for Source)   920   Wron. Mix. Pdts.
240   Fly. Part. (Flying
        Particles)   <u>40   Security</u>
260   Fumes   160   Illegal Act.
280   Hvy. Wx. (Heavy Weather)   320   Larceny
300   Implosion   480   Liab. Clm. (Liability Claim)
320   Ingestion   640   Sabotage (Known or Suspected
340   Inhalation   or Hostile Act)
360   Ioniz. Rad. (Radiation)   800   Vandalism
380   Irradiation
400   Lifting   <u>60   Survey & Insp.</u>
420   Liquid (Contact With)   160   Comp. Carrier
430   Material Move. (Movement)   320   Consultant
440   Mist   480   Fire Carrier
460   Nat. Disas. (Flood,   640   Govt. Agency
        Earthquake)   800   Indust. Hyg. (Industrial)
480   Noise   960   Safety
500   Non-Ioniz. Rad.   975   Systems Safety (Analysis)
        (Radiation)
520   Off The Job   <u>80   Fire Source</u>
540   Over-Exert.   070   Auto Ignition
560   Pdt. Contam.   140   Chem. Reaction
580   Poll. - Air (Pollution)   210   Compres. Ignit. (Compression)
600   Poll. - Chem.   Ignition
620   Pres. Extreme (Pressure   280   Diesel Engine
        Extreme)

12. <u>Type of Case</u> (Continued)

<u>Code</u>

350	Ele. Equip. D/D (Electrical Equip. Defective/Damaged)
420	Ele. Equip. N/D (Not Defective)
490	Friction
560	Lighting
630	Match (or Lighter)
700	Open Flame (Furnace, etc.)
770	Overheat
840	Spon. Ignit.
910	Static Spark
980	Welding

25	<u>Near Miss (No property damage/no injury)</u>
020	Breakdown (Machine or Equipment)
040	Caught (In or Between)
060	Chemicals
100	Dust
120	Elect. Sh. (Electrical Shock)
140	Explosion (See 80 for Source)
160	Fall DL (Different Level)
180	Fall SL (Same Level)
200	Fal. Obj. (Falling Object)
220	Fire (See 80 for Source)
240	Fly. Part. (Flying Particles)
260	Fumes
280	Hvy. Wx. (Heavy Weather)
300	Implosion
320	Ingestion
340	Inhalation
360	Ioniz/ Rad. (Radiation)
380	Irradiation
400	Lifting
420	Liquid (Contact With)
430	Material Move. (Movement)
440	Mist
460	Nat. Disas. (Flood, Earthquake)
480	Noise
500	Non-Ioniz. Rad. (Radiation)
520	Off The Job
540	Over-Exert.

<u>Code</u>

560	Pdt. Contam.
580	Poll. - Air (Pollution)
600	Poll. - Chem.
620	Pres. Extreme (Pressure Extreme)
640	Qual. Prob. (Quality)
660	Reaching
700	Slip (or Twist)
720	Spill (Splash)
740	Sports (Company Sponsored)
760	Strk. Against (Struck)
780	Struck By
800	Temp. Extr. (Extremes)
840	Traf. Col.
860	Vapor
880	Veh. Damage (Vehicle)
900	Vibration
920	Wrong Mix. Pdts.

30	<u>Prop. Damage (Property)</u>
050	Breakdown (Machiner or Equipment)
150	Explosion
200	Fire
250	Heavy Wx. (Heavy Weather)
300	Implosion
350	Material Move. (Movement)
400	Nat. Disaster (Natural)
450	Pdt. Contam. (Product Contamination)
475	Pdt. Destroyed (Unintentional)
500	Quality Problem
550	Strk. Against (Struck)
600	Struck By
650	Traffic Coll. (Collision)
700	Veh. Damage (Vehicle)
750	Wron. Mix. Pdts. (Wrong Mixed Products)

<u>Section A Identity</u>   Items 13 to 16   Page 4

13. <u>Type of Property</u> Code where the incident occurred.

<u>Code</u>   <u>Code</u>

999 No Prop.   172 Mtr. Veh.
004 Aircraft   180 Off. Bldg.
012 Airport   188 Other Mfg.
020 Animal Area   196 Pkg. Opn. (Packaging)
028 Barge   204 Pkg. Lot (Parking)
236 Barn   212 Pharm. Mfg.
044 Boiler Hse   220 Pilot Plant
052 Cafeteria   228 Port Fac.
060 Chem. Mfg.   236 Recrea. Fac.
068 Cust. Loc (Customers)   244 Res. Bldg. (research)
076 Drum Shed   252 Sidewalk
084 Grass   260 Silo
092 Guard Fac.   264 Stairway
100 Kennel   268 Sterile Pdts.
108 Lab.   276 Storage
116 Lab. Bldg.   284 Store. Shed
124 Leased Fac.   288 Street (Roadway)
132 Leased Veh.   292 Tank Car
140 Maint. Bldg.   300 Tank Farm
148 Mech. Eq. Rm. (Mechanical   304 Transp. Facil. (Transpor-
       Equip.)   tation Facility)
156 Med. Fac. (Medical)   308 Vessel
164 Motel (or Restaurant)   316 Warehse.

14. <u>Property Ownership</u>

<u>Code</u>   <u>Code</u>

999 No Prop.   040 Empl. Own.
008 Co.Owned   048 Leased
016 Co. Subs.   056 Priv. Own.
024 Concess. (Concessionaire)   064 Other
032 Contr.

15. <u>Phase of Operation at Time of Incident</u>

<u>Code</u>   <u>Code</u>

99 No Factor   32 Overtime
01 More Data   40 Shift, Day
08 Gerry-Rigged (Operation)   48 Shift, Midn. (Midnight)
16 Idle (Out of Service,   56 Shift, N. (Evening, Night)
      During Non-Occupancy)
24 Maint. (Repair)

16. <u>Months Since Last Inspection</u> - usually refers to equipment but
     can be used to identify a poor inspection schedule, i.e.,
     housekeeping or safety.

<u>Code</u>

00 More Data
99 No Factor

17. Weather Ccnd.

Code		Code	
001	More Data	450	Hot, Rain
999	No Factor	480	Hot, R., Fog (Rain)
030	Cold, Dry	510	Warm, Dry
060	Cold, Rain	540	Warm, Humid
090	Cold, R., Fog (Rain)	570	Warm, H., Fog
120	Cold, Snow	600	Warm, Rain
150	Cool, Dry	780	Warm, R., Fog
180	Cool, Humid		
210	Cool, Rain		Severe Weather
240	Cool, R. Fog (Rain)	810	Earthquake
270	Freeze, Dry	840	Elec. Stm. (Electrical Storm)
300	Freeze, Snow	880	Flood
330	Freeze, Rain	900	Hurricane
360	Hot, Dry	930	Tornado
390	Hot, Humid	960	Trop. Stm. (Tropical Storm)
420	Hot, H. Fog		

18. Physical Movements or Operations (This file may be used in
       place of Specific Operation or may be used as an added file.)

Code		Code	
999	No Factor	510	Pushing
001	More Data	540	Reaching
		570	Running
030	Bending	600	Sitting
060	Carrying	630	Slipping/Tripping
090	Catching	660	Squatting
120	Climbing	690	Standing
150	Crawling	720	Stepping (up or down)
180	Falling	750	Stooping
210	Grabbing	780	Striking (with a tool)
240	Jerking	810	Sudden chg. in posture
270	Jumping	840	Throwing
300	Kneeling	870	Turning
330	Lifting-knuckle & lower	900	Twisting
360	Lifting-knuckle to shoulder	930	Walking
390	Lifting-shoulder & above		
420	Lowering		
450	Lying		
480	Pulling		

18. <u>Specific Operation Involved</u>

<u>Code</u>

99999	No Factor
00001	More Data

<u>06   Admin./Off.</u>
108   Data Proc.
416   Library
624   Sec./Cler.
823   Other (All)

<u>12   Biolog. Test.</u>
208   Assay
416   Gen'l. Microb.
624   Microb. Sys.
832   Toxicology

<u>18   Chem. Mfg.</u>
008   Acetyla.
050   Acidica.
075   Alkyation
100   Amidation
125   Bromina.
150   Carbethoxy.
175   Cathyl.
200   Charging
225   Chlorin.
250   Chlorosulfon.
275   Condens.
300   Conglomer.
325   Cyanation
350   Decarboxyl.
375   Decyanogen.
400   Dehydrobrom.
425   Dehydromesy.
450   Distall.
475   Elution
500   Esterifica.
525   Extraction
550   Fluorination
575   Fractiona.
600   Freidel Craf.
      (Reaction)
625   Grogmard
650   Hydrogena.
675   Hydrolysis
700   Iodination
725   Maleate Form.
750   Oxidation
775   Quaterniza.

<u>Code</u>

800   Racemiza.
825   Recrystal.
850   Reduction
875   Reflux
900   Resolution
925   Sampling
950   Saponifica.
975   Thiona.
990   Williamson (Reaction)

<u>24   Comm. Oper.   (Common)</u>
108   Cleaning
116   Compound.
224   Equip. Prep. (Setup)
228   Fastening
332   Glass. Wash.
436   Heating Mtl. (Material)
540   Inspection
648   Mtl. Move. (Movement &
      Transfer)
752   Sterilization
854   Traying
956   Waste Proc. (Disposal)

<u>27   Non-Wk/Rel/Act. (Non-Work
    Related Activities)</u>
110   Break Time
350   Meals
560   On-site Tvl. (Travel)
770   Sports Act (Activity)

<u>30   Maint.</u>
108   Carpentry
216   Electrical
324   Gen'l Labor.
432   HVAC (Heat. Vent. Air.
      Cond.)
540   Insulating
600   Pack. Blachive
648   Pipe Fit.
756   Reg. Mech.
864   Sheet Met.
972   Utility (Oiler/Insp.)
980   Welding

<u>36   Pkg. (Packaging)</u>
008   Blister Ln. (Line)
116   Finished Gds.

Section A Identity   Items 18 to 19   Page 7
                                                      Changed 11/27/XX

18. Specific Operation Involved (Cont'd)

Code   Code

    324 Liquid Ln.   732 Microfilm.
    532 Pkg. Supply   840 Printing
    740 Tablet Lne.   848 Rec. Store.
    948 Tube Lne.   948 Wd. Proc.

     42 Pharma. Mfg.   70 Sterile
    008 Aerosol   208 Bact. Monit.
    110 Blend   416 Fill & Pack.
    200 Brand   624 Steriliz.
    290 Capsule (encapsulation)   832 Wtr. Syst. (Mtaintenance)
    380 Coating
    470 Compress   76 Warehse.
    560 Creams (Ointment)   708 Rcvg. (Receiving)
    650 Dry   816 Ship.
    740 Granulate
    830 Milling   82 Res. Opn.
                                          007 Analytical
     48 Q.C. Chem. Cntl.   114 Animal Cr. (Care)
    208 Liq. & Cream   221 Biol. Res.
    416 Parenterals   328 Chem. Res.
    624 Raw Mtls.   435 Drug Metab.
    832 Tablet Test.   442 Elect'ics.
                                          549 Med. Res.
                                          556 Microbio.
     52 Clinical Mfg.   563 Microbio. Dev.
                                          670 Pharma. R & D
     58 Secur. Svs.   777 Reg. Aff.
    208 Emer. Resp.   784 Screen Lab
    416 Mtr. Patrol   791 Stock Handl.
    624 On Post   898 Toxicol.
    832 Tours

     64 Site Services
    408 Cafet. (Cafeteria)
    516 Duplic. (Duplicating)
    020 Forms Control
    624 Mail

19. Specific Component

Code   Code
                                       40 Boiler, Lo. (Low Pressure)
  99999 No Component   48 Compressor
  00001 More Data   56 Heater (Hot Water)
                                       64 Steam
    008 Air Press.
         (Pressure Vessels)   016 Burning Equip.
     08 Air Press. Ln.   20 Blow Torch
     16 Air Tank (Receivers)   40 Cut. Equip.
     24 Autoclave   60 Solder Iron
     36 Boiler, Hi. (High   80 Weld. Equip.
         Pressure)

19. <u>Specific Component</u> (Cont'd)

<u>Code</u>   <u>Code</u>

024   <u>Clean./Cook.</u>   60   Roller
 15   Brooms (Brush, Mop)   80   Pwr. Screw
 30   Bucket (Pail)
 45   Can Opener   064   <u>Dryers</u>
 60   Cutlery   15   Aeromatic
 75   Pots, Pans   30   Fluid Bed.
                                        45   Spray
032   <u>Clothing</u>   60   Stokes Vac.
 15   Cloth   75   Tray
 30   Coat
 45   Glove   072   <u>Earth Moving</u>
 60   Hat   20   Cart (Wheelbarrow)
 75   Pants/Dress (Shirt,   40   Earth Scraper (Hand)
        Overalls, etc.)   60   Pick
 90   Shoes (Boots)   80   Shovel (Hoe, Rake)

040   <u>Constr. Equip.</u>   080   <u>Elect. Equip.</u>
 08   Batching Plant   05   A/C Equip.
 16   Cable Ways   10   Batteries (Charging)
 24   Concrete Vibr.   20   Capacitors
        (Vibrators)   22   Cords (Electrical)
 36   Drill. Equip.   25   Fans (Blowers)
 38   Lumber (Durage)   30   Insulation
 40 Rock Crusher   35   Motors (Relays)
 48   Screen. Plant   40   Motors (Generators)
 56   Tampers   45   Pwr. Bkr. (Power Breaker)
                                        50   Reactor
048   <u>Containers</u>   55   Rectifier
 04   Basket, Wire   60   Rotary Cond.
 05   Bottle   65   Solenoid
 07   Box   70   Sw. Bd. (Switchboard)
 14   Cylinder (Compressed   75   Switches
        Gas)   80   Transfor. (Transformer)
 21   Drum   85   Trans. Line
 35   Sack & Bag
 42   Tank, Feed   088   <u>Enclosure</u>
 49   Tank, Holding   20   Fence
 56   Tank, Precip.   40   Gate
 63   Tank, Recvr. (Receiver)   60   Guardrail
 67   Tote   80   Wall
 70   Trash Cntr. (Container,
        Box, Can, Bin)   096   <u>Food Equip.</u>
 74   Tray   15   Bread Cutter
 77   Rectr, Glass (Reactor)   30   Bottle Mach.
 84   Rectr, Monel   45   Dough Brake
 91   Rectr, S/S   60   Dough Mixer
                                        75   Meat Grinder
056   <u>Conveyor (No Pass.)</u>
 20   Belt (or Chain Link)   104   <u>Furniture</u>
 40   Gravity Slide   08   Ashtray

19. <u>Specific Component</u> (Cont'd)

<u>Code</u>   <u>Code</u>

    16 Bed (Bunk, Cot)   152 Kitchen Equip.
    24 Chair   07 Cup
    32 Desk (Table)   15 Dishwasher
    40 Door   22 Fork
    48 Elec. Apply   30 Freezer
    56 Ext. Cord   45 Fryer
    64 File (Bookcase, Lockers,   60 Mixer
          Shelves, Safe, etc.)   75 Stove (Oven)
    72 Floor (Floor Covering)   85 Urn (Coffee Maker)
    80 Window
                                       160 Laboratory
   112 Glass/Ceramic   06 Bench/Cabnet
    20 Dishes   12 Cage (Animal)
    40 Flat Glass   18 Chromatogr. (Chromato-
    60 Light Bulb (Fluorescent,   graph)
          Incandescent)   24 Compres. Gas (Cylinder)
    80 Tubing (Flasks, Beakers)   30 Evaporator
                                        36 Extract. Bal. (Extractor
   116 Guards   Balance)
    20 Barrier   42 Fume Hood
    40 Enclosure   48 Lg. Animal
                                        52 Lg. Glasw.
   120 Hack/Punch   60 Mtr./Inst. (Meter/
    15 Adze   Instrument)
    30 Awl, (Prost Pin)   63 Pipette
    45 Chisel   66 Refrig./Freez.
    60 Crowbar   70 Sm. Animal
    75 Drills   76 Sm. Glasw.
    90 Hatchet (Axe)   79 Syringe
    95 Knife (Razor)   82 Vac/Pump

   128 Heat Equip.   168 Ladders
    30 Fireplace   20 Extension
    60 Heating Tape   40 Fixed (Permanent)
    90 Stove (Electric, Gas)   60 Single
                                        80 Step
   136 Hoist Equip.   83 Stool, Step
    15 Air Hoist
    30 Chain Hoist   176 Laundry
    45 Escalators   20 Dryer
    60 Hydraul. Lft. (Lift)   40 Ironer
    75 Life Lines   60 Tumbler
                                        80 Washer
   144 Ice Maker
                                       184 Machine Shop
   148 Instruments   10 Drillpress
    85 Temp. Probe   20 Grinder (Stationary)
                                        30 Lathe
                                        40 Mill. Machine
                                        50 Pipe Thread
                                        60 Pwr. Press

19. <u>Specific Component</u> (Cont'd)

<u>Code</u>

188	Mtl. Handl. (Material Handling)
05	Boiler
07	Crusher
10	Dockboard
20	Handtruck
30	Lift Truck
40	Pallet (Skid)
50	Push Cart
60	Stacker Crane
80	Walkie

192	Metal Form. (Large)

200	Metal Items
07	Cable (Wire)
14	Chain
21	Fittings (Elbows, Trees & Flanges)
28	Heavy Metal
35	Molten Metal
42	Nails (Nails in Boards)
49	Nuts, Bolts
56	Particles (Slivers, Chips)
63	Pipe
70	Plate, Sheet
77	Rebar (Reinforcing Steel)
84	Scrap
88	Spring (All Types)
91	Struct. Steel

208	Mills
10	Chemicol (Chemicolloid)
20	Chilson. (Chilsonator)
30	Cottom
40	Dry Mill.
50	Eppenbach (Colloid)
60	Fitzmill
72	Gifford (Wood Colloid)
80	Wet Mill.

224	Misc. Equip.
08	Alsop
16	Balance, (Torsion etc.)
24	Centifuge
32	Condensers, (Glass, S/S)

<u>Code</u>

40	Evaporator
48	Evaporator (Rodney Hunt, etc.)
56	Filter, (Fluid Dynamics etc.)
64	Filter Press
72	Nutsche
80	Sparkler

232	Mixers
15	AMF
30	Gemco
45	Groen
60	Natua
75	Paterson (Blender)

240	Office Equip.
10	ADP Equip.
20	Calculator
30	Dupli. Equip.
32	Punches, All
35	Telephone
40	Teletype
50	Terminal
60	Typewriter
70	Xerox

248	Office Types
20	Paper
25	Pencil (Pen, Drafting Tools)
50	Scissors
75	Stapler

256	Pharma. Mach.
08	Blower (Ointment Tube, Plastic Cap, etc.)
16	42" Pans (Coating)
24	Pellegr. Pans
32	Sterilizer (Steam & Gas)
40	Sort. Mach.
48	Tablt. Brands
56	Wash, Bottle
64	Wash, Prosp.
72	Washer, Rota.

<u>Section A Identity</u>   Item 19   <u>Page 11</u>
                                                    Changed 11/27/XX

19. <u>Specific Component</u> (Cont'd)

<u>Code</u>   <u>Code</u>

264 <u>Pkg. Equip.</u>   312 <u>Pwr. Trans. (Mechanical)</u>
 07 Blister Pack.   20 Belts (Chains, Ropes,
 10 Bottle Cleaner   Cables)
 14 Bundler   40 Drums (Pulleys, Sheaves)
 21 Capper   60 Fan Blades (Spokes)
 28 Cartoner   80 Gears (Sprockets)
 35 Case Sealer
 42 Cleaner   316 <u>Surfaces</u>
 49 Cottoner   10 Curb
 53 Elev. Hopper (Elevator)   -- Floor (See 104)
 56 Feeder   30 Grass
 63 Filler   40 Gravel
 70 Gluer   70 Road
 77 Labeler (Pony)   75 Step
 84 Plugger   80 Sidewalk

268 <u>Piping/Hose</u>   320 <u>Saw/Strike</u>
 -- Metal (see 200)   20 Hammer (Mallet)
 20 Glass   40 Hand Saw
 60 Plastic (Tygon, etc.)   60 Post Driver
 70 Rubber   80 Sledge (Maul)
 80 Stainless Flex
                                  328 <u>Tablet Press</u>
272 <u>PPE (Protective</u>   25 Manesty
    <u>Equipment</u>   75 Stokes
 15 Apron
 20 Boots   336 <u>Tools Port Pw.</u>
 30 Chem. Suit   06 Buffing Mach.
 45 Face Shield   12 Circular Saw (Electric)
 60 Glasses (Safety)   18 Chain Saw (Gas or Electric)
 75 Goggles   24 Drain Clean
 90 Respirator   30 Drill (Electric)
                                   36 Edger (Powered)
280 <u>Printing</u>   42 Floor Waxer (Polisher)
 20 Corner Cutter   50 Grinder
 40 Papercutter   56 Hedge Trim (Electric)
 60 Press   62 Hole Auger
 80 Slitter   68 Jackhammer (Air)
                                   74 Lawnmower
288 <u>Pull/Scrape</u>   80 Router
 20 Bolt Cutter   86 Sander
 30 File Rasp   92 Steam Cleaner
 45 Hand Planer
 60 Nail Puller   344 <u>Turn/Lift</u>
 75 Pliers   15 Jack
 85 Tweezers   30 Pry Bar
                                   45 Screw Driver
296 <u>Pulverizer</u>   60 Tire Iron
 50 Micronizer   75 Vise
                                   90 Wrench

Section A Identity   Items 19 to 21   Page 12
                                                       Changed 11/27/XX

19. Specific Component (Cont'd)

    Code   Code

      348   Vehicles   30   Planer
       20   Car   45   Sander
       80   Truck   60   Shaper
                                        75   Table Saws (All Types)
      352   Woodwork
       15   Jointer

20. Model Year (Model or Construction)

    If vehicle state model, year - e.g., 75.
    If process unit or other equipment, state estimated purchase
    or construction year. It is not important unless it is vital
    to the case.

    Code

       99   No Equip. (or does not apply)

21. Contaminant (Based on Registry of Toxic Effects of Chemical
                  Substantces)
    Code   Code
              Substances   BB35000   Allyl Glycidyl Ether
    ZZ99999   No Contam.   BC36750   Allyl Propyl Disulfide
    AA00001   More Data   BC85740   Alpha Naphythyl
    AA36750   Abate   Thiourea (ANTU)
    AB19250   Acetaldehyde   BG56000   m-Aminobenzotriflouride
    AF12250   Acetic Acid   BI49000   4-Aminodiphenyl
    Ak19250   Acetic Anhydride   BI70000   2-Aminoethanol
    AL31500   Acetone   BM59500   2-Aminopyridine
    AL77000   Acetonitrile   BM98000   2-Aminothiazole
    AM40250   Acetophenazine Maleate   BN77000   Amitryptyline HCL
    AO29750   2-Acetylaminofluorene   BN89250   Ammate (Ammonium
    A064750   Acetyl Chloride   Sulfamate)
    AP08750   Acetylene Chloride   B009750   Ammonia
              (A,2-dichloroethylene)   BP45500   Ammonium Chloride
    AO96250   Acetylene   BQ96250   Ammonium Hydroxide
    AP12250   Acetylene Tetrabromide   BS43750   Ammonium Sulfamate
    AR71750   Acridine   (Ammate)
    AS10500   Acrolein   BU57750   n-Amyl Acetate
    AS33250   Acrylamide   BU59500   sec-Amyl Acetate
    AS43750   Acrylic Acid   BU63000   Amyl Alcohol
    AT52500   Acrylonitrile   BW66500   Aniline
              (Vinylcyanide)   BZ54250   Anisidine (o,p-isomers)
    AZ01750   Alcohol, Anhydrous 2B   CC40250   Antimony
    AZ08750   Alcohol, Denatured (All)   ---   ANTU see (Alpha
    AZ40250   Aldrin   Naphtyl Thiourea)
    BA50750   Allyl Alcohol   CG05250   Arsenic and Compounds
    BA98000   Allyl Chloride   CG64750   Arsine
                                       CJ03500   Aspirin

21. Contaminant (Continued)

Code		Code	
CM59500	Azinphos. Methyl (Guthion)	---	n-Butyl Glycidyl Ether (BGE)
CN59500	Azodrin (Monocrotophos)	EQ77000	Butyl Mercaptan (Butanethiol)
CQ85750	Barium Carbonate	ER77000	p-tert-Butyl Toluene
CU43750	Benzaldehyde	EU98000	Cadmium (Metal Dust and Soluble Salts)
CX14000	Benzene (Benzol)		
DC96250	Benzidine	EV19250	Cadmium Oxide (Fume)
DG08750	Benzoic Acid	EV94500	Calcium Arsenate
---	Benzol (Benzene)	EWO5250	Calcium Cyanamide
DK26250	p-Benzoquinone (Quinone)	EW31000	Calcium Oxide
DM85750	Benzoyl Peroxide	EX12250	Camphor
DN31500	Benzyl Alcohol	FE17500	Carbaryl (Sevin)
DP82250	Benzyl Benzoate	FF58600	Carbon Black
DQ03500	Benzyl Chloride	FF64000	Carbon Dioxide
DS17500	Beryllium	FF66500	Carbon Disulfide
DT19250	BGE (n-butyl glycidyl ether)	FG35000	Carbon Monoxide
		FG49000	Carbon Tracholoride (Tetrachloromethane)
DU80500	Biphenyl (Diphenyl)		
DY01750	Bis (Chlormethyl) Ether	FG50750	Carbonyl Chloride (Phosgene)
ED79000	Boron Oxide		
ED82250	Boron Triflouride	FG66500	Carbophenothion (Thithion)
EF91000	Bromine		
EG26250	p-Bromobenzyl Cyanide	FJ43750	Cellosolve Acetate (Othoxyethyl Acetate)
EG73500	Bromoform		
EH28000	n Bromosuccinimide	FN59500	Chlordane
EH89250	Brucine	FQ31500	Chlorobutanol
EI91000	Butadiene (1,3-Butadiene)	FQ61250	4-Chloro-m-cresol
		FO21000	Chlorine
EJ42000	Butane	FO22750	Chlorine Dioxide
EK63000	Butanethiol (Butyl Mercaptan)	FO57750	Chloroacetaldehyde
		FO80500	a-Chloroacetophenone (Phenacylchloride)
EL64750	2-Butanone (MEK) (Methyl Ethyl Ketone)		
		ZN20025	m-Chloroaniline
EN50750	2-Butoxy Ethanol (Butyl Cellosolve)	FP24500	p-Chloroaniline
		FP36750	Chlorobenzene (Mono-chlorbenzene)
E007000	n-Butyl Acetate		
E007350	sec-Butyl Acetate	FP92750	o-Chlorobenzylidene Malonitrile (OCEM)
E007550	tert-Butyl Acetate		
E014000	Butyl Alchol	ZN20030	p-Chlorobenzylcyanide
E015750	n-Butyl Alcohol	ZN20032	2(p-Chlorobenzyl Pyridine)
E017500	sec-Butyl Alcohol		
E019250	ter-Butyl Alcohol	FQ14000	Chlorobromomethane
E031500	n-Butylamine	FQ22750	2-Chloro-1,3-butadiene (Chloroprene)
E094500	Butylated Hydroxy Toluene		
EP36750	Butyl Cellosolve (2-butosky ethanol)	FR87500	Chlorodiphenyl (54% CL)
		ZN21006	Chlorodiphenylamine
EP48050	tert-Butyl Chromate	FR96250	1-Chloro-2,3-epoxypropane (Epichlorohydrin)
EQ22750	n-Butyl Ether		

<u>Section A Identity</u>   Item 21   <u>Page 14</u>

21. <u>Contaminant</u> (Cont'd)

Code

Code		Code	
FS08750	2-Chloroethanol (Ethyl-ene Chlorohydrin)	HG78750	Demeton (Systox)
KU96250	Chloroethylene (Vinyl Chloride)	HI10500	DGE (Diglycidyl Ether)
FS91000	Chloroform (Trichloro-methane)	HI56000	Diacetone Alcohol (4-methyl-2-pentanone)
ZN37429	7-Chloro-3 Methyl-2H-1,2,4-Benzothiadiazine 1,1-dioxide (Hyperstat I.V.)	HK57750	1,2-Diaminoethane (Ethylenediamine)
		HM80500	Diazomethane
		---	Dibrome see (Dimethyl-1,2-dibromo-2,2-dichloroethyl phosphate)
FU19250	Chloromethyl Ether (Methyl-Chloromethyl Ether)		
FU96250	1-Chloro-1-nitropropane	HR26250	1,2-Dibromoethane (Ethylene Dibromide)
ZN21007	4-Chlor-n-methyl Piperi-dine HCL	HS64750	Dibutyl Phosphate
		HS66500	Dibutylphthalate
ZN91080	10-(3-Chloropropyl)-2-Chlorophenothiazine (PTP-H)	HT50750	Dichloracetylene
		ZN29018	Dichloroacetaldehyde-Ethyl-Hemi-Acetal
FW68250	Chloropicrin (Nitrotri-chloromethane)	HT70000	o-Dichlorobenzene
		HT71750	p-Dichlorobenzene
FW71750	Chloroprene (2-chloro-1,3-butadiene)	HT78750	Dichlorobenzidine
		HU34000	Dichlorodifluoromethane
ZN21020	2-Chloropyridine	HU49000	1,3-Dichloro-5,5-dimethyl hydantoin
ZN21040	Chlorosulfonic Acid		
GB24500	Chromic Acid	HU61250	1,1-Dichloroethane (Ethylidine Chloride)
GF87500	Cobalt		
ZN23601	Cobalt Chloride Hexa-hydrate	HU63000	1,2-Dichloroethane (Ethylene Dichloride)
GG71750	Codeine Phosphate	HV03500	1,2-Dichloroethylene (Acetylene Dichloride)
GO57750	Creosote		
GO59500	Cresol	HV05250	Dichloroethyl Ether
GP50750	Cresylic Acid	HV56000	Dichloromethane (Methylene Chloride)
GP96250	Crotonaldehyde		
GR85750	Cumene	HV77800	Dichloromonofluoro-methane
GS26250	Cypric Sulphate		
GS71750	Cyanide	HV96250	1,1-Dichloro-1-nitroethane
ZN26048	Cyanoacetamide		
GT19250	Cyanogen	---	2,4-Dichlorophenoxy-acetic Acid (see 2,4-D)
GU63000	Cyclohexane		
GW08750	Cyclohexanol		
GW10500	Cyclohexanone	HX10500	1,2-Dichloropropane (Propylene Dichloride)
GW25000	Cyclohexene		
GY09500	Cyclopentadiene	HX39050	Dichlorotetrafluor-ethane
HW26250	2,4-D (2,4-dichloro-phenoxyacetic acid)	HX87500	Dieldrin
HC80500	DDT	ZN99102	Dienestrol Diacetate
HC84000	DDVP	HZ87500	Diethylamine
HD12250	Decaborane	IA73500	2-Diethylaminoethanol
ZN27301	Decyl Oleate	---	1,4-Diethylene Dioxide see (p-Dioxane)

<u>Section A Identity</u>   Item 21   <u>Page 15</u>

21. <u>Contaminant</u> (Cont'd)

<u>Code</u>

Code		Code	
IE12250	Diethylene Triamine	JG03500	Dioctyl Phthalate
IE31500	Diethyl Ether (Ethyl Ether) (Ethyl Oxide)	JG82250	p-Dioxane (1,4,Diethylene Dioxide) (1,4,Dioxane)
IE78750	Di(2-ethylhexyl)phthalate (di-sec-octyl-phthalate)	JJ78750	Diphenyl Amine, Amino-
		---	Diphenyl see (Biphenyl)
IG85750	Difluorodibromomethane	JK84000	p,pl-Diphenylmethane Diisocyanate (Methylene Bis(pheny isocyanate) (MDI)
IH35000	Diglycidyl Ether (DGE)		
IJ70000	Dihydroxy Acetone		
ZN31002	Dihydrostreptomycin Sulfate	JM14000	Dipropylene Glycol Methyl Ether
IJ78750	Dihydroxybenzene (Hydroquinone)	J001750	Disulfoton (Di-Syston)
		JZ17500	Fndosulfan (Thiodan)
IM24500	Diisobutyl Ketone (2,6-dimethylheptanone)	JZ40250	Endrin
		KB38500	Epichlorohydrin (1-chloro-2,3-epoxypropane)
IM38500	Diisopropyl Adipate		
IN40250	Diisopropylamine	KC12250	EPN
IN66500	Dimecron (Phosphamidon)	KD24500	1,2-Epoxypropane (Propylene-Oxide)
IP15750	Dimethoxymethane (Methylal)	KD26250	2,3-Epoxy-1-Propanol (Glycidol)
IP70000	n,n-Dimethylacetamide		
IP87500	Dimethylamine	KI96250	Ethanethiol (Ethyl Mercaptan)
IQ35000	4-Dimethylaminoazobenzene	---	2-Ethoxyethanol see (Ethylene Glycol Monethyl Ether)
ZN31007	b-Dimethylamino Ethyl Chloride HCL		
IU73500	2,4-Dimethylaniline (2,4-xylidene)	KP66500	Ethoxyethyl-Acetate (Cellosolve Acetate)
IV42000	Dimethylbenzene (xylene)	KQ49000	Ethyl Acetate
		KQ59500	Ethyl Acrylate
IX2275C	Dimethyl-1,2-dibromo-2,2-dichloroethyl phosphate (dibrome)	KQ63000	Ethyl Alcohol
		KQ70000	Ethyl Amine
		KS36750	Ethyl sec-amyl ketone (5-methyl-3-heptanone)
IZ05250	n,n-Dimethylformamide		
IZ24500	2,6-Dimethylheptanone (Dilsobutyl ketone)	KS65000	Ethyl Benzene
		KS80500	Ethyl Bromide
IZ43750	1,1 Dimethylhydrazine	KT08750	Ethyl Butyl Ketone (3-heptanone)
JA84000	Dimethylnitrosoamine (n-nitrosodimethyl-amine)		
		KT43750	Ethyl Chloride
		KT54250	Ethyl Chloroformate
JC42000	Dimethylphthalate	KU92750	Ethylene Chlorohydrin (2-chloroethanol)
JD05250	Dimethylsulfate		
ZN31021	Dimethyl Sulfoxide (DMSO)	---	Ethylenediamine see (1,2-diaminoethane)
JE47250	m-Dinitrobenzene		
JE77000	2,4-Dinitrochlorobenzene	KV77000	Ethylene Dibromide (1,2-dibromoethane)
JE82250	Dinitro-o-cresol		
JF17500	Dinitrophenol	KV91000	Ethylene Dichloride (1,2-dichloroethane)
JF54250	2,3-Dinitrotoluene		
		KW56000	Ethylene Glycol Dinitrate

21. <u>Contaminant</u> (Cont'd)

Code

		Code	
KW78750	Ethylene Glycol Mono- ethyl Ether (2-ethoxyethanol)	---	3-Heptanone (Ethyl Butyl Ketone)
KW85750	Ethylene Glycol Mono- methyl Ether Acetate (Methyl Cellosolve Acetate)	ML35000 ML51400 MN92750	Hexochloroethane Hexachloronaphthalene Hexane (n-hexane)
KX15750	Ethylene Imine	MP14000	2-Hexanone (Methyl n-Butyl Ketone)
KX24500	Ethylene Oxide	MQ28000	Hexone (Methyl Isobutyl
---	Ethyl Ether (Ethyl Oxide) see (Diethyl Ether)	MQ36750 MQ85750	Ketone) Hexyl Acetate Hexylresorcinol
KX89500	Ethyl Formate	MU71750	Hydrazine
KZ14000	Ethylidene Chloride (1,1-dichlorethane)	MW38500 MW40250	Hydrobromic Acid Hydrochloric Acid (Muriatic Acid)
---	Ethyl Oxide (Ethyl Ether) see (Diethyl Ether)	MW92750 MW94500	Hydrogen Azide Hydrogen Bromide
---	Ethyl Mercaptan see (Ethanethiol)	MW96250 MW98000	Hydrogen Chloride Hydrogen Cyanide
LB87500	Ethyl Paration (Parthion)	MX01750	Hydrogen Fluoride (Hydrofluroic Acid)
LA94500	4-Ethylmorpholine	MX09000	Hydrogen Peroxide (90%)
LC66500	Ethyl Silicate	MX10500	Hydrogen Selenide
ZN37030	Euprocin Hydrochloride	MX12250	Hydrogen Sulfide
LJ61250	Ferbam	MX35000	Hydroquinone
LJ91000	Ferric Chloride	NB38500	Hydroxyethyl Piperazine
LJ98000	Ferric Nitrate	---	Hyperstat I.V. see
LK14000	Ferrovanadium Dust		(7-Chloro-3 Methyl-
---	Fluoride, Sodium		2H-1,2,4-benzothiadia-
LM64750	Flourine		zine 1,1-dioxide)
LP18700	Fluorotrichloromethane	NH98800	IGE (Isopropyl Glycidyl Ether)
LP35300	Fluphenazine (FPH-F-II)	NK82250	Indene
LP39500	Fluphenazine Dihydro- chloride	NL10500 NN15750	Indium Iodine
LP89250	Formaldehyde	NP38500	Isoamyl Acetate
LQ49000	Formic Acid	NP40250	Isoamyl Alcohol
LS62000	Freon II	NP91000	Isobutyl Acetate
LU91000	Furfuryl Alcohol	NP96250	Isobutyl Alcohol
LY49000	Germanium	NQ10500	Isobutylcarbonol
LZ01750	Gitalin	NT15750	Isophorone
MA69050	Glycerin	NT75250	Isopropyl Acetate
---	Glycidol (2,3-epoxy- 1-propanol)	NT80500 NT84000	Isopropyl Alcohol Isopropylamine
MG47250	Hafnium Chloride Oxide	NV08750	Isopropylether
MI17500	Heptachlor	---	Isopropyl-Glycidyl Ether see (IGE)
MI77000	Heptane (n-Heptane)		
MJ50750	2-Heptanone (Methyl (n-amyl) Ketone)	OF12250 OF75250 OI70800	Lauric Diethanolamide Lead Lime, Hydrated

21. <u>Contaminant</u> (Cont'd)

Code

Code		Code	
OI82250	Lindane	---	Methyl Cellosolve (2-Methoxyethanol)
ZN50950	Lithium Amide		
ZN51001	Lithium Bromide	---	Methyl Cellosolve see (Ethylene Glycol Mono-methyl Ether Acetate)
OJ57750	Lithium Carbonate		
ZN51020	Lithium		
OJ67300	Lithium Hydride	PJ01750	Methyl Chloride
OK27050	L.P.G. (Liquified Petroleum Gas (Natural Gas)	PJ17500	Methyl Chloroform 1,1,1-trichloroethane)
		---	Methyl Chloromethyl Ether see (Chloromethyl Ether)
OM21000	Magnesium		
OM38500	Magnesium Oxide	PJ70000	Methyl 2-cyanoacrylate
OM85750	Malathion	PJ78750	Methylcyclohexane
ON36750	Maleic Anhydride	PJ80500	Methylcyclohexanol
0092750	Manganese and Compounds	PJ84000	o-Methylicyclohexanone
---	MDI (Methylene Bis (phenylene Isocya-nate)) see Diphenyl-methane Diisocyanate)	---	Methylene Bis (Phenylene Isocyanate) (MDI) see (Diphenylmethane Diisocyanate)
---	MEK see (2-Butanone) (Methyl Ethyl Ketone)	PL28000	4,4-Methylene (Bis) Chloroaniline
OT03500	Menthol	---	Methylene Chloride see (Dichloromethane)
OV45500	Mercury	PN19250	Methyl Formate
OX71750	Mesityl Oxide	---	5-Methyl-3-heptanone see (Ethyl sec-amyl Ketone)
OY68250	Meta-Systox R (Oxydeme-ton Methyl)	PO50750	Methyl Iodide
ZN53030	Methane Sulfonyl Chloride (Mesyl Chloride)	---	Methyl Isobutyl Carbinol see (Methyl Amyl Alcohol)
PC14000	Methanol (Methyl Alcohol)	---	Methyl Isobutyl Ketone see (Hexone)
PD83100	Methoxychlor	PO66500	Methyl Isocyanate
PE07000	2-Methoxyethanol (Methyl Cellosolve)	PP71750	Methyl Methacrylate
		PR80500	Methyl Parathion
ZN53070	Methoxyphenylacetic Acid	PU54250	Methyl Propyl Ketone (2-pentanone)
PF21000	Methyl Acetate	PV35000	Methyl Salicylate
PF49000	Methyl Acrylate	PV36750	Methyl Silicate
---	Methylal see (Dimethoxy-methane)	PV38500	a-Methyl Styrene
		PX92750	Mevinphos (Phosdrin)
---	Methyl Alcohol see (Methanol)	PY80300	Mineral Oil
		ZN55015	Monoamyl Amine
PF63000	Methylamine	---	Monocrotophos see (Azodrin)
PG80500	Methyl Amyl Alcohol (Methyl Isobutyl Carbinol)	QB82250	Monomethyl Aniline
		QB89250	Monomethyl Hydrazine
---	Methyl (n-amyl) Ketone see (2-heptanone)	---	Muriatic Acid see Hydrochloric Acid)
		QI94500	Naphtha (Coal Tar)
PI84000	Methyl Carbitol	QJ05250	Naphthalene

21. <u>Contaminant</u> (Cont'd)

<u>Code</u>   <u>Code</u>

Code	Name	Code	Name
---	Natural Gas see (L.P.G.) (Liquified Petroleum Gas)	---	2-Pentanone see (Methyl Propyl Ketone)
QR59500	Nickel	SC64750	Peracetic Acid
QR63000	Nickel Carbonyl	ZN61025	Perchloric Acid
QS52500	Nicotine	SD10500	Perchloroethylene (Tetrachloroethylene)
QV57750	Nitric Acid		
QV31500	Nitric Oxide	SD15750	Perchloromethyl Mercaptan
QV66500	p-Nitroaniline	SD19250	Perchloryl Fluoride
QV75250	Nitrobenzene	---	Petroleum Ether see (Naphtha)
QV80500	p-Nitrobenzoic Acid	---	Phenacylchloride see (a-chloraceto-phenone)
QW15750	p-Nitrochlorobenzene		
QW31500	p-Nitrodiphenyl		
ZN91004	Nitrofurathiazide	---	PGE see (Phenyl Glycidyl Ether)
ZN57500	5-Nitrofurural Diacetate		
QW36750	Nitroethane	SE12250	Perphenazine
QW98000	Nitrogen Dioxide	SJ33250	Phenol Perphenazine
QX19250	Nitrogen Trifluoride	SQ72250	Phenylactonitrile
QX21000	Nitroglycerine	SS80500	p-Phenylene Diamine
QX64750	Nitromethane	SR87500	Phenyl Carbinol
QY15750	o-Nitrophenol	ST35200	Phenylephrine
QY81350	1-Nitropropane	ST57750	Pheylephrine Hydrochloride
---	N-Nitrosodimethylamine see (Dimethyl Nitrosoamine)	ST80500	Phenyl Ether
		SU35000	Phenylethylene (Styrene Monomer)
QZ04500	p-Nitrotoluene	SU89250	Phenyl Glycidyl Ether (PGE)
---	Nitrotrichloromethane see (Chloropicrin)		
---	OCBM (o-chlorobenzyli-dene Malonitrile)	SV01750	Phenylhydrazine
		SV70000	Phenylmercuric Acetate
ZN59005	2-Octadecanol	SV77800	Phenylmercuric Nitrate
RF91900	Octachloronaphthalene	SW64750	Phenylpropanolamine Hydrochloride
RG84000	Octane		
RN11400	Osmium Tetroxide	ZN62600	Phenylsuccinic Acid
RO24500	Oxalic Acid	---	Phosdrin see (Mevinphos)
---	Oxydemetonmethyl see (Meta-Systox R)	---	Phosgene see (Carbonyl Chloride)
RS21000	Oxygen Difluoride	---	Phosphamidon see (Dimecron)
RS82250	Ozone		
RV03500	Paraffin	SV75250	Phosphine
ZN60010	Paraformaldehyde	TB63000	Phosphoric Acid
RV66500	Paraquat	TH35000	Phosphorous
---	Parthion see (Ethyl Parathion)	ZN63003	Phosphorous Oxychloride
		TH40250	Phosphorous Pentachloride
RV87500	Pentaborane	ZN63005	Phosphorous Pentoxide
RZ25400	Pentaerythritol Chlorol	TH41200	Phosphorous Pentasulfide
RZ06400	Pentachlornaphthalene	TH46700	Phosphorous Trichloride
RZ26750	Pentachlorophenol	TH98000	Phthalic Acid Anhydride
RZ94500	Pentane	ZN63007	3-picoline-n-oxide

21. <u>Contaminant</u> (Cont'd)

Code		Code	
TJ78750	Picric Acid (2,4,6-trinitrophenol)	VZ28000	Sodium Borohydrite
TO75250	Pival (2-Pivoloyl-1,3-indandione)	VZ40250	Sodium Carbonate
		VA75250	Sodium Cyanide
TP22750	Platinum Chloride	WA92750	Sodium Fluoroacetate
ZN64007	Polyphosphoric Acid	WB03500	Sodium Fluoride
TR77400	Polytetrafluoroethylene	ZN77601	Sodium Formaldehyde Sulfoxylate
ZN65007	Potassium-T-Butoxide	WB49000	Sodium Hydroxide
ZN65040	Potassium Ethyl Xanthate	WB64750	Sodium Iodide
TT21000	Potassium Hydroxide	WB78750	Sodium Lauryl Sulfate
TT29750	Potassium Iodide	ZN79030	Sodium Methylate
TX22750	Propane	WE19250	Sodium Sulfide
UC43750	Propargyl Alcohol	WE21000	Sodium Sulfite
UD91000	Propiolactone	WI28000	Stearic Acid
UE59500	Propionic Acid	WJ07000	Stibine
UE91000	Propionic Anhydride	WJ89250	Stoddard Solvent
UH77000	n-Propyl Acetate	WL22750	Strychnine
UH82250	Propyl Alcohol	---	Styrene Monomer see (Phenylethylene)
---	Propylene Dichloride see (1,2-dichloropropane)	WN08750	Succinic Anhydride
		WO59500	Sulfamic Acid
UJ14000	Propylene Glycol	ZN87015	Sulfur
UJ36750	Propylene Imine	WS45500	Sulfur Dioxide
---	Propylene Oxide see 1,2-epoxypropane)	WS52500	Sulfur Hexafluoride
		WS56000	Sulfuric Acid
UK03500	n-propyl Nitrate	WT22750	Sulfur Monochloride
UR42000	Pyrethrum	WT45500	Sulfur Pentafluoride
UR8400	Pyridine	WT50750	Sulfuryl Fluoride
ZN69020	Pyridine-2-Aldehyde	---	Systox see (Demeton)
ZN69004	Pyrophosphryl Chloride	WY92750	2,4,5-T (2,4,5-trichloro phenoxyacetic acid)
---	Quinone see (p-benzoquinone)	WW55050	Tantalum
		WX92750	TEDP
VF07000	RDX	WX99100	Teflon
VG66500	Reserpine	WY26250	Tellurium
VI91000	Rhodium Chloride	WY28000	Tellurium Hexafluoride
VK98000	Ronnel	WY85750	TEPP
VL10500	Rotenone	WZ64750	p-Terphenyl
VN64750	Salicylamide	XA78750	Tetracaine Hydrochloride
VO05250	Salicylcic Acid	XB05250	1,1,2,2-Tetrachloro-1,2-difluorethane
VS77000	Selenium		
VS96250	Selenium Hexaflouride	XB15750	1,1,2,2-Tetrachloro-ethane
---	Sevin see (Carbaryl)		
VW07000	Silicon Carbide	---	Tetrachloroethylene see (Perchloroethylene)
VV73500	Silicon Dioxide		
VW35000	Silver	---	Tetrachloromethane see Carbon Tetrachloride)
ZN79020	Sodium		
VY08750	Sodium Acetate	XB29750	Tetrachloronaphthalene
VZ19250	Sodium Bisulfite	XC28000	Tetraethyl Lead

21. Contaminant (Cont'd)

Code		Code	
XC77000	Tetrahydrofuran	YJ66500	Trinitrotoluene
ZN90035	1,1,3,3 Tetramethoxy-propane	YJ91000	Triorthocresyl Phosphate
XE59500	Tetramethyl Succinonitrile	YK35000	Triphenyl Phosphate
XF01750	Tetranitromethane	---	Trithion see (Carbophenothion)
XG02500	Tetryl (2,4,6-trinitrophenyl-methinitramine)	YM05250	Triton X-100
XG35000	Thallium	YO71750	Tungsten
XK15750	Thimersol	YO84000	Turpentine
XK50750	Thiocarbamide	YQ32500	Undecylenic Acid
---	Thiodan see (Endosulfan)	---	Vinyl Chloride see (Chloroethylene)
XM73500	Thiophene	---	Vinylcyanide see (Acrylonitrile)
---	Thiouria see (Thiocarbamide)	ZA02500	Vinyl Toluene
XO28000	Thiram	---	Warfarin
XP22750	Thymol	---	Xylene see (Dimethylbenzene)
XR21000	Titanium Dioxide	ZE89250	2,4-Xylidine (2,4-Dimethylaniline)
XS52500	Toluene (Toluol)		
XT08750	Toluene-2,4-diisocyanate	ZG19250	Yittrium
ZN60015	p-Toluenesulfonyl Chloride	ZH14000	Zinc Chloride
XT63000	p-Toluenesulfonic Acid	ZH45500	Zinc Oxide
XU29750	o-Toluidine	ZH47250	Zinc Oxide (Fume)
---	Toluol see (Toluene)		
YA15750	Tributyl Phosphate		**Unidentified Chemicals**
---	1,1,1-Trichloroethane see (Methyl Chloroform)	ZP00100	Bleach
YA50750	1,1,2-Trichloroethane	ZP00200	Caustic Cleaner Mist
YA52500	Trichloroethylene	ZP00250	Coffee
---	Trichloromethane see (Chloroform)	ZP00300	Cosmetic Solvent Vapor
XC17500	Trichloronaphthalene	ZP00400	Degreaser
---	2,4,5-Trichlorophenoxyacetic Acid see (2,4,5-T)	ZP00500	Detergent
YC54250	1,2,3-Trichloropropane	ZP00600	Duplicator Vapors
YD82250	Triethanolamine	ZP00700	Dye
YE01750	Triethylamine	ZP00800	Dye Solvent Vapor
ZN37024	Triethyl Orthopropionate	ZP00900	Enzyme
YF31500	Trifluoromonobromomethane	ZP01000	Explosives
ZN92030	Trimethylene Chlorobromide	ZP01100	Flavoring Agents
---	2,4,6-Trinitrophenol see (Picric Acid)	ZP01200	Flux
---	2,4,6-Trinitrophenyl-methylnitramine see (Tetryl)	ZP01300	General Solvent Vapor
		ZP01400	Glue
		ZP01500	Glue Vapors
		ZP01600	Grease
		ZP01700	Ink
		ZP01800	Ink Mist
		ZP01900	Ink Solvent Vapor
		ZP02000	Metal Fume
		ZP02100	Microbiological Pathogens
		ZP02200	Oil
		ZP02300	Oil Mist
		ZP02400	Paint and Varnish Thermal Decomposition Products

21. <u>Contaminant</u> (Cont'd)

<u>Code</u>

ZP02500	Paint Mist
ZP02600	Paint Thinner
ZP02700	Pesticide, Unspecified
ZP02800	Photographic Developer Vapor
ZP02900	Plant and Food Products
ZP03000	Rubber Accelerators and Antioxidants
ZP03100	Soap
ZP03200	Stainless Steel (Composition Unknown or Unspecified)
ZP03250	Starch
ZP03275	Sugar (Syrup, Granular)
ZP03300	Tar
ZP03600	Welding Gasses, Unspec. (Ozone, Nox, CO_1, CO_2)
ZP03700	Other Alcohols
ZP03800	Other Solvents
ZP03900	Other and Unidentified Chemicals

<u>Physical Hazards</u>

ZP04000	Noise, Continuous
ZP04100	Noise, Intermittent
ZP04200	Noist, Impact
ZP04300	Noise, Ultrasonic
ZP04400	Vibration
ZP04500	Alpha Radiation
ZP04600	Beta Radiation
ZP04700	Gamma Radiation - Sealed Source
ZP04900	X-Radiation
ZP05000	Other Ionizing Radiation
ZP05100	Exposure to Cold
ZP05200	Heat Stress - Dry
ZP05300	Heat Stress - Humid
ZP05325	Hot Water
ZP05350	Steam
ZP05400	Improper Illumination
ZP05500	Infrared
ZP05600	Long Wave Radio - Frequency
ZP05700	Microwave
ZP05800	Sunlight
ZP05900	Ultra-Violet
ZP06000	Lasers - Infrared
ZP06100	Lasers - Microwave and RF

<u>Code</u>

ZP06200	Lasers - Visible
ZP06300	Lasers - X-Radiation
ZP06400	Low Pressure
ZP06500	High Pressure
ZP06600	Electrical Shock
ZP06700	All Other Physical Hazards

<u>Dust</u>

ZP06800	Cristobalite
ZP06900	Crystalline Free Silica
ZP07000	Silica (Silicon Dioxide)
ZP07100	Tridymite
ZP07200	Animal dander
ZP07300	Amorphous Silica, including Diatomaceous Earth
ZP07400	Asbestos
ZP07500	Brass Dust (Composition Unknown or Unspecified)
ZP07600	Brick Dust
ZP07700	Clay Dust
ZP07800	Coal
ZP07900	Dust Containing Enzymes
ZP08000	Flour Dust
ZP08100	Graphite
ZP08200	Leather
ZP08300	Mica
ZP08400	Mortar Dust
ZP08500	Organic Dust
ZP08600	Plastic Dust
ZP08700	Portland Cement
ZP08800	Rubber Dust
ZP08900	Soapstone
ZP09000	Talc
ZP09100	Tremolite
ZP09200	Wood - Sawdust
ZP09300	Inert or Nuisance Dust, NEC

22. <u>Recommendations Outstanding</u>

   a. Record under "Original" the total number of recommendations
      made in the report.

   b. Under the word "Remain" indicate only those recommendations
      which are outstanding or not yet completed; i.e., the item
      is either fully complete or it is not complete.

   c. When the recommended action has been taken, recode the recom-
      mendations outstanding section in the remain block by the
      number left to be resolved.

23. <u>Batch</u>

   <u>Code</u>

   999999 No Factor

24. <u>Work Order/AR Number</u>

   a. Enter work order on A/R Number as applicable.

   b. If no number is used, enter:

   <u>Code</u>

   999999 No Factor

25. <u>Fatality Date</u> Enter Date. If not a factor enter 99/99/99.

Section B People   Items 30 to 34   Page 23
                                                              Changed 11/27/XX

30. <u>Social Security Number</u>   No need to code - Provided by
                                      E.I.S.

31. <u>Employee Number</u> Code for Beisa and Manati. Also code for those
                        employees not on present status. Do not code
                        non-employees.

32. <u>Age in Years</u>   No need to code - provided by E.I.S.

33. <u>Employment Status</u>

    <u>Code</u>   <u>Code</u>

       10 Contr. (Contractor)   50 Temp. Empl. (Temporary)
       20 Empl. Farm. (Employee Family) 60 Vendor
       30 Perm. Empl. (Permanent)   70 Visitor
       40 Public O. (Other)

34. <u>Injury/Illness</u> If two types, select most severe.

    <u>Code</u>   <u>Code</u>
       999 None   440 Heart Disease
       020 Abrasion   460 Hernia
       040 Alcohol   480 Infection
       060 Amputation   490 Ingestion
       080 Animal Disease- (Insect)   500 Internal Inj.
       100 Asphyziation (Not Drowning)   520 Lacer. (Laceration)
       120 Avulsion (Loss of flesh by   540 Multi. Inj.
              shearing, tearing)   560 Part. in Eye (Particle)
       140 Bites (Trauma)   580 Phys. Agents (Disorder
       160 Blinded (Eye Injury)   due to)
       180 Burns C. (Chemical)   600 Poisoning (Systematic
       200 Burns T. (Thermal)   effects of toxic
       220 Concussion (Head)   materials) (Include
       240 Contagious (Disease)   food poisoning)
       260 Contusion (Bruised but   610 Puncture
              surface intact)   620 Repeat. Trauma (Disorders
       280 Crush   due to)
       300 Cut   640 Respir. Cond. (Due to
       320 Dermatitis (Occ. Skin   toxic agents)
              Disease)   660 Scratches
       340 Dislocation   680 Shock (Electric)
       360 Drug Abuse   700 Smoke Inhal.
       380 Dust Disease (of the   720 Sprain
              lungs)   740 Strain Bk. (Back)
       390 Eye Irr. (Irritation   760 Strain No. Bk. (Not Back)
              from Chem.)   780 Stress (Emotional)
       400 Fracture   800 Stroke
       420 Fungus   820 Unknown

35. Parts of Body Injured

Code

9999	No Injury		3200	Lower Extrem.
			15	Ankle
0800	Head		30	Foot - Feet (Not Toes)
10	Ear(s)		45	Knee(s)
20	Eye(s)		60	Leg(s)
30	Face		75	Thigh(s)
40	Jaw		90	Toe(s)
45	Mouth			
50	Neck		4000	Body System
60	Nose - Throat		10	Circulatory
70	Skull (& Scalp)		20	Digestive
80	Teeth (& Gums)		30	Excretory
			40	Genitor - (Urinary)
1600	Arm - Hand		50	Nervous
10	Elbow		60	Reproductive
20	Fingers		70	Respiratory
30	Hand(s) - (Not Fingers)			
40	Lower Arm (Below Elbow)		4800	NEL (Not Elsewhere Listed
50	Upper Arm (Above Elbow)		25	All Skin (Wide area-
60	Wrist			Poison Ivy, Burns, etc.)
			75	All Body Inj. (Multiple
2400	Trunk			Body Injuries)
10	Abdomen			
20	Back (Spine)			
30	Buttocks			
40	Chest, Rib(s)			
50	Groin			
60	Hip(s)			
70	Shoulders			

36. Severity of Injury

Note: For a fatality resulting from Injury Code 48. For occupational illness related fatality, code the specific type;
i.e., dust disease. Enter fatality data.

Code

06	No Injury		66	Poisoning (Syst. eff.
12	No Aid Req. (Bruises)			Toxic Mtls.)
18	1st Aid		72	Repeat Trauma (Disorders
24	All Other (Occ. Illness)			due to)
30	Dust Disease (of the Lungs)		78	Respir. Cond. (Due to
--	Fatality (Indicate by date			Toxic Agents)
	on form)		84	Restr. Act. (Restricted
42	L/T & Restr. Act.			Activity)
48	L/T Only (Lost Time)		90	Skin Disease (Occ.
54	Med. Tmt.			Disorder)
60	Physical Ag. (Disorders			
	due to			

37. <u>Days Off Work</u> (4 digits)

This is the best estimate of time loss away from work. When
actual days off are determined, submit a new form.

Code   Description

    A   Actual Days
    E   Estimated Days

Coding: Prefix E or A then four digit code for estimate or
            actual days away from work.

Not a Factor Code A0000

40. <u>Restr. Activity</u>

This is the best estimate of days of restricted activity. When
actual days are determined, submit a new form.

Code   Description

    A   Actual   Not a Factor - Code A00
    E   Estimated

Coding: Prefix E or A then a two digit code for estimate or actual
            days of restricted activity.

41. <u>Relation to Work</u>

<u>Code</u>

    99 No Factor
    01 More Data
    25 Co. Med. Pln.
    50 Wkr. Comp. (but not work related)
    75 Work Related (Occurred in course of work)

41a. <u>Occupation Code</u> - These codes are obtained from a listing of Job
     Codes in the EIS System and are entered next to Work Relation on
     the input document. Use 4 digits (xxxx) in brackets. The next
     change to the input form will include this change.

42. <u>Employee Name and Initials</u>: Write in last name and up to three
    initials.

43. <u>Supervisor's Name</u>: Write last name and up to three initials.

44. <u>Supervisor's Employee Number</u>:

<u>Code</u>

99999 No Factor

50. <u>Freq. Pot</u>. The expected rate (number of times) this event will
    occur.

    <u>Code</u>

    08   Freq. - once per week to twice per month
    16   Occas. - once per month to twice per year
    24   Rare - has been known to occur

51. <u>Severe Pot</u>. The most probable result of the <u>next</u> mishap event.
    Note: Medical treatment can be put in either minor or severe
    since consideration is <u>next</u> event.

    <u>Code</u>

    08   Maj. - (Major) - Amputation, fatality, property damage
              above $75,000.
    16   Serious - Disabling injury, property damage $1,000 -
              $75,000.
    24   Minor - 1st aid, property damage to $1,000

52. <u>Human Factor</u>

    <u>Code</u>   <u>Code</u>

    0001 More Data   4560 Oper. Too Fast
    9999 None   4575 Wrong Place (Improper
                                             Placement, Mix, Combine)
    1500 D/N Recog. (Failed to
          Recog. Risk Cond.)   6000 Wrong Decis. (Decision)
    1525 Body Pos. LTA (Body   6025 D/N Check (With Supervisor)
          Position)   6050 D/N Use (Procedure)
    1550 Equip. Use LTA   6075 Veh. Opr. Err.
    1565 Knowledge LTA
    1575 Meter Use LTA (Instrument   7500 Wrng. Funct. (Did Function
          Reading)   Not Required)
                                       7515 Altercation
    3000 DN Respon. (Response to   7530 Distract. (Tease, Abuse,
          Contingency LTA)   Scaring)
    3012 Act N/R (Act Not   7545 D/N do Co. Pol. (D/N Follow
          Required)   Co. Policy)
    3024 D/N Get Help   7560 Viol. of Law
    3036 D/N Warn   7575 X Safe Dev. (Made Safety
    3048 Expr. LTA (Experience)   Device Inoperative)
    3060 F/T Act (Failed to Act)
    3072 Wrong Sequ. (Wrong
          Sequence)

    4500 Omission (Failed to
          Perform Proper Function)
    4515 D/N Shut Dn. (Shut Down)
    4530 Omitted PPE
    4545 Omitted Step

53. Condition Factor

Code		Code	
001	More Data	460	Rd. Block. (Road Blocked – Storm, etc.)
999	None	480	Rough Gnd.
020	Clothing	500	Rough Sur.
040	Comm. LTA (Communication)	520	Secure. LTA (Securing)
060	Cntl. Fail. (Control Failure)	540	Shoring LTA
070	Equipment LTA	560	Shield LTA
080	Fire Fite LTA (Fighting)	580	Slippery Sur.
100	Fire Prot. LTA	600	Steep Sur.
120	Ground LTA (Electrical)	620	Supply LTA
140	Guard LTA	660	Surface LTA
160	Guard R. (Removed)	680	Too Bulky (For Manual Handling)
180	High Heat/Cold	700	Too Heavy (For Manual Handling)
200	Hi Noise (Excessive)	720	Tools LTA
220	Hi Smoke	740	Ventl. LTA
240	Hsekeep. LTA	760	Ver. Spa. LTA (Vertile Space)
260	Label LTA	780	Vis. LTA
280	Lift Fac. LTA		
300	Light. LTA		Security
320	Method LTA	800	Disord. CD (Disorderly Crowd)
340	Miss. Equip. (Missing)	850	Disord. Per. (Disorderly Person)
360	Nar. Spa. (Narrow Space)	900	Riot
380	No Energ. Bar. (No Energy Barrier)	950	Traf. Cntl. LTA (Traffic Control)
400	Piling LTA		
420	Placmnt. LTA		
440	PPE LTA		

54. Root Cause Factors

Code this section by selecting from the fields within each area as
they apply. One selection must be made.
   Note: Three selections are available for developing cause data.

			Page
	Unwanted Energy Flow		
Maintenance Codes	08	Design Factor	28
	16	Maintenance/Environ.	28
9999 – No Factor	24	Inspection	28
0001 – More Data	32	Supervision Factor	28
	Management System		
	40	Risk Assessment	29
	48	Implementation	29
	Security Factor		
	56	Security	29
	64	Problem Not Related	29

54. <u>Root Cause Factors</u> (Cont'd)

Unwanted Energy Flow

Code

Code		Code	
0800	Design/Factor	0855	Inst. LTA (Instrumentation)
0805	No Haz. Rev. (No Hazard Review)	0860	Man. Cntl LTA (Manual Control LTA)
0810	Arrange. LTA (Arrangement)	0865	Specs. LTA (Specifications
0815	Auto. Cntl. LTA (Automatic Control LTA)	0870	Struct. LTA (Structure)
0820	D/N Cntrl Ener. (D/N Control Energy	0875	Under Des. (Under Designed)
0825	D/N Insp. (Inspect)	0880	Vis. LTA (Visibility)
0835	Drain LTA (Drainage)	0885	Warn. LTA (Warnings LTA)
0840	Energ. Bar. LTA (Energy Barriers LTA)		
0845	Exp. Haz. (Exposed Hazard)		
0850	Fire Pro. LTA (Fire Protection)		

Code		Code	
1600	Maint./Env. Fact.		
1608	Cleaning LTA	1648	D/N Use Per (D/N User Permit)
1616	Disposal LTA	1656	Exhaust LTA
1624	D/N Anal. Fail. (D/N Analyze Failure for Cause)	1664	Hsekeep LTA
1632	D/N Insp. (Inspect)	1672	Sched. LTA (Scheduling)
1640	D/N Test	1680	Vent. LTA (Ventilation)
		1688	Wx Factor (Weather Severe)

Code

2400   Insp. Factor
2420   D/N Spec. (Specify)
2440   Fail. Anal. LTA (D/N Analyze for Failure)
2460   Per. In. LTA (Periodic Inspection)
2480   Sched. LTA (Inspection Schedule)

Code		Code	
3200	Super. Factor	3240	Emer. Pln. LTA (Emergency Planning)
3208	Coord. LTA (Coordination Inter Dept.)	3248	Instr. LTA (Verbal or Written)
3216	D/N Cor. Rsk. (Correct Risk)	3256	JSA LTA (Job Safety Analysis)
3224	D/N Det. Haz. (Detect Hazard)	3264	Resp. Unclr. (Responsibility Unclear)
3232	E. Tng. LTA (Employee Training)	3272	OJT LTA (On Job Training)

Management System

Code

4000	Risk Assessment
4012	Budget LTA
4024	D/N Id. Rsk. (Identify Risk)
4036	ID Rsk. N. Act. (Ident. Risk No Action)

Code

4048	Kn. Crit. Inc. (Known Critical Incident)
4060	Kn. Prec. (Known Precedent)
4072	Plnd. Chg. (Planned Change)
4084	Proc. LTA (Procedure)

Code

5600	Security Factor
5620	Arson
5640	Burglary

Code

5660	Theft
5680	Vandalism

Code

6400	Prob N/R
6412	Attacks
6424	Embezzlement
6436	Fraud

Code

6448	Murder
6460	Suicide
4672	Vagrancy

SAMIS INPUT ERROR RULES

I. <u>General</u>
   A. Case Number
      1. must be numeric
      2. if transaction code 'A' - must be unique
      3. if transaction code 'C' or 'D' - must be on file

   B. Transaction Code - must be 'A,' 'C,' or 'D'

   C. Card - must be 1-5.

II. <u>Card 1</u>
   A. Corp/Div
      1. must be numeric or blank
      2. must be on table #2 (personnel DIV)

   B. Org. Unit
      1. must be numeric or blank
      2. must be on table ORG (personnel)

   C. Expense Center
      1. must be numeric or blank
      2. must be on table EXP (personnel)

   D. Location
      1. must be numeric
      2. must be on table #5 (personnel)

   E. Incident Date - will validate
      1. must be numeric
         a. day - cannot exceed for month
         b. month - must be 1-12
         c. year - if leap year can have 0229
      2. cannot be GT current date

   F. Time
      1. must be numeric
      2. not greater than 2400
      3. must have leading zero if less than 1000

      G. A/R
         1. must be 'A,' 'R' or blank

      H. Property Cost
         1. first position must be 'E' or 'A'
         2. following positions must be numeric

      I. Medical Cost Same as II, H

      J. Compensation Cost Same as II, H

      K. Other Cost Same as II, H

      L. Type of Case
         1. must be numeric
         2. must be on table #12
         3. must have primary (secondary & other can be blank)

      M. Type of Property
         1. must be numeric
         2. must be on table #13

      N. Property Ownership
         1. must be numeric
         2. must be on table #14

III. Card 2
      A. Phase of Operation
         1. must be numeric
         2. must be on table #15

      B. Last Inspection
         1. must be numeric
         2. see code manual, #16

      C. Weather
         1. must be numeric
         2. must be on table #17

      D. Specific Operation
         1. must be numeric
         2. must be on table #18

III. <u>Card 2</u> (Cont'd)

    E. Specific Component

       1. must be numeric

       2. must be on table #19

    F. Model Year

       1. must be numeric

       2. see code manual, #20

    G. Contminant #1 & #2

       1. first two positions must be alphabetic

       2. following 5 positions must be numeric

       3. must be on table #21

       4. Contaminant #2 may be left blank

    H. Recommendations Outstanding

       1. must be numeric

       2. see code manual, #22

    I. Batch Number

       1. must be numeric

       2. see code manual, #23

    J. Work Order A/R Number

       1. must be numeric

       2. see code manual, #24

    K. Fatality Date

       1. must be numeric

       2. if not 9's will validate same as Incident Date (II, E)

IV. <u>Card 3</u>

    A. Social Security Number

       1. must be numeric or blank

       2. see code manual, #30

    B. Employee Number

       1. must be numeric or blank

       2. see code manual, #31

IV. <u>Card 3</u> (Cont'd)

    C. Age

       1. must be numeric or blank

       2. see code manual, #33

    D. Employee Status

       1. must be numeric

       2. must be on table #33

       3. If employee status does not = 30 then employee # must
          be blank otherwise must be numeric

    E. Nature of Injury/Illness

       1. must be numeric

       2. must be on table #34

    F. Part of Body

       1. must be numeric

       2. must be on table #35

    G. Injury Severity

       1. must be numeric

       2. must be on table #36

    H. A/R

       1. must be 'A' or 'R' or blank

    I. Days Off

       1. first position must be 'E' or 'A'

       2. following 4 positions must be numeric

       3. see code manual, #37

    J. Restricted Activity

       1. first position must be 'E' or 'A'

       2. following positions must be numeric

       3. see code manual, #40

    K. Work Relation

       1. must be numeric

       2. must be on table #41

          Note: For OSHA Report + Inj. Analysis Rpts. Work Relation
          must be coded 75

IV. <u>Card 4</u>

 A. Employee Name - will accept any characters or blank

 B. Supervisor Name - will accept any characters or blank

 C. Supervisor Employee Number
  1. must be numeric or blank
  2. see code manual, #44

V. <u>Card 5</u>

 A. Frequency Potential
  1. must be numeric
  2. must be on table #50

 B. Severity Potential
  1. must be numeric
  2. must be on table #51

 C. Human Factor
  1. must be numeric
  2. must be on table #52
  3. must have first (second is optional)

 D. Condition Factors
  1. must be numeric
  2. must be on table #53
  3. must have first (second is optional)

 E. Root Cause
  1. must be numeric
  2. must be on table #53
  3. must have first (second & third optional)

VI. <u>Extras</u>

 A. If type of case = 60xxx, card #5 may be omitted.

 B. Once a field (cost fields, days off, and restricted activity fields) has an 'A' (actual) in front of it, it can no longer be changed to 'E' (estimate).

VI. <u>Extras</u> (Cont'd)

    C. If location is outside the United States, Employee Number must be coded.

    D. For addition of case, must have cards 1, 2, and 3; 4 and 5 are optional.

APPENDIX C

A CONSTRUCTION ASSOCIATION'S SAFETY DATA SYSTEM

The Construction Safety Association of Ontario developed this safety data system for their members. The goal of the association, through the use of this system, is to encourage construction management and labor to develop programs in safe working practices. The Association provides consulting services for identification and measurement of hazards and acts as a catalyst in defining and documenting corrective practices using all known resources.

The system is designed to make reporting as simple as possible. The system is flexible enough to allow the Association to change data that have already been stored. In addition, the system allows alphabetic coding of all storage information. This enables the system to store "Bulldozer" as "BLDZ" or any other combination of letters, rather than "04," which would have to be looked up in a code book. The system is also adaptive enough to accept the entry of dates as 1/12/83, 01/12/83, or 01/12/1983.

Output from the system includes complete printout of a firm's injuries, research studies concerning potential problems, and identification of serious industrywide problems. An example of an output report, the data system's user's manual (*Accident Causal Data: 1983 Code Book*), and a copy of an accident investigation form follow.

1983 CODEBOOK

Prepared by:

Research & Development Department
Construction Safety Association of Ontario
74 Victoria Street
10th Floor
Toronto, Ontario
M5C 2A5

July 1983

I N D E X

<u>LOST TIME INJURY INPUT</u>

CLAIM-NUMBER	12345678
FIRM-NUMBER	123456AB
RATE-NUMBER	123
COUNTY	01 (SEE COUNTY, p. 3)
NAME	DOE J (15 characters)
DATE-OF-INJURY	24FEB83
TIME-OF-INJURY	1630 (24 hour clock)
AGE	34
OCCUPATION	ABCD (SEE OCCUPATION, p. 10-11)
MAJOR-PART-OF-BODY	ABCD (SEE PART-OF-BODY, p. 12)
MINOR-PART-OF-BODY	ABCD (SEE PART-OF-BODY, p. 12)
PROJECT-TYPE	ABCD (SEE PROJECT-TYPE, p. 13)
CONSTRUCTION-TYPE	ABCD (SEE CONSTRUCTION-TYPE, p. 14-15)
EMPLOYEE-ACTIVITY	ABCD (SEE EMPLOYEE-ACTIVITY, p. 16)
MAJOR-ACTED-ON	ABCD (SEE EQUIPMENT & MATERIALS, p. 17-32)
MINOR-ACTED-ON	ABCD (SEE EQUIPMENT & MATERIALS, p. 17-32)
MAJOR-ACTED-WITH	ABCD (SEE EQUIPMENT & MATERIALS, p. 17-32)
MINOR-ACTED-WITH	ABCD (SEE EQUIPMENT & MATERIALS, p. 17-32)
MAJOR-TYPE-OF-ACCIDENT	ABCD (SEE TYPE-OF-ACCIDENT, p. 33-35)
MINOR-TYPE-OF-ACCIDENT	ABCD (SEE TYPE-OF-ACCIDENT, p. 33-35)
MAJOR-OTHER-INVOLVEMENT	ABCD (SEE EQUIPMENT & MATERIALS, p. 17-32)
MINOR-OTHER-INVOLVEMENT	ABCD (SEE EQUIPMENT & MATERIALS, p. 17-32)
DISTANCE	8 (FEET)
POUNDS	8 (POUNDS)
WORK-SURFACE	ABCD (SEE WORK-SURFACE, p. 36)
CONDITION-OF-WORK-SURFACE	ABCD (SEE CONDITION-OF-WORK-SURFACE, p. 38)
MAJOR-NATURE-OF-INJURY	ABCD (SEE NATURE-OF-INJURY, p. 39-40)
MINOR-NATURE-OF-INJURY	ABCD (SEE NATURE-OF-INJURY, p. 39-40)
MAJOR-CAUSES	ABCD (SEE CAUSE, p. 41-43)
MINOR-CAUSES	ABCD (SEE CAUSE, p. 41-43)
MAJOR-CAUSE-INVOLVEMENT	ABCD (SEE EQUIPMENT & MATERIALS, p. 17-32)
MINOR-CAUSE-INVOLVEMENT	ABCD (SEE EQUIPMENT & MATERIALS, p. 17-32)

<u>RATE – NUMBER</u>   <u>-1-</u>

<u>(Major Type of Construction)</u>

<u>RATE 736</u>

1   Culverts or Small Bridges

2   Road Construction & Airports

<u>RATE 744</u>

3   Asphalt and Paving Material
4   Sidewalk Construction

<u>RATE 753</u>

5   Sewer & Watermain Installation
6   High Rise Concrete Forming
7   Reinforcing Steel
8   Rental & Operation of
      Construction Equipment
9   Utility (Gas, Bell Telephone)
10 Pipe Lines
11 Land Clearing and Grubbing
12 Excavating
13 Blasting
14 Large Bridges
15 Water Works
33 Crushing Plants & Gravel Pits
54 Soil/Rock Testing

<u>RATE 761</u>

16 Tunnelling

<u>RATE 809</u>

17 Prefabricated Steel or Concrete Erection

<u>RATE 827</u>

18 Metal Work or Machinery/
      Equipment Installation
19 Welding

<u>RATE 836</u>

20 Canal or Dam Construction
21 Railways
22 Caisson-Work and Pile
      Driving
23 Dredging
24 Diving

<u>RATE 854</u>

25 Building Construction/
      General Contractors
26 Bricklaying and Mason Work
27 Blast Furnaces & Coke
      Ovens, etc.
28 Cement or Concrete Work
29 Lathing and Plastering,
30 Roofing
31 Carpentry
32 Restoration
34 Window Cleaning
35 Supplying of Labour

<u>RATE 859</u>

36 Demolition

<u>RATE-NUMBER</u>, cont'd.   <u>-2-</u>

<u>(Major Type of Construction)</u>

<u>RATE 864</u>

37 Electrical Work
38 Mechanical Work
39 Sheet Metal Work/Eavestroughing/Siding
40 Refrigeration & Airconditioning
41 Floor Laying, Carpets, Tiles
42 Mechanical Insulation
43 Ceramic and Wall Tile Installation
44 Floor Finishing & Carpet Cleaning
45 Inspection Services
46 Acoustic & Ceiling Tiling
47 Overhead Door Work
29 Plastering & Drywall

<u>RATE 873</u>   <u>ALL RATES</u>

48 Insulating   52 Other
49 Painting & Decorating   53 N/A
50 Steeple-Jack Work
51 Window Work

Construction Safety Data System

<u>COUNTY</u>   -3-

01	–	ALGOMA	28	–	MUSKOKA
02	–	BRANT	29	–	NIPISSING
03	–	BRUCE	30	–	NORFOLK
04	–	CARLETON	31	–	NORTHUMBERLAND
47	–	COCHRANE	33	–	OXFORD
05	–	DUFFERIN	34	–	PARRY SOUND
06	–	DUNDAS	36	–	PEEL
07	–	DURHAM	37	–	PERTH
08	–	ELGIN	38	–	PETERBOROUGH
09	–	ESSEX	39	–	PRESCOTT
10	–	FRONTENAC	40	–	PRINCE EDWARD
11	–	GLENGARRY	41	–	RAINY RIVER
12	–	GRENVILLE	42	–	RENFREW
13	–	GREY	43	–	RUSSELL
14	–	HALDIMAND	44	–	SIMCOE
15	–	HALIBURTON	45	–	STORMONT
16	–	HALTON	46	–	SUDBURY
17	–	HASTINGS	47	–	TIMISKAMING (COCHRANE)
18	–	HURON	48	–	THUNDER BAY
19	–	KENORA	49	–	VICTORIA
20	–	KENT	50	–	WATERLOO
21	–	LAMBTON	52	–	WELLINGTON
22	–	LANARK	53	–	WENTWORTH
23	–	LEEDS	54	–	YORK
24	–	LENNOX & ADDINGTON	55	–	OUTSIDE ONTARIO
25	–	NIAGARA	56	–	ONTARIO WATERS
26	–	MANITOULIN	98	–	OUT OF BUSINESS
27	–	MIDDLESEX	99	–	TEMPORARILY UNKNOWN

<u>TOWN WHERE INJURY OCCURRED</u>   -4-

TOWN	COUNTY CODE	TOWN	COUNTY CODE	TOWN	COUNTY CODE
ACTON	16	BATTERSEA	10	CAINSVILLE	02
AGINCOURT	54	BAYSVILLE	28	CALEDON	36
AILSA CRAIG	27	BEACHBURG	42	CALEDON EAST	36
AJAX	07	BEAMSVILLE	25	CALEDONIA	14
ALDERSHOT	53	BEARDMORE	48	CALLANDER	34
ALEXANDRIA	11	BEAVERTON	07	CALSTOCK	47
ALFRED	39	BEETON	44	CAMBRAY	49
ALGONQUIN	01	BELLE RIVER	09	CAMBRIDGE	50
ALLISTON	44	BELLEVILLE	17	CAMLACHIE	21
ALMA	52	BELLS CORNERS	04	CAMPBELLFORD	31
ALMONTE	22	BELMONT	08	CAMBELLVILLE	16
ALVINSTON	21	BETHANY	07	CAMP BORDEN	44
AMELIASBURG	40	BINBROOK	53	CANFIELD	14
AMHERSTBURG	09	BLACKSTOCK	07	CAPREOL	46
ANCASTER	53	BLENHEIM	20	CARAMAT	48
ANGUS	44	BLIND RIVER	01	CARDINAL	12
ANSONVILLE	47	BLOOMFIELD	40	CARGILL	03
APPLE HILL	11	BLYTH	18	CARLETON PLACE	22
APSLEY	38	BOBCAYGEON	49	CARLISLE	53
ARDOCH	10	BOLTON	36	CARP	04
ARKONA	21	BOND HEAD	44	CARTIER	46
ARMSTRONG STATION	48	BONFIELD	29	CASSELMAN	43
ARNPRIOR	42	BOTHWELL	20	CASUMMIT LAKE	19
ARTHUR	52	BOURGET	43	CAYUGA	14
ARVA	27	BOWMANVILLE	07	CENTRALIA	18
ASHTON	04	BRACEBRIDGE	28	CHALK RIVER	42
ATHENS	23	BRADFORD	44	CHAPLEAU	46
ATIKOKAN	41	BRAESIDE	42	CHATHAM	20
ATWOOD	37	BRAMALEA	36	CHATSWORTH	13
AUBURN	18	BRAMPTON	36	CHELMSFORD	46
AUDEN	48	BRANTFORD	02	CHESLEY	03
AULTSVILLE	45	BRECHIN	07	CHESTERVILLE	06
AURORA	54	BRESLAU	50	CHIPPAWA	25
AVONMORE	45	BRIGDEN	21	CHRYSLER	45
AYLMER	08	BRIGHTON	31	CHURCHILL	44
AYR	50	BRINSTON	06	CLAREMONT	07
AYTON	13	BRITT	34	CLARKSBURG	13
AZILDA	46	BROCKVILLE	23	CLARKSON	36
BADEN	50	BRONTE	16	CLIFFORD	52
BAILIEBORO	31	BROOKLIN	07	CLINTON	18
BALA	28	BROWNSVILLE	33	COBALT	47
BALMERTOWN	19	BRUCE MINES	01	COBDEN	42
BANCROFT	17	BRUSSELS	18	COBOCONK	49
BARNICH	41	BURFORD	02	COBOURG	31
BARRIE	44	BURGESSVILLE	33	COCHRANE	47
BARRY'S BAY	42	BURK'S FALLS	34	COCHENOUR	19
BATAWA	17	BURLINGTON	16	COLBORNE	31
BATH	24	BURWASH	46	COLDWATER	44

TOWN WHERE INJURY OCCURRED, CONT'D.   -5-

TOWN	COUNTY CODE	TOWN	COUNTY CODE	TOWN	COUNTY CODE
COLE	19	DUNDALK	13	FONTHILL	25
COLLINGWOOD	44	DUNDAS	53	FORDWICH	18
COLLINS BAY	10	DUNGANNON	18	FOREST	21
COMBER	09	DUNNVILLE	14	FORT ERIE	25
COMBERMERE	42	DURHAM	13	FORT FRANCES	41
CONCORD	54	DUTTON	08	FORT WILLIAM	48
CONISTON	46	DWIGHT	28	FOURNIER	39
CONSECON	40	EADES	47	FOXBORO	17
COOKSTOWN	44	EAR FALLS	19	FRANKFORD	17
COOKSVILLE	36	EASTVIEW CENTRE	04	FRANKVILLE	23
COPPER CLIFF	46	EDGAR	44	FRANZ	01
CORBYVILLE	17	EGANVILLE	42	FRASERDALE	47
CORNWALL	45	ELDORADO	17	FREELTON	53
CORUNNA	21	ELGIN	23	FREEMAN	29 & 16
COURTRIGHT	21	ELK LAKE	47	FRUITLAND	53
CREDITON	18	ELLIOT LAKE	01	GALT	50
CREEMORE	44	ELMIRA	50	GANANOQUE	23
CREIGHTON MINE	46	ELMVALE	44	GARDEN HILL	07
CUTLER	01	ELMWOOD	03	GEORGETOWN	16
CYRVILLE	04	ELORA	52	GERALDTON	48
DALHOUSIE MILLS	11	EMBRO	33	GLENCOE	27
DALKEITH	11	EMBRUN	43	GODERICH	18
DALTON	01	EMO	41	GODFREY	10
DANE	47	EMPIRE	48	GOGAMA	46
DASHWOOD	18	ENGLEHART	47	GOLDEN VALLEY	34
DEEP RIVER	42	ENNISKILLEN	07	GOLD PARK	01
DELAWARE	27	ENTERPRISE	24	GOODWOOD	07
DELHI	30	ERIEAU	20	GORDON LAKE	19 & 01
DELORO	17	ERIN	52	GORE BAY	26
DELTA	23	ERINDALE	36	GORRIE	18
DEMORESTVILLE	40	ESPANOLA	46	GRAFTON	31
DESBARATS	01	ESSEX	09	GRAND BEND	21
DESERONTO	17	ETHEL	18	GRAND VALLEY	05
DIXIE	36	EVERETT	44	GRANTON	27
DOFASCO	53	EXETER	18	GRAVENHURST	28
DORCHESTER	27	FALCONBRIDGE	46	GRIMSBY	25
DORNOCH	13	FENELON FALLS	49	GUELPH	52
DOUGLAS	42	FENWICK	25	HAGERSVILLE	14
DOUGLAS POINT	03	FERGUS	52	HALEY	42
DOWNSVIEW	54	FERRIS	29	HAILEYBURY	47
DRAYTON	52	FIELD	29	HALIBURTON	15
DRESDEN	20	FINCH	45	HALL'S GLEN	38
DRUMBO	33	FINGAL	08	HAMILTON	53
DRYDEN	19	FISHERVILLE	14	HANMER	46
DUART	20	FLESHERTON	13	HANNON	53
DUBLIN	37	FLINTON	24	HANOVER	13
DUBREUILVILLE	01	FLORENCE	21	HARRIETSVILLE	27
DUNBARTON	07	FOLEYET	46	HARRISTON	52

<u>TOWN WHERE INJURY OCCURRED</u>, CONT'D.   -6-

TOWN	COUNTY CODE	TOWN	COUNTY CODE	TOWN	COUNTY CODE
HARROW	09	KERWOOD	27	LONG BRANCH	54
HARROWSMITH	10	KESWICK	54	LONGLAC	48
HASTINGS.	31	KILBRIDE	16	LONG SAULT	45
HAVELOCK	38	KILLALOE STATION	42	L'ORIGNAL	39
HAWKESBURY	39	KILLARNEY	26	LORING	34
HAWK JUNCTION	01	KINBURN	04	LORNE PARK	36
HEARST	47	KINCARDINE	03	LUCAN	27
HENSALL	18	KING CITY	54	LUCKNOW	03
HEPWORTH	03	KING KIRKLAND	47	LYN	23
HERON BAY	48	KINGSTON	10	LYNDEN	53
HESPELER	50	KINGSVILLE	09	MCDONALD'S CORNERS	22
HICKSON	33	KINMOUNT	49	MCKENZIE ISLAND	19
HIGHGATE	20	KINTAIL	18	MACTIER	28
HILLSBURGH	52	KINTORE	33	MADAWASKA	29
HILLSDALE	44	KIRKFIELD	49	MADOC	17
HOLLAND CENTRE	13	KIRKLAND LAKE	47	MADSEN	19
HOLSTEIN	13	KIRKTON	18	MAGNETAWAN	34
HONEY HARBOUR	28	KITCHENER	50	MAITLAND	12
HONEYWOOD	05	KOMOKA	27	MALLORYTOWN	23
HORNBY	16	LAFONTAINE	44	MALTON	36
HORNEPAYNE	01	LAKEFIELD	38	MANITOUWADGE	48
HUDSON	19	LAKESIDE	33	MANITOWANING	26
HUMBERSTONE	25	LAKEVIEW	08 & 36	MANOTICK	04
HUNTSVILLE	28	LAMBETH	27	MAPLE	54
HYDE PARK	27	LAMBTON MILLS	54	MARATHON	48
IGNACE	19	LANARK	22	MARKDALE	13
INDIAN RIVER	38	LANCASTER	11	MARKHAM	54
INGERSOLL	33	LANGSTAFF	54	MARLBANK	17
INGLEWOOD	36	LANGTON	30	MARMORA	17
INNERKIP	33	LANSDOWNE	23	MARTINTOWN	11
INVERARY	10	LANSING	54	MASSEY	46
INWOOD	21	LARDER LAKE	47	MATACHEWAN	47
IROQUOIS	06	LASALLE	09	MATHESON	47
IROQUOIS FALLS	47	LEAMINGTON	09	MATTAWA	29
ISLINGTON	54	LEASIDE	54	MATTICE	47
JACKSON'S POINT	54	LEFAIVRE	39	MAXVILLE	11
JAMESTOWN	18 & 01	LEVACK	46	MAYNOOTH	17
JARVIS	14	LIMEHOUSE	16	MEAFORD	13
JASPER	12	LINDSAY	49	MELBOURNE	27
JELLICOE	48	LINWOOD	50	MERLIN	20
JORDON	25	LION'S HEAD	03	MERRICKVILLE	12
KANATA	04	LISTOWEL	37	MERRITTON	25
KAPUSKASING	47	LITTLE BRITAIN	49	METCALFE	04
KEARNEY	34	LITTLE CURRENT	26	MIDDLEVILLE	22
KEENE	38	LITTLE LONG RAPIDS	47	MIDHURST	44
KEEWATIN	19	LIVELY	46	MIDLAND	44
KEMPTVILLE	12	LOBO	27	MILDMAY	03
KENORA	19	LONDON	27	MILLBROOK	07

TOWN WHERE INJURY OCCURRED, CONT'D.   -7-

TOWN	COUNTY CODE	TOWN	COUNTY CODE	TOWN	COUNTY CODE
MILLEROCHES	45	NORTH AUGUSTA	12	POINTE AU BARIL	34
MILLHAVEN	24	NORTH BAY	29	POINTE AUX ROCHES	09
MILLIKEN	54	NORTH GOWER	04	PONTYPOOL	07
MILTON	16	NORTHWOOD	20	PORT ARTHUR	48
MILVERTON	37	NORVAL	16	PORT BURWELL	08
MIMICO	54	NORWICH	33	PORT CARLING	28
MINDEMOYA	26	NORWOOD	38	PORT COLBORNE	25
MINDEN	15	OAKVILLE	16	PORT COLDWELL	48
MISSANABIE	01	OAKWOOD	49	PORT CREDIT	36
MISSISSAUGA	36	O'BRIEN	47 & 48	PORT DALHOUSIE	25
MITCHELL	37	ODENBACH	29	PORT DOVER	30
MONKTON	37	ODESSA	24	PORT ELGIN	03
MOONBEAM	47	OHSWEKEN	02	PORT HOPE	31
MOOREFIELD	52	OIL SPRINGS	21	PORT LAMBTON	21
MOOSE CREEK	45	OJIBWAY	09	PORTLAND	23
MOOSE FACTORY	47	OMEMEE	49	PORT LORING	34
MOOSONEE	47	ONAPING	46	PORT MCNICOLL	44
MORRISBURG	06	ORANGEVILLE	05	PORT PERRY	07
MORRISTON	52	ORILLIA	44	PORT ROBINSON	25
MOSHER	01	ORLEANS	04	PORT ROWAN	30
MOUNT ALBERT	54	ORONO	07	PORT STANLEY	08
MOUNT BRYDGES	27	OSGOODE	04	PORT UNION	07
MOUNT DENIS	54	OSHAWA	07	POWASSAN	34
MOUNT ELGIN	33	OTTAWA	04	PRESCOTT	12
MOUNT FOREST	52	OTTERVILLE	33	PRESTON	50
MOUNT HOPE	53	OWEN SOUND	13	PRINCETON	33
MOUNT PLEASANT	02	PAISLEY	03	QUIRKE LAKE	01
MUNCEY	27	PAKENHAM	22	RAINY RIVER	41
NAKINA	48	PALERMO	16	RAMSEY	46
NANTICOKE	14	PALGRAVE	36	REDBRIDGE	29
NAPANEE	24	PALMERSTON	52	RED LAKE	19
NAVAN	43	PARIS	02	RED ROCK	48
NEPEAN	04	PARKHILL	27	REGAN	48
NEUSTADT	13	PARRY SOUND	34	RENABIE	01
NEWBORO	23	PELEE ISLAND	09	RENFREW	42
NEWBURY	27	PEMBROKE	42	REXDALE	54
NEWCASTLE	07	PENETANGUISHENE	44	RICEVILLE	39
NEW DUNDEE	50	PERTH	22	RICHARD'S LANDING	01
NEW HAMBURG	50	PETAWAWA	42	RICHMOND	08 & 04
NEWINGTON	45	PETERBOROUGH	38	RICHMOND HILL	54
NEW LISKEARD	47	PETERSBURGH	50	RICHVALE	54
NEWMARKET	54	PETROLIA	21	RIDGETOWN	20
NEWTONVILLE	07	PICKERING	07	RIDGEWAY	25
NEW TORONTO	54	PICKLECROW	19	RIPLEY	03
NIAGARA FALLS	25	PICTON	40	RIVERSIDE	09
NIAGARA-ON-THE-LAKE	25	PLANTAGENET	39	ROCKLAND	43
NIPIGON	48	PLATTSVILLE	33	ROCKWOOD	52
NOELVILLE	46	POINT EDWARD	21	RODNEY	08

TOWN WHERE INJURY OCCURRED, CONT'D.   -8-

TOWN	COUNTY CODE	TOWN	COUNTY CODE	TOWN	COUNTY CODE
ROSENEATH	31	SPRAGGE	01	TOLEDO	23
ROSLIN	17	SPRINGFIELD	08	TORONTO	54
ROSSEAU	34	SPRUCEDALE	34	TOTTENHAM	44
RUSSELL	43	STAMFORD CENTRE	25	TRAFALGAR	16
ST. ALBERT	43	STAYNER	44	TRENTON	17
ST. CATHARINES	25	STEVENS	48	TROUT CREEK	34
ST. CHARLES	46	STEVENSVILLE	25	TWEED	17
ST. DAVIDS	25	STIRLING	17	UNIONVILLE	54
ST. EUGENE	39	STONEY CREEK	53	UPSALA	48
ST. GEORGE	02	STOUFFVILLE	54	UXBRIDGE	07
ST. JACOBS	50	STRAFFORDVILLE	08	VAL CARON	46
ST. MARY'S	37	STRATFORD	37	VALORA	19
ST. THOMAS	08	STRATHCONA	24	VANKLEEK HILL	39
ST. WILLIAMS	30	STRATHROY	27	VANIER	04
SANDWICH	09	STREETSVILLE	36	VARS	43
SANITORIUM	28	STROUD	30	VERNER	29
SAPAWE	41	STURGEON FALLS	29	VERONA	10
SARNIA	21	SUDBURY	46	VICTORIA HARBOUR	44
SAULT STE. MARIE	01	SULTAN	46	VINELAND	25
SAVANT LAKE	48	SUNDERLAND	07	VIRGINIATOWN	47
SCARBORO JUNCTION	54	SUNDRIDGE	34	VITTORIA	30
SCHOMBERG	54	SUTTON	54	WADE	19
SCHUMACHER	47	SWANSEA	54	WAINFLEET	25
SCHREIBER	48	SWASTIKA	47	WALES	45
SCOTLAND	02	SYDENHAM	10	WALKERTON	03
SEAFORTH	18	TAMWORTH	24	WALKERVILLE	09
SEBRINGVILLE	37	TARA	03	WALLACEBURG	20
SEELEY'S BAY	23	TAVISTOCK	33	WALLACETOWN	09
SELKIRK	14	TECUMSEH	09	WALTER'S FALLS	13
SHAKESPEARE	37	TEESWATER	03	WARDSVILLE	27
SHALLOW LAKE	13	TEMAGAMI	29	WARKWORTH	31
SHANNONVILLE	17	TERRACE BAY	48	WARREN	46
SHARBOT LAKE	10	THAMESFORD	33	WARSAW	38
SHEDDEN	08	THAMESVILLE	20	WASAGA BEACH	44
SHEFFIELD	53	THEDFORD	21	WASHAGO	07
SHELBURNE	05	THESSALON	01	WATERFORD	30
SIMCOE	30	THORNBURY	13	WATERLOO	50
SIOUX LOOKOUT	19	THORNDALE	27	WATERDOWN	53
SKEAD	46	THORNHILL	54	WATFORD	21
SMITH'S FALLS	22	THORNTON	44	WAUBAUSHENE	44
SMITHVILLE	25	THOROLD	25	WAWA	01
SMOOTH ROCK FALLS	47	THUNDER BAY	48	WEBBWOOD	46
SOUTHAMPTON	03	TILBURY	20	WELLAND	25
SOUTH MOUNTAIN	06	TILLSONBURG	33	WELLANDPORT	25
SOUTH PORCUPINE	47	TIMMINS	47	WELLESLEY	50
SOUTH RIVER	34	TIVERTON	03	WELLINGTON	40
SOUTH WOODSLEE	09	TOBERMORY	03	WESTBORO	04
SPENCERVILLE	12	TODMORDEN	54	WEST HILL	54

TOWN WHERE INJURY OCCURRED, CONT'D.   -9-

TOWN	COUNTY CODE	TOWN	COUNTY CODE	TOWN	COUNTY CODE
WEST LORNE	08	WHITNEY	29	WOODBRIDGE	54
WESTMEATH	42	WIARTON	03	WOODHAM	37
WESTON	54	WILBERFORCE	15	WOODSTOCK	33
WEST PORT	23	WILLIAMSBURG	06	WOODVILLE	49
WESTREE	46	WILLIAMSTOWN	11	WOOLER	31
WHEATLEY	20	WILLOWDALE	54	WROXETER	18
WHITBY	07	WINCHESTER	06	WYOMING	21
WHITE DOG FALLS	19	WINDSOR	09	YARKER	24
WHITE LAKE	42	WINGHAM	18	ZURICH	18
WHITE RIVER	01	WINONA	53		

OUTSIDE ONTARIO....55

ONTARIO WATERS.....56
(MARITIME ACCIDENTS)

OCCUPATION

CODE	DESCRIPTION
ANTI	Antenna Installer/Repairman
APPL	Appliance Repairman
BOIL	Boilermaker
BRIA	Bricklayer's Apprentice
BRIC	Bricklayer/Stone Mason
CARA	Carpenter's Apprentice
CARP	Carpenter/Wood Worker
CEMT	Concrete & Cement Finisher
CHIM	Chimney Worker/Steeplejack
CHMA	Chimney Apprentice
CONC	Concrete and Cement Worker (Not Finisher)
DIVE	Diver
DOOR	Overhead Door Serviceman
DRIV	Truck Driver
ELEA	Electrician's Apprentice
ELEC	Electrician
ELEV	Elevator Tradesman
ENGI	Engineer
EREC	Precast Concrete Erector
EXEC	Executive/Self-Employed
FACT	Factory/Plant/Shopworker
FLAG	Flagman
FLOR	Flooring Tradesman
FLRA	Flooring Apprentice
FORM	Formsetter
FURN	Furnace Serviceman
GLAA	Glazier's Apprentice
GLAZ	Glazier
HEQA	Mechanic's Apprentice
HEQP	Heavy Equipment or Vehicle Mechanic
INSA	Insulator's Apprentice
INST	Instrument/Piping Mechanic
INSU	Insulator
IRON	Ironworker/Ironworker Apprentice
JANI	Janitor/Cleaner
LABO	Labourer
LANS	Landscaper/Gardener
LUMB	Lumberjack
MILA	Millwright's Apprentice
MILW	Millwright
MINR	Miner/Driller/Powderman
MISC	Miscellaneous
OFWO	Office Worker
OPER	Operator - Equipment or Machine
PAIA	Painter's Apprentice
PAIN	Painter/Paper Hanger

-11-

OCCUPATION

CODE	DESCRIPTION
PIPA	Pipelayer's Apprentice
PIPE	Pipelayer/Drainlayer
PLAA	Plasterer's Apprentice
PLAS	Plasterer/Lather/Drywall/Taper
PLUA	Plumber's Apprentice
PLUM	Plumber
RADI	Radiographer
RAKE	Concrete or Asphalt Raker
REFA	Refrigeration Apprentice
REFR	Refrigeration/Airconditioning Mechanic
RODA	Rodbuster's Apprentice
RODB	Rodbuster/Rodman – Reinforcing Steel
ROOF	Roofer (Flat Roofing)
SFTA	Steamfitter's Apprentice
SFTR	Steamfitter/Pipefitter
SHNG	Shingler (Sloped Roofing)
SIDE	Siding Installer/Eavestrough Installer
SMTA	Sheet Metal Worker's Apprentice
SMTL	Sheet Metal Worker
SNBL	Sandblaster/High Pressure Water Blaster
SUPR	Superintendent/Foreman/Manager
SURV	Surveyor
WASH	Window Washer
WELA	Welder's Apprentice
WELD	Welder/Flame Cutter

PART-OF-BODY

MAJOR CODE	MAJOR DESCRIPTION	MINOR CODE	MINOR DESCRIPTION
HEAD	Head Area	EARS	Ears
		EYES	Eyes
		FACE	Face/Lips/Chin
		GUMS	Teeth & Gums
		JAWS	Jaw
		NECK	Neck
		NOSE	Nose & Throat
		SKLL	Skull & Scalp
TRNK	Trunk Area	ABDO	Abdomen
		BACK	Back/Spine
		BUTT	Buttocks
		CHST	Chest/Ribs
		GROI	Groin
		HIPS	Hips/Pelvis
		SHOU	Shoulders
ARMS	Arm/Hand	ARMS	Arm
		FNGR	Fingers
		ELBO	Elbow
		HAND	Hands
		WRST	Wrist
		THUM	Thumbs
LOWR	Lower Body	ANKL	Ankle
		FOOT	Foot
		HEEL	Heel
		KNEE	Knee
		LEGS	Leg/Thigh
		META	Metatarsal (Upper Foot)
		SOLE	Sole
		TOES	Toes
BODY	Body System	BODY	General Body Injury (e.g. Electric Shock)
		CIRC	Circulatory (e.g. Heart Attack)
		DGST	Digestive
		EXCR	Excretory
		GENI	Genital/Urinary
		INTE	Internal Body Organ
		NERV	Nervous
		REPR	Reproductive
		RESP	Respiratory

-13-

PROJECT-TYPE

CODE	DESCRIPTION
ASPP	Asphalt Plant
BILD	Other Building (Barn, Garage)
BRID	Bridge
BUSH	Bush Clearing
CBHR	Commercial Building – High Rise
CBLR	Commercial Building – Low Rise
COMM	Commercial Building – Height Unknown
CRSH	Crushing Plants & Gravel Pits
DAMS	Dams/Piers/Breakwater
DRIV	Driveways and Parking Lots
EXCA	Other Excavation (Swimming Pool, Parks)
FLTR	Filtration Plant
HIND	Heavy Industrial
HYDR	Hydro
INST	Institutional (Schools, Prisons, Hospitals, Army Bases, Church, Fire & Police Stations, Gov't Bldgs)
LIND	Light Industrial
MINE	Mining
MSUR	Marine (Surface)
NONC	Misc. Non–Construction, Activities
NUCL	Nuclear
PIPE	Pipeline (Natural Gas/Oil/Water)
PTRO	Petrochemical
RAIL	Railroad/Track Work
RBHR	Residential Building – High Rise
RBLR	Residential Building – Low Rise
RESI	Residential Building – Height Unknown
ROAD	Road/Highway/Street
SEWG	Sewage Treatment Plant
SEWR	Sewer & Watermain
SHOP	Shop/Yard/Factory/Suppliers Premises
SILO	Silo
SUBW	Subway
TEMP	Temporary Agency Work
UTIL	Utility other than Hydro

CONSTRUCTION-TYPE

CODE	DESCRIPTION
ACOU	Acoustic/Suspended Ceiling Work
CARP	Carpentry
CEMT	Concrete & Cement Finishing
CHIM	Chimney Work
CONC	Concrete/Cement Work/Core Drilling
DECO	Decorating/Painting
DEMO	Demolition
DRED	Dredging
DRIL	Drilling and Blasting
ELEC	Electrical
ELEV	Elevator Work
ENGS	Engineering Services (Inspection, Testing, Surveying)
EREC	Erection (Concrete, Steel)
EREO	Erection (Other, i.e. Prefabs)
EXCA	Excavation
FLOR	Flooring Installation (Carpet or Tile)
FLRS	Floor Sanding/Finishing
FORM	Forming
GARB	Garbage Collection
GENL	General Contracting
GLAZ	Glazing/Window Construction/Repair
GRAV	Gravel & Rock Work
GRUB	Land Clearing/Grubbing/Tree Removal
HAUL	Hauling Materials/Earthmoving
INSU	Insulation/Waterproofing/Fireproofing
JANI	Janitorial
LANS	Landscaping
MACH	Machinery & Equipment Installation
MAIN	Maintenance of Building & Equip. therein
MANA	Fabrication/Manufacturer in Shop/Yard/Factory
MASO	Masonry
MECH	Mechanical (Plumbing, Steamfitting, Ductwork)
NONC	Misc. Non-Construction Activities
PAVE	Paving/Grading/Hot Laid Asphalt
PILE	Piledriving/Foundation Shoring
PIPE	Oil & Gas Pipe Installation
PLAS	Plastering/Drywall
RAIL	Railway Construction/Maintenance
RDMN	Road Maint./Line Painting/Catch Basin/Traffic Light Repair
REIN	Reinforcing Steel
REPA	Construction Equip. and Vehicle Repair
REFR	Refractory

-15-

<u>CONSTRUCTION-TYPE</u>, cont'd.

<u>CODE</u>	<u>DESCRIPTION</u>
REST	Restoration (Sandblasting/Renovating)/Exterior Cleaning
ROAD	Road Construction
ROFF	Flat Roofing
ROFS	Sloped Roofing
ROOF	Roofing – Type Unknown
SEWR	Sewer Installation
SIDE	Siding/Eavestroughing/Sheet Metal
SIDW	Sidewalk Work/Curb/Gutter
SNOR	Snow Removal/Sanding & Salting Roads
TRAN	In Transit (To or from a site)
TUNL	Tunnelling
UTIL	Utilities
WASH	Window Cleaning
WELD	Welding

EMPLOYEE-ACTIVITY

CODE	DESCRIPTION	CODE	DESCRIPTION
AJST	Adjusting	LOWR	Lowering
APPL	Applying	MEAS	Measuring
BEND	Bending	MIX	Mixing
BOOS	Boosting	MOUN	Mounting (Of Equip.)
BREA	Breaking	MOVE	Moving
CARY	Carrying	OPEN	Opening
CHNG	Changing	OPER	Operating
CHIP	Chipping	PAIN	Painting
CHSL	Chiseling	PLAC	Placing
CLEN	Cleaning	PLAS	Plastering
CLMB	Climbing	POUR	Pouring
CLOS	Closing	PRY	Prying
CONN	Connecting	PULL	Pulling
CRAW	Crawling	PUSH	Pushing
CUT	Cutting	RAIS	Raising
DEMO	Demolishing	REAC	Reaching
DISC	Disconnecting	REMO	Removing
DISM	Dismantling	REPA	Repairing
DMNT	Dismounting	REPL	Replacing
	(From Equipment)	ROLL	Rolling
DSND	Descending	SCRA	Scraping
DRIL	Drilling	SECU	Securing
DRIV	Driving	SGNL	Signalling
DUMP	Dumping	SHOV	Shovelling
ERCT	Erecting	SHRP	Sharpening
FINI	Finishing Concrete/Parging	SIT	Sitting
GRND	Grinding	SMOO	Smoothing
GUID	Guiding	SOLD	Soldering
HAMR	Hammering	STAC	Stacking
HANG	Hanging	STAN	Standing
HNDL	Handling	STRI	Stripping
HEAT	Heating	STRT	Starting
HOLD	Holding	TAPE	Taping
HOOK	Hooking	TGHT	Tightening
HSKP	Housekeeping	THRW	Throwing
INSP	Inspecting	TURN	Turning
INST	Installing	TYE	Tying
JUMP	Jumping	UNHK	Unhooking
KNEE	Kneeling	UNLD	Unloading
LAY	Laying	WALK	Walking
LGHT	Lighting	WELD	Welding
LIFT	Lifting		
LOAD	Loading		
LOOS	Loosening		

-17-

EQUIPMENT AND MATERIALS

The codes located in the equipment and materials section of this book are used
in the following fields when coding:

A)	MAJOR-ACTED-ON	E)	MAJOR-OTHER-INVOLVEMENT
B)	ACTED-ON	F)	OTHER-INVOLVEMENT
C)	MAJOR-ACTED-WITH	G)	MAJOR-CAUSE-INVOLVEMENT
D)	ACTED-WITH	H)	CAUSE-INVOLVEMENT

MAJOR CODE	MAJOR DESCRIPTION	MINOR CODE	MINOR DESCRIPTION
ANIM	Animals, Insects, Birds, Reptiles,	DOG	Dog
		HUMN	Human
		INSC	Insect/Bee
		OTHR	Other
BILD	Building Parts	BILD	Building Parts, nec.
		CEIL	Ceiling
		CHIM	Chimney
		DFRM	Door Frame
		DOOR	Door
		ELED	Elevator Door
		FLOR	Floor
		GDOR	Garage/Metal Door
		ROOF	Roof
		STAI	Stairs
		STRU	Other Structures/Entire Building (i.e. Sheds)
		TOPS	Counter Tops
		WALL	Wall
		WINP	Window Parts
		WPAR	Wall Partition
		WWAL	Frame Wall Under Construction
BOIL	Boilers, Pressure Vessels	BOIL	Boilers
		CALH	Compressed Air Line (Hose)
		CMGC	Compressed Gas Container/Air
		HYDH	Hydraulic Hose(Line)/ Steam Hose
		PRES	Pressure Vessel, nec.
CHEM	Chemicals (Gas, Liquid or Solid)	ACET	Acetylene

-18-
<u>EQUIPMENT AND MATERIALS</u>, cont'd.

<u>MAJOR CODE</u>	<u>MAJOR DESCRIPTION</u>	<u>MINOR CODE</u>	<u>MINOR DESCRIPTION</u>
Chemicals continued		AMMO	Ammonia
		BIOL	Biological Agents (Infections & Viruses)
		CAIR	Compressed Air
		CHEM	Chemical, nec.
		CHLN	Chlorine
		CORR	Corrosives (Acids, Caustics)
		EPOX	Epoxies
		EXH	Exhaust Fumes
		EXPL	Flammable/Explosive Gases
		FUME	Fumes nec
		FUMP	Pitch/Asphalt Fumes
		GAS	Gasoline
		GASN	Gas (Natural)
		GLUE	Glue/Adhesive
		LEAD	Lead
		OIL	Oil/Grease/Lubricants
		OXYG	Oxygen
		PAIN	Paint/Coatings
		RADS	Radiating Substances
		REFR	Refrigerant
		SMOK	Smoke from fires
		URET	Urethane Insulation/Urea Formaldehyde
		WATL	Water
		WFUM	Welding/Cutting Fumes
		WPRE	Wood Preservatives
CONT	Containers (Empty or Full)	BARR	Barrel/Drum
		BCKT	Bucket/Pail
		BOX	Box
		CAN	Can
		CBUC	Concrete Bucket
		CONT	Container, nec.
		CRAT	Crate
		REEL	Reel (For Wire Or Cable)
		TANK	Tank
		TRSH	Trash Boxes/Cans/Bins

-19-

<u>EQUIPMENT AND MATERIALS</u>, cont'd.

<u>MAJOR CODE</u>	<u>MAJOR DESCRIPTION</u>	<u>MINOR CODE</u>	<u>MINOR DESCRIPTION</u>
DUST	Dust	CEME	Cement Dust/Mortar Dust/ Brick
		DUST	Dust, (Not Available)
		METL	Metallic
		MINR	Mineral Dust, nec.
		SAWD	Sawdust/Wood Dust
ELEC	Electrical Equipment and Hardware	AIRC	Airconditioning Equipment
		ANTE	Antenna
		ATTP	Attachment Plug, Other Connector
		BATT	Batteries/Battery Charger
		EBOX	Electrical Junction Box/ Outlet Box
		BUSB	Bus Bar/Bus Duct
		BULB	Light Bulb/Fluorescent Bulb
		CABL	Electrical Cable/Wire/ Extension Cords
		COND	Conduit
		ELEC	Electrical Equipment & Materials, nec.
		FIXT	Light Fixture/Traffic Lights
		FUSE	Fuse, nec.
		GENR	Generator, nec.
		GROU	Ground Rod
		METR	Meters/Relays
		MOTR	Motor
		RECE	Receptacle
		SBOX	Switch Box
		SWGR	Disconnect Switches & Switchgear/Panel/ Cubicle, nec.
		TRAN	Transformer, nec.

-20-

EQUIPMENT AND MATERIALS, cont'd.

MAJOR CODE	MAJOR DESCRIPTION	MINOR CODE	MINOR DESCRIPTION
ENCL	Enclosures	FENM	Fence (Metal/Wire)
		FENW	Fence (Wood)
		GARD	Guardrails
		OPEG	Gate Door
		RETW	Retaining Wall/Gabion
EQUP	Other Equipment	BOXT	Trench Box
		HORS	Saw Horse
		PILE	Steel, Cement Or Wood Pile
		RAMP	Ramp
		SCAF	Scaffold
		SHOR	Shore
		STAG	Swing Stage
GLAS	Glass & Ceramic Items	BROK	Broken Glass
		CERA	Ceramic Item, nec.
		GLSO	Glass Item, nec.
		GRIL	Glazed Tile (Decorative Wall/Mosaic)
		SHET	Sheet of Glass/Window Glass
HEAT	Heating Equipment	AHTR	Asphalt Heater
		BURN	Oil & Propane Burner
		CHTE	Construction Heater (Electric)
		CHTO	Construction Heater (Other)
		FURN	Furnace
		HEAT	Heating Equipment, nec.
		KETL	Roofers Kettle/Tar Pot
		RADI	Radiator
		STOV	Wood/Airtight Stoves
		WATR	Water Heater

-21-

<u>EQUIPMENT AND MATERIALS</u>, cont'd.

<u>MAJOR CODE</u>	<u>MAJOR DESCRIPTION</u>	<u>MINOR CODE</u>	<u>MINOR DESCRIPTION</u>
HAND	Hand Tools-Portable	AIRG	Air Gun
		AUGP	Auger
		AXE	Axe/Hatchet
		BROM	Broom/Mop
		BRSN	Hand Brush (i.e., Wire Brush)
		CART	Cart/Dolly
		CAUK	Caulking Gun
		CHAB	Chain Binder/Chain Tightener/Bear Trap
		CHIP	Chipper
		CHSL	Chisel
		CLMP	Clamp
		CLEN	Cleanout Rod
		CRWB	Crowbar/Prybar/Tire Iron
		CUT	Cutting Equipment, nec.
		DRIL	Drill − Non-Powered
		DRLA	Drill (Air)/Air Hammer
		DRLE	Drill (Electric)
		EXPT	Explosive Actuated Tool
		FILE	File
		FISH	Fish Wire/Tape
		GRGN	Grease Gun
		GRND	Grinder (Wheel/Disc)
		HAMR	Hammer/Mallet
		JHAM	Jackhammer
		JUMP	Jumping Jack Compactor/Tamper
		KNEE	Knee Kicker
		KNIF	Knife
		NALR	Nailer
		PAIB	Paint Brush/Roller
		PICK	Pick/Pick Axe
		PIPB	Pipebender
		PIPC	Pipe Cutter
		PIPT	Pipethreader
		PLIE	Pliers/Wirecutters/Strippers
		PLNE	Electric Plane
		PLNN	Plane
		PMPN	Pump
		PMPP	Pump-Powered

-22-

EQUIPMENT AND MATERIALS, cont'd.

MAJOR CODE	MAJOR DESCRIPTION	MINOR CODE	MINOR DESCRIPTION
Hand tools continued		POSH	Post Hole Digger
		POST	Post Driver
		PUNC	Punch
		RAKE	Rake
		REAM	Reamer
		ROLL	Roller
		ROUT	Router
		SAW	Saw
		SAWC	Saw (Chain)
		SAWE	Saw (Electric)
		SAWH	Hacksaw
		SAWJ	Jigsaw
		SAWR	Saw (Skilsaw/Circular)
		SAWQ	Quick-cut/Cutoff Saw
		SCRD	Screwdriver (Powered)
		SCRN	Scraper
		SCRW	Screwdriver
		SCSS	Scissors/Shears/Snips
		SHOV	Shovel/Spade
		SLDG	Sledgehammer
		SNDB	Sand/Water Blaster
		SNDP	Sander-Powered
		SOLD	Soldering/Seaming Iron
		SPRN	Sprayer
		SPUD	Spud Wrench (Ironworkers)
		STAP	Stapler/Staple Gun
		STMC	Steam Cleaner
		STUD	Stud Gun (Welds Bolts & Metal)
		TAPE	Tape Measure
		TBOX	Tool Box
		TEST	Testing Equipment
		TOLP	Tool Part
		TOOL	Tool, nec
		TORC	Acetylene/Blow Torch
		TRWL	Trowel
		VACU	Vacuum Cleaner
		VIBC	Concrete Vibrator
		VISE	Vise
		VISG	Vise Grips
		VISP	Power Vise
		WBAR	Winding Bar

-23-
<u>EQUIPMENT AND MATERIALS</u>, cont'd.

<u>MAJOR CODE</u>	<u>MAJOR DESCRIPTION</u>	<u>MINOR CODE</u>	<u>MINOR DESCRIPTION</u>
Hand Tools continued		WELD	Welding Equipment (Electric)
		WELG	Welding Equipment (Gas)
		WHEB	Wheelbarrow/Buggy
		WRNC	Wrench
		WRNP	Impact Wrench/Spud Gun
HMEM	Heavy Mobile Equipment and Machinery	ASPV	Asphalt Paver/Spreader
		BCKH	Backhoe/Shovel
		BUCK	Loader Bucket
		BULC	Bulldozer
		COMP	Compactor
		CONF	Concrete Finisher
		DRLR	Rock Drill/Air Trac
		EQPP	Equipment Part
		EXCA	Excavating Equipment,nec
		FLTR	Fork Lift Truck/ Towmotor
		GRAD	Grader
		LOAD	Front End Loader/Track Loader
		PILD	Pile Driver
		ROLM	Roller
		SCRP	Scraper
		SNOW	Snow Removal Equipment
		HMEM	Mobile Equipment, nec.
HOT	Hot Substances	ASPH	Asphalt (Hot Only)
		FIRE	Open Flame (Fire)
		FLAR	Flare
		HEAF	Heat from Fire
		HMTL	Molten/Hot Metal
		HSBS	Hot Substance, nec.
		MATC	Match
		SLAG	Slag (Welded On)
		SPRK	Sparks
		STM	Steam
		TAR	Pitch/Tar (Hot Only)
		WATH	Hot Water

-24-

EQUIPMENT AND MATERIALS, cont'd.

MAJOR CODE	MAJOR DESCRIPTION	MINOR CODE	MINOR DESCRIPTION
LADD	Ladders	EXTL	Extension Ladder
		FXDL	Fixed Ladder (Permanent)
		LADD	Ladder
		STEP	Step Ladder
LIFT	Rigging, Hoisting & Lifting Equipment	BLTA	Block and Tackle
		BOOM	Crane Boom
		BTRK	Boom Truck
		COME	Come Along
		CRNB	Crane (Bridge/Monorail/ (Overhead)
		CRNM	Crane (Mobile)
		CRNT	Crane (Tower)
		DRUM	Drum/Pulley/Sheave
		ELEV	Elevator
		HOIS	Hoist, nec.
		HSTC	Hoist (Chain/Chain Fall)
		HSTR	Hoist (Roofers)
		JACK	Jacks
		LIFP	Lifting Equipment Part
		MANL	Man Lift
		RIG	Rigging Equipment, nec.
		SLNG	Slings/Chocker
		SPRD	Spreader Beam/Equalizer
		WNCH	Winch (Manually Powered)
MACH	Powered Machinery (Excluding Portable Hand Tools)	AUGR	Auger
		BEND	Bending Press/Brake
		BLWR	Blower/Fan
		CMPA	Compressor (Air)
		CONM	Concrete Mixer (Not Truck)
		CONP	Concrete Pump
		CONV	Conveyor
		CRSH	Crusher
		DRLP	Drill Press

-25-

<u>EQUIPMENT AND MATERIALS</u>, cont'd.

<u>MAJOR CODE</u>	<u>MAJOR DESCRIPTION</u>	<u>MINOR CODE</u>	<u>MINOR DESCRIPTION</u>
Powered Machinery, continued		GRNS	Grinder (Stationary)
		INSM	Insulation Machinery
		JOIN	Jointer
		LATH	Lathe
		LAWN	Lawnmower
		MACH	Machinery, nec.
		MACP	Machine Part
		PLNM	Stationary Plane
		PMPM	Pump
		SAWA	Saw (Radial Arm)
		SAWB	Bandsaw
		SAWM	Saw (Masonry)
		SAWT	Saw (Table)
		SNDF	Floor Sander/Polisher
		SPRY	Spray Machine
		TRWM	Trowelling Machine
		VISM	Power Vise
		WLDR	Welding Equipment
METL	Metal Items	ANGL	Angle Iron
		BAND	Steel Band for Tying
		BAR	Bar
		BEAM	Beam
		BOLT	Bolt
		CHAI	Chain
		CHAN	Channel
		CHUT	Chute
		COVR	Manhole Cover
		DROD	Drill Rod
		DUCT	Duct/Ductwork
		FAST	Metal Fastener
		FLUX	Soldering Flux
		FRAM	Manhole or Catchbasin Frame
		FTTG	Flanges/Elbows/Tees/Valves/Fittings/Faucets
		GALV	Galvanized Metal/Zinc Coated
		GRAT	Grate
		HOOK	Hook

-26-

EQUIPMENT AND MATERIALS, cont'd.

MAJOR CODE	MAJOR DESCRIPTION	MINOR CODE	MINOR DESCRIPTION
Metal Items, continued		METI	Metal Items, nec.
		MFOR	Form
		NAIL	Nail
		NUT	Nut/Screw
		PANL	Panel
		PART	Metal Particles
		PIN	Pin
		PIPE	Pipe/Piping
		PLAT	Plate
		POLE	Pole
		RACK	Rack
		RAIL	Rail
		REIN	Reinforcing Steel
		SMET	Sheet Metal/Siding/ Aluminum/Flashing
		SPRG	Spring
		STDM	Upright Column Stud
		STST	Structural Steel
		WIRE	Wire (Not Electrical)
		WIRM	Wire Mesh
		WROD	Welding Rod
		TRAK	Door Track/T-Bar
		WIRR	Wire Rope/Cable
MINR	Mineral Items	ASPC	Asphalt
		BRIC	Brick
		CATC	Catch Basin/Manhole
		CAUC	Caulking Compound
		CEMT	Cement
		CLAY	Clay
		CONB	Concrete (Blocks)
		CONC	Concrete (Chips)
		CONL	Liquid Concrete
		CONS	Solid Concrete, nec.
		CURB	Curb
		DIRT	Dirt/Sand
		DRYW	Drywall
		DTAP	Drywall Tape

-27-

<u>EQUIPMENT AND MATERIALS</u>, cont'd.

<u>MAJOR CODE</u>	<u>MAJOR DESCRIPTION</u>	<u>MINOR CODE</u>	<u>MINOR DESCRIPTION</u>
Mineral Items, continued		FELT	Asphalt Rolls/Felt Paper
		FOOT	Footings
		GRAV	Gravel
		GROT	Grout
		INSU	Insulation (Fiberglass, Asbestos, etc.)
		LIME	Lime
		MORT	Mortar
		MUD	Mud
		MSNR	Masonry Products, nec.
		PATI	Patio Stone
		PITC	Pitch/Tar
		PLST	Plaster
		ROCK	Stone/Rock
		SEWR	Sewer Pipe or Watermain
		SHNG	Asphalt Shingling
		SPAP	Sandpaper/Sanding Block
		TERR	Terrazzo
		TILE	Ceiling Tile
		TREN	Trench
MISC	Miscellaneous Items	APPL	Appliances (Stove, etc.)
		BATH	Washroom Items (Tub/Sink/Toilet)
		CABI	Cabinets (Filing/Cupboard, etc.)
		CLTH	Clothing
		FURT	Furniture, nec.
		HOLE	Hole in Surface
		MFIX	Fixture (Not electrical)
		MISC	Miscellaneous Items, nec.
		PAPE	Paper/Cardboard
		SIGN	Sign
PLAS	Plastic & Rubber Items	HOSE	Hose
		LAMI	Laminates

-28-

EQUIPMENT AND MATERIALS, cont'd.

MAJOR CODE	MAJOR DESCRIPTION	MINOR CODE	MINOR DESCRIPTION
Plastic & Rubber Items, continued		MOUL	Mouldings
		PCAB	Plastic Cable
		PIPP	Pipe
		PLAS	Plastic Item, nec.
		RUBB	Rubber Item, nec.
		TILF	Floor Tile
		TIRE	Rubber Tire
		VNYL	Vinyl Flooring/Linoleum
		WEEP	Weeping/Drainage Tile
		WELC	Welding Cable
PROT	Protective Equipment	BOOT	Boots/Shoes
		CLTP	Protective Clothing
		EAR	Ear Protector
		FACE	Face Shield
		GOGG	Glasses/Goggles
		GLOV	Gloves
		HHAT	Hard Hat
		KNEP	Knee Pad
		LANY	Safety Belt/Lanyard
		LIFE	Life Preserver
		PROT	Protective Equipment, nec.
		RESP	Respirator
		WELH	Welding Helmet

-29-

<u>EQUIPMENT AND MATERIALS</u>, cont'd.

MAJOR CODE	MAJOR DESCRIPTION	MINOR CODE	MINOR DESCRIPTION
SCRP	Scrap, Debris, Waste Materials	DEBR	Scrap/Debris/Waste Material, nec.
TEXT	Textile Items	CARP	Carpet
		ROPE	Rope
		STRG	String
		TARP	Tarp
		TEXT	Textile, nec.
		TWAL	Wallcoverings
TREE	Plants, Trees, Vegetation	BRSH	Brush
		LIMB	Branch/Limb/Twig/Thorn
		PIVY	Poison Ivy/Poison Oak
		SOD	Sod
		STRA	Straw
		TREE	Tree/Log/Stump
		VEG	Vegetation, nec.
VEHS	Vehicles	AUTO	Automobile
		BOAT	Barge/Boat
		BMPR	Truck Bumper/Fender
		BUS	Bus
		CTRK	Cement Truck
		CYCL	Motorcycle
		DUMP	Dump Truck
		FLOT	Float/Low-Bed Trailer
		LOCO	Locomotive
		MCAR	Muck Car (Skip)
		SMOB	Snowmobile
		TAIL	Tailgate
		TOWM	Stand-on Towmotor
		TRAC	Farm Tractor
		TRAI	Trailer Unit (Not Low-Bed)
		TRCK	Truck, nec.
		VDOR	Vehicle Door
		VEHP	Vehicle Parts
		VRMP	Vehicle/Equip. Ramp

-30-

EQUIPMENT AND MATERIALS, cont'd.

MAJOR CODE	MAJOR DESCRIPTION	MINOR CODE	MINOR DESCRIPTION
WOOD	Wood Items	BLCK	Block
		FORM	Form
		LUMB	Lumber/Timber/Lagging/Board
		MOLW	Moulding/Baseboard
		PLAN	Plank
		PLYW	Panel/Plywood
		PUSH	Push Stick
		RAFT	Rafters/Joists
		SHAK	Shingle/Shake
		SKID	Skid/Pallet
		SPLN	Chip/Splinter
		STAK	Stake
		STDW	Studs
		WBEA	Beam
		WEDG	Wedge
		WOOD	Wood Products, nec.
		WPOL	Pole
		WRAI	Rail
		TRUS	Trusses

-31-

TYPE-OF-ACCIDENT

MAJOR CODE	MAJOR DESCRIPTION	MINOR CODE	MINOR DESCRIPTION
ATTA	Heart Attack	HERT	Heart Attack
		ATTO	Attack N.E.C.
*BITE	Bite or Sting	ANIM	Animal Bite
		INSE	Insect Bite or Sting
*CGHT	Caught In, Under or Between	CGHT	Caught In, Under or Between, nec.
		COLL	Collapsing Structure
		MASO	A Moving and A Stationary Object
		MOVE	Moving Object(s) (Meshing and Non-Meshing)
		SLID	Land Slides/Cave-In
		STAO	Stationary Object(s)
*CONT	Contact with, Exposure to Dust, Fumes, Caustic, Toxic, Noxious or Other Hazardous Substances	ABSO	By Absorption (Body Contact, Eyes, etc.)
		INGE	By Ingestion
		INHA	By Inhalation
*ELEC	Contact with Electric Current	ELEC	Contact With Electric Current
		ELEF	Electrical Flash/ Explosion
*EXPL	Explosion	EXPL	Explosion (Commercial Explosives)
		GAS	Gas Explosion
		OEXP	Other Explosions
FALL	Falls	FALD	Fall – Different Level
		FALS	Fall – Same Level
*NOIS	Contact with, Exposure to Noise (Ears only)	NOIS	Contact With, Exposure to Noise, nec.
		PRSN	Air Pressure Exposure
		WATR	Water Pressure

-32-

TYPE-OF-ACCIDENT, cont'd.

MAJOR CODE	MAJOR DESCRIPTION	MINOR CODE	MINOR DESCRIPTION
OVER	Overexertion	OVEO	Other Overexertions
		OVER	Overexertion Due To Lifting/Pushing/ Pulling
		OVRT	Overexertion twist
RADS	Contact with, Exposure to Radiations	OTHR	Other
		WELD	Welding Flash
*RUBD	Rubbed or Abraded	HAND	By Objects Being Handled (Not Vibrating)
		PRES	By Repetition Of Pressure
		RUBD	Rubbed or Abraded, nec.
		VIBR	By Vibrating Objects
*SAIG	Struck Against/ Thrown Against	MOVO	Moving Object(s)
		STOB	Stationary Object(s)
SLIP	Slip (Not a Fall)	SLPD	Slip (Different Level)
		SLPS	Slip (Same Level)
		SLPT	Slip Twist
*STEP	Stepped On/ Knelt On	MOVS	Moving Object(s)
		STTS	Stationary Object(s)
*STRK	Struck By	EYES	Particle In Eye
		FLYO	Flying Object(s)
		FODL	Falling Object(s) - (From Level Above Injured Person)
		FOIP	Falling Object(s) - (During Handling By Injured Person)
		FONC	Falling Object(s) - (Same Level, Injured Person Not Controlling)

-33-
TYPE-OF-ACCIDENT, cont'd.

MAJOR CODE	MAJOR DESCRIPTION	MINOR CODE	MINOR DESCRIPTION
		MOIP	Moving Object(s) - (During Handling By Injured Person)
		MONC	Moving Object(s) - (Where Injured Person Is Not Controlling)
		REVR	Struck By Reversing Vehicle Or Equipment
		STRK	Struck By, nec.
TEMP	Contact with Temperature Extremes	*COBJ	Cold Objects Or Substances
		COLD	General Cold - Atmosphere Or Environment
		HEAT	General Heat - Atmosphere Or Environment
		*HOBJ	Hot Objects Or Substances
*TRIP	Trip Over	MOBJ	Moving Object
		OBJS	Stationary Object
*VEHS	Vehicle or Equipment Accidents	JACK	Jack knife
		MCOL	Collision Or Sideswipe With Another Vehicle - Both Vehicles In Motion.
		NCAO	Non-Collision - Other
		NCAR	Non-Collision - Ran Off Roadway (Out of control)
		NCAS	Non-Collision - Sudden Stop Or Start
		NCAT	Non-Collision - Overturned
		SCOL	Collision Or Sideswipe With A Standing Vehicle Or Stationary Object.
		VSTR	Vehicle Standing, Struck By
		VEHS	Vehicle Accidents, nec.

-34-

WORK-SURFACE

CODE	DESCRIPTION
BOAT	Boat/Barge/Vessel
BCHR	Boatswain's Chair/Hoist
BENC	Bench, Chair, etc.
BFUC	Building Floor (Under Construction)
BRID	Bridge
BMPR	Truck Bumper
CONC	Concrete Structure
CONW	Wet Concrete
CRAW	Crawl Space (e.g., Basements, False Ceilings, Underneath a House, etc.)
DOCK	Loading Dock
ELES	Elevator Shaft
ELEV	Elevating Device (Workman or Personnel)
EXCA	Excavation (Pit)
FLOR	Floor
FORM	Formwork
GROU	Ground
HMEM	Heavy Mobile Equipment & Machinery
HOLE	Hole
ICE	Ice (Frozen Lake/Artifical Ice)
LADF	Ladder (Fixed)
LADP	Ladder (Portable)
MACH	Stationary Machinery
MANH	Manhole
MATL	Piled Material/Hand Tools
MISC	Miscellaneous
MVEH	Motor Vehicle
PATH	Sidewalk/Path/Walkway
PIER	Pier/Wharf/Breakwater/Dam
PIPE	Pipe
PLAN	Plank
PLAT	Platform
POLE	Pole
POOL	Swimming Pool
RAMP	Ramp/Catwalk
REIN	Reinforcing Steel
ROAD	Road/Street/Driveway/Parking Lot
ROFF	Roof (Flat)
ROFS	Roof (Sloped)
ROOF	Roof (type unknown)
SCAF	Scaffold (Supported from below)
SILO	Silo
SLOP	Slope
STAG	Staging (Suspended from above)

-35-

<u>WORK-SURFACE</u>, cont'd

<u>CODE</u>   <u>DESCRIPTION</u>

STEP   Stairs/Steps
STRU   Steel Skeleton Structure (Structural Steel
                          Only)
TANK   Tank/Drum
TRAI   Trailer or Truck Platform
TREE   Tree
TREN   Trench/Ditch
TUNL   Tunnel
UNDW   Underwater
UNDR   Under equipment or vehicle
VSTP   Vehicle Step
WALL   Top of Wall
WOOD   Wood Skeleton Structure

CONDITION-OF-WORK-SURFACE

CODE   DESCRIPTION

CLEN   Clean
CONF   Confined
DARK   Dark
DEFE   Defective
DUST   Dusty
LITT   Littered
MUD   Muddy
NARR   Narrow
ROUG   Rough/Uneven
SAND   Sandy
SLIP   Slippery (Not Ice)
SNOW   Ice or Snow Covered
STEE   Steep
UNPR   Unprotected
UNSH   Unshored - Trench
UNST   Unstable
WET   Wet

-37-

NATURE-OF-INJURY

MAJOR CODE	MAJOR DESCRIPTION	MINOR CODE	MINOR DESCRIPTION
OINJ	Occupational Injury	ACHE	Ache, Pain
		AMP	Amputation
		BITE	Bites/Stings
		BLOP	Blood Poisoning (from Injury)
		BLND	Blinded
		BRNC	Burn (Chemical)
		BRNE	Burn (Electrical)
		BRNF	Burn (Friction)
		BRNS	Burn/Scald (Hot Substances)
		CONC	Concussion
		CONT	Contusion (Bruised, but Surface Intact)
		CRSH	Crush
		CUT	Cut/Laceration (Includes Loss of Teeth)
		DRMA	Poison Ivy
		DISL	Dislocation/Locked Joint
		DRWN	Drowning
		ELES	Electrical shock
		EYES	Irritation or Particle in Eye
		FRAC	Fracture
		FRST	Frost Bite/Freezing
		HEAR	Hearing Loss (Rapid)
		HEMA	Hematoma (Int. Bleeding)
		HERN	Hernia/Rupture
		INFE	Infection
		INTE	Internal Injuries
		JNTS	Stiff Joints
		MULT	Multiple Injuries
		NERV	Pinched Nerve
		NUMB	Numbness Due to Direct Injury
		PARA	Paralysis
		PUNC	Puncture
		SCRA	Scratches/Abrasions/ Superficial Wounds
		SHCK	Shock (Not Electrical)
		SPRN	Sprain/Strain
		TENO	Tendonitis or Teno- synovitis

-38-

NATURE-OF-INJURY, cont'd

MAJOR CODE	MAJOR DESCRIPTION	MINOR CODE	MINOR DESCRIPTION
Occupational Injury continued		TORN	Torn Ligaments or Cartilage
		WATR	Water on the Knee, Elbow, etc.
		WELD	Welding Flash/Electrical Flash
OILL	Occupational Illness	ASBE	Asbestosis
		ATTA	Heart Attack
		BLDP	Blood Poisoning (Working With chemicals, etc.)
		BURS	Bursitis (Knees, Elbows, Shoulders)
		COLO	Collapse
		CRMP	Cramp
		DERM	Dermatitis/Rash
		DIZZ	Dizziness
		HERL	Hearing Loss (Long Term)
		HEAT	Heat Stroke/Sun Stroke
		NAUS	Nausea
		NUMO	Numbness
		OILL	Occupational illness n.e.c.
		POIS	Poisoning
		RESP	Respiratory Conditions
		SILI	Silicosis
		STRS	Stress (Emotional)
NOCC	Non-Occupational Illness	BREA	Breathing
		COLL	Collapse
		CONV	Convulsion (Epilepsy, etc.)
		NAUN	Nausea
		NJNT	Stiff Joints/Muscles
		NDIZ	Dizziness
		NUMN	Numbness
		ULCR	Ulcer

-39-

CAUSE

MAJOR CODE	MAJOR DESCRIPTION	MINOR CODE	MINOR DESCRIPTION
ENVR	Environmental	*ANIM	Animal/Insect Attack
		BARR	Barricade Missing or Inadequate
		COLD	Cold
		CONF	Confined Space
		DUST	Dust
		EXPL	Explosion
		FIRE	Fire - Flying Sparks, Heating Equipment Failure, Open Flame, Burning Trash
		*FUME	Noxious Fumes/Poor Ventilation
		HEAT	Excessive Heat
		*HSKP	Bad Housekeeping
		MUDD	Muddy Boots
		*NOIS	Noise
		NOSI	No Warning Sign
		OCCH	Occupational Hazard
		ROUG	Rough Surface
		SHOR	Shoring Inadequate or Missing/Improperly Sloped
		SLIP	Slippery Surface
		SMOK	Smoke
		STEE	Steep Slope
		TRAF	Traffic Control
		VISI	Poor Visibility/Dark
		WEAT	Poor Weather
		WIND	Strong Wind
EQUP	Equipment	*DEFE	Defective Equipment (Not Vehicle)
		*DVEH	Defective Vehicle
		*ELEC	Electrical Short
		*GARD	Guard Missing or Uncovered
		*GRND	Ungrounded/Poorly Grounded

-40-
CAUSE, cont'd

MAJOR CODE	MAJOR DESCRIPTION	MINOR CODE	MINOR DESCRIPTION
Equipment continued		*LABL	Inadequate Labelling
		ILFT	Inadequate Lifting Facilities
		*FPRO	Failure of Personal Protection
		*SLPE	Equipment Slipped or Jammed
HUMN	Human	AIM	Incorrect Aim (Hammering, etc.)
		ALCO	Alcohol/Drugs
		ALLE	Allergy
		*CLTH	Unsafe Clothing
		COMM	Communication Difficulty
		EMOT	Emotional Problem
		FAID	Failed to Get First Aid
		*FAIL	Failed to Use Equipment
		FATI	Fatigue
		FIRH	Human Error Producing Fire
		FOOT	Lost Footing (No Apparent Slippery Surface)
		GARL	Missing or Inadequate Guardrails on 8' Scaffold or No Fall Arrest Protection
		GRIP	Improper Grip, Object Slipped
		HELP	Failure to Get Help
		*IEQP	Improper Use of Equipment or Vehicle
		ILLS	Illness
		INAT	Inattention To Activity
		*INSP	Faulty Inspection
		LIFT	Lifted or Moved Object Incorrectly
		*MISS	Equipment Not Provided
		MOVE	Moved Body Incorrectly
		NSIG	No Signalman
		OTHR	Other Vehicle at Fault
		OVEO	Repetitive Work
		OVER	Lifted Excess Weight
		*PILE	Material Improperly Piled
		POST	Unsafe Working Position
		*PROF	Improper Fit

-41-

<u>CAUSE</u>, cont'd

MAJOR CODE	MAJOR DESCRIPTION	MINOR CODE	MINOR DESCRIPTION
Human continued		*PROT	Not Wearing Protective Equipment
		*PROW	Used Wrong or Inadequate Protective Equipment
		REOC	Recurrence of Old Injury
		RUSH	Rushing, Hurrying, etc.
		*SDEV	Disconnected Safety Device
		*SECU	Secured Poorly
		*SHUT	Failure to Shut Down
		SPEE	Unsafe Speed
		STRS	Working under Stress
		SUPR	Supervision Problem (Poor Planning, Organization, Directing, Controlling)
		TEAS	Distraction, Abuse, Teasing
		*TOOL	Used Wrong Tool
		TRAI	Poor Training
		UNSF	Unsafe Action
		*WEQP	Used Wrong Equipment
		WIPE	Wiped Eyes
MATL	Material	*DEFM	Defective Material
		*IMAT	Improper Material
		*SHRP	Sharp Edges on Material
		*SPLL	Material Spilled
		*SLPM	Material Slipped or Jammed

APPENDIX D

AN OIL COMPANY'S SAFETY DATA SYSTEM

The Loss Management Information System (LOMIS) is a safety data system designed by Robert Wright of Gulf Oil Canada, Ltd., and patterned after the system designed for the U.S. Department of the Interior by William Pope. LOMIS is the second-generation system from which the pharmaceutical company's system was developed. It is included in this book because it provides a different approach to file segments, and is representative of another industry.

Within the following code manual there are excellent descriptions concerning the uses of this system and the methods for input of data. Reportedly, the system has proved to be an excellent accident-prevention tool for this company. A sample of the incident report form is reproduced as well.

LOMIS

**Loss Management Information
System**

**Incident Report
Code Manual**

GULF OIL CANADA LIMITED

IDENTITY - ITEMS 1 - 17

1. Organization Code

    From table of Company organization codes. The organization codes and
    report number uniquely identify all reports.

2. Report Number

    The Supervisor will assign numbers serially in his Department.

3. Location of Incident

    Use the common name of the location. Include service station code number
    where applicable.

    e.g. Edmonton Refinery, Burnaby Terminal, Varennes Petro Works

4. Date of Incident

    Express in numerals - year, month, day.

5. Time of Incident

    24 Hour Clock - 0100 (1 a.m.)
                  - 1200 (noon)
                  - 1700 (5 p.m.)
                  - 2400 (midnight)

6. Total Cost of Incident

    Estimate all extra costs to the operation to closest five dollars.

7. Liability Claim Expected

    Where the Company is liable, use estimate in dollars, of liability incurred.

8. Amount Awarded

    In dollars (show minus sign if in Company favour).

9. <u>Type of Incident</u> - Primary must be completed. Secondary and Other are
   important but optional.

Code   Code

01	Absenteeism	46	Solid
02	Barometric Pressure Extremes	47	Stretching to Reach
03	Caught In or Between	48	Struck Against
04	Chemicals	49	Struck by
05	Crude or Product Contamination	50	Temperature Extremes
06	Customer Complaint	51	Tissue Corrosion
07	Delivery Problem	52	Vapour
08	Dust	53	Vehicle Damage
09	Electrical Shock	54	Vessel Collision with Fixed or Floating Object
10	Explosion (See 71-90 for source)	55	Vessel Collision with Other Vessel
11	Fall - Different Level		
12	Fall - Same Level	56	Vessel Standing/Grounding
13	Falling, Flying Particles	57	Damage to Fixed Marine Structures
14	Falling Object		
15	Fire (See 71-90 for Source)	58	Hull Damage Incl. Damage from Wash
16	Flood, Earthquake, or Natural Disaster		
17	Force Majeure	59	Pier Damage
18	Fumes	60	Vibration
19	Illegal Activity	61	Well Blow Out
20	Illness	62	Wind or Heavy Weather
21	Ingestion	63	Wrongly Mixed Product
22	Inhalation		
23	Ionizing Radiation	70	Source of Ignition (71-90) if Fire or Explosion
24	Irradiation		
25	Known or Suspected Sabotage or Hostile Act	71	Auto Ignition on Release
26	Larceny	72	Chemical Reaction
27	Liquid	73	Compression Ignition
28	Loss of Sale	74	Diesel Engine
29	Machinery, Equipment Breakdown or Loss	75	Electrical Equipment (Defective/Damaged)
30	Mist	76	Electrical Equipment (Not Defective)
31	Muscular Over-Exertion		
32	Noise	77	Flare
33	Non-Ionizing Radiation	78	Friction Overheating
34	Off-the-Job Incident	79	Gasoline Engine
35	Out of Product	80	Hot Line or Equipment
36	Physical Agents	81	Iron Sulphide Oxidation
37	Pollution - Air	82	Lightning
38	Pollution - Chemical	83	Match or Cigarette Lighter
39	Product Error Incl. Off-Spec.	84	Open Flame (Furnace, etc.)
40	Product Spill	85	Other Overheating
41	Public Liability Claim	86	Spontaneous Ignition
42	Quality Problem	87	Spread from Outside Sources
43	Skin Absorption	88	Static Spark
44	Skin Disorder - Dermatitis	89	Vandalism, Sabotage, Military Action
45	Slip or Twist (Not Fall)	90	Welding, Cutting, Brazing

10. Type of Property

    Code   Code

        00   No Property Involved   35   Gas Plant
        01   Airport Facility   36   Offshore Platform
        02   Barge   37   Onshore Storage Tank/Tank
        03   Bulk Plant   Battery
        04   Car Care Centre   38   Onshore Well
        05   Chemical Plant   39   Other Leased Facility
        06   Customers' Facility   40   Other Production
        07   Grease Plant   41   Salt Water Disposal Facility
        08   LNG Plant   42   Satellite Station
        09   Lube Packaging   43   Sub-Surface Production Equip-
        10   Lube Plant   ment
        11   Motel or Restaurant   44   Water Flood Plant
        12   Motor Vehicle   45   Well Servicing Unit
        13   Other Manufacturing
        14   Refinery   50   Other
        15   Service Station   51   Aircraft
        16   Tank Car   52   Contracted Vehicle
        17   Terminal   53   Leased Storage
                                           54   Leased Vehicle
        20   Pipeline   55   Mining Facility
        21   Gas Pipeline   56   Miscellaneous (Describe)
        22   Oil Pipeline   57   Office Building
        23   Other Pipeline   58   Port Facility
        24   Pipeline Terminal or   59   Research Facility
                Pumping Station   60   Vessel
        25   Product Pipeline   61
                                           62
        30   Production   63
        31   Compressor Station   64
        32   Computer Control Unit   65
        33   Drilling Rig - Offshore   66
        34   Drilling Rig - Onshore   67

11. Property Ownership

    Code   Code

        00   No Property Involved   08   Other
        01   Company Owned   09
        02   Company Subsidiary   10
        03   Concessionaire   11
        04   Contractor   12
        05   Employee Owned   13
        06   Leased   14
        07   Privately Owned

12. Specific Unit Involved

Code		Code	
	Process Units	108	Linear Olefin Production
000		109	Propylene Hydrofiner
001	Alkylation	110	Steam Cracker
002	Amine Unit	111	Other Olefins
003	Asphalt Plant		
004	Atmospheric Distillation		Alcohols
005	Butamer Unit	120	
006	C B & G Plant	121	Ethanol Dehydration
007	Caustic Treating	122	Ethanol Finishing
008	Clay Treating	123	Ethanol Manufacture
009	Cumene Plant	124	Isopropanol Dehydration
010	D D S	125	Isopropanol Manufacture
011	Carbonizing Unit	126	Methanol Finishing
012	Delayed Coking Plant	127	Methanol Manufacture
013	Desalter (Separate Unit)	128	Oxo Alcohol Feed Preparation
014	Dewaxing Unit	129	Oxo Alcohol Manufacturing
015	Fluid Catalytic Cracking Unit	130	Secondary Butyl Alcohol Finishing
016	Furfural Unit		
017	H D S	131	Secondary Butyl Alcohol Manufacturing
018	Hydrar Unit		
019	Hydrogen Plant	132	Other Alcohol Finishing
020	L C O Hydrotreating Unit	133	Other Alcohol Manufacturing
021	Merox Plant		
022	Mercapfining Unit	140	Elastomers
023	Polymerication	141	Butyl/Chlorobutyl Rubber Manufacture
024	Propane Deasphalting Unit		
025	Reformer	142	Elastomer Finishing
026	Saturate Gas Plant	143	Polybutadiene Manufacture
027	Solvent	144	Polyisobutylene Manufacture
028	Sulfur Plant	145	S B R Manufacture
029	Toluene Hydrodealkylation Unit (T H D)	146	Vistalon Rubber Manufacture
		147	Other Elastomer/Rubber
030	U D E X Unit		
031	Unsaturate Gas Plant	150	Specialty Chemicals
032	Vacuum Distillation	151	Additives Manufacture
033	Other Process Units	152	Oil Field Chemicals Manufacture
034		153	Other Specialty Chemicals
035			
036		160	Commodity Intermediates & Solvents
100	Olefins	161	Acetone Manufacture
101	Acetylene Removal	162	Acrylonitrile Manufacture
102	Butadiene and other Diolefin Extraction	163	Benzene Manufacture
		164	Caprolactam Manufacture
103	Cyclopentadiene Purification	165	Cyclohexane Manufacture
104	Ethylene Purification	166	Cyclohexanol Manufacture
105	Isoprene Extraction	167	Cyclohexanone Manufacture
106	Isobutylene Extraction	168	Methyl Ethyl Ketone Manufacture
107	Isobutylene Purification	169	Methyl Isobutyl Ketone Manufacture

12. <u>Specific Unit Involved - Continued</u>

<u>Code</u>

170	Phthalic Anhydride Plasticizer (Phythalates Manufacture)	216	Polyester Fibers Plant
171	Phthalic Anydride Manufacture	217	Polyethylene Finishing
172	Toluene Manufacture	218	Polyethylene Manufacture
173	Xylene Extraction	219	Polyolefin Unit
174	Other Commodity Intermediates	220	Polypropylene Finishing
175	Other Solvents	221	Polypropylene Manufacture
		222	Polyvinyl Chloride Manufacture
180	Other Industrial Chemicals & Inorganic Acids	223	Other Fibers and Yarns
181	Carbon Dioxide Purification	224	Other Plastic Finishing
182	Caustic Soda/Chlorine Manufacture	225	Other Plastic Manufacture
183	Gypsum Plant	230	Other Fabricated Products
184	Hydrochloric Acid	231	Blown Bottle Plant
185	Hydrogen Sulfide Manufacture	232	Building Materials Plant
186	Nitric Acid Plant	233	Cordage Plant
187	Phosphoric Acid	234	Experimental Process Unit
188	Sulfuric Acid Plant	236	Other Fabricated Products Manufacture
189	Other Industrial Chemicals		
		340	Tankage
190	Fertilizers & Fertilizer Intermediates	341	Concrete Tank
191	Ammonia Plant	342	Cone or Dome Roof Tank
192	Ammonium Nitrate Manufacture	343	Diaphragm Type Tank
193	Ammonium Phosphate Manufacture	344	Ethylene Refrigerated Storage
194	Calcium Ammonium Nitrate Manufacture	345	Floating Roof Tank
195	N P K Complex Manufacture	346	High Pressure Tank (100 kPa and Over)
196	Sulfur Plant	347	Internal Floating Roof (Hard Hat)
197	U R E A Manufacture	348	L N G Refrigerated Storage
198	Other Fertilizer	349	L P G Refrigerated Storage
		350	Lifer Roof Tank
200	Other Chemicals & Resins	351	Low Pressure Tank (Less Than 100 kPa)
201	Buton Resin Manufacture	352	Near but not Involving Tank
202	Coatings Manufacture	353	Open Reservoir or Pit
203	Neo-Acids Manufacture	354	Open Top Tank
204	Petroleum Resin Manufacture	355	Reservoir or Pit with Fixed Roof
205	Other Hydrocarbon Resin Manufacture	356	Reservoir or Pit with Floating Roof
206	Other Resins	357	Storage Heated Above 100°C
210	Plastics, Fibers & Yarns	358	Underground Cavern
211	Combination Fibers Plant	359	Underground Steel Tank
212	Film & Sheet Packaging Plant	360	Wood Roof Tank
213	Nylon Fibers Plant	361	Other Refrigerated Storage
214	Plastic Film Plant	362	Other Tank
215	Plastic Fibers Plant		

12. Specific Unit Involved - Continued

Code   Code

370	Buildings, Utilities & Offsites		426	Truck Loading Rack
371	Boiler House		427	Other Loading or Filling Facility
372	CO Boiler			
373	Circuit Breaker, Switch		428	Other Waste Disposal Facility
374	Commissary			
375	Compressor House		430	Vehicles
376	Control House		431	Aircraft
377	Cooling Tower		432	Automobile
378	Dining Room		433	5 Axle Tractor Trailer
379	Electrical Sub-Station		434	Hovercraft
380	Gas or Fuel Line		435	6 Axle Tractor Trailer
381	Ground		436	Tandem Axle Straight Truck
382	Hospital		437	Tank Car
383	Laboratory		438	Truck Train
384	Living Quarters, House Trailer		439	Other Vehicles
385	Manufacturing Building		440	
386	Office Building		441	
387	Pier or Wharf		442	
388	Power House		443	
389	Power Plant			
390	Pumphouse		750	Field Facilities
391	Recreation Building		751	Automatic Custody Transfer Facility
392	Refrigeration Plant		752	Dead Oil System
393	School		753	Drilling Rig Offshore Platform
394	Shop		754	Filed Pumping Equipment
395	Steam Station		755	Flow Station
396	Temporary Construction Building		756	Gas Compression Facility
397	Utility Pumping Station		757	Gas Flow Line
398	Warehouse		758	Liquid Hydrocarbon Unit
399	Other Building		759	Living Quarters
400	Other Utility		760	Oil Field Treater, Separator
			761	Oil Flow Line
410	Loading Facilities		762	Onshore Gas Compression Facility
411	Barrell or Drum Type Filling Facility		763	Platform Drilling Rig
412	Catch Basin		764	Platform Well Servicing Unit
413	Centrifuge		765	Salt Water Disposal Well
414	Coagulator		766	Salt Water Line
415	Emulsion Breaker		767	Well
416	Filter		768	Other Production Facility
417	Flare, Vent Stack			
418	Incinerator		770	Pipelines
419	Refuse Dump		771	Delivery Terminal
420	Separator		772	Gathering Lines
421	Sewer		773	Main Line
422	Solid Waste Collector		774	Pump Station
423	Sour Water Stripper		775	Receiving Station
424	Tank Car Loading Rack		776	Transmission Line
425	Tanker or Barge Loading Facility		777	Truck Line

12. <u>Specific Unit Involved - Continued</u>

<u>Code</u>   <u>Code</u>

778   Other Pipeline   782   Data Storage Area
                                            783   Outdoor Storage Area
780   <u>Miscellaneous - Other</u>   784   Service Station
781   <u>Computer Equipment</u>   785   Other Mining

13. <u>Model or Construction Year</u>

If Vehicle, state model year - eg. 1975

If Process Unit or other Equipment, state estimated purchase or
construction year.

14. <u>Specific Component Involved</u>

<u>Code</u>   <u>Code</u>

010   <u>Air Pressure</u>   048   Tanks, Vats
                                            049   Trash Boxes, Cans, Bins
020   <u>Animals, Insects, Birds,</u>
         <u>Reptiles</u>   050   <u>Conveyor (non passenger type)</u>
021   Bears   051   Belt or Chain Link
022   Birds   052   Gravity Slide
023   Deer   053   Power Screw
024   Dogs   054   Roller
025   Fish   055
026   Horses
027   Insects   060   <u>Clothing/Cloth</u>
028   Reptiles   061   Boots, Shoes
029   Mammals (other)   062   Cloth
                                            063   Gloves, Mittens
030   <u>Boilers, Pressure Vessels</u>   064   Hats
031   Air Pressure Lines   065   Outer Coats, Rain Wear
032   Air Tanks, Receivers   066   Shifts, Sweaters, Inner Coats
033   Autoclaves   067   Stockings, Leggings, Socks
034   Boilers, high pressure   068   Suits, Pants, Overalls,
035   Boilers, low pressure   Dresses
036   Compressor (all types)   069   Underwear
037   Heater, Hot Water
038   Steam   070   <u>Electrical Equipment</u>
039   071   Air Conditioning Equipment
                                            072   Batteries, Battery Charging
040   <u>Containers (metal, wood, cloth</u>   073   Bus Bar
041   Barrels, Drums   074   Capacitors
042   Boxes   075   Conductors Wires
043   Carboys (acid)   076   Disconnect Switches & Switch
044   Cylinders, Compressed Gas   Gear
045   Hoses (fire and water)   077   Fans, Blowers (large)
046   Nets and Slings   078   Heating Appliances (major types)
047   Sacks and Bags   079   Magnetic, Electrolytic

14. Specific Component Involved - Continued

Code   Code

080	Meters and Relays	140	Food Products
081	Motors, Generators	141	Coffee Spills
082	Power Circuit Breaker	142	Hot Grease
083	Power Line Insulations	143	
084	Reactors		
085	Rectifier	150	Furniture, Fixtures
086	Rotary Condensers	151	Ashtrays
087	Solenoid	152	Beds, Bunks, Cots
088	Station Service Equipment	153	Cash Registers
089	Switchboard	154	Chair, Bench, Stool
090	Transformers, Convertors	155	Desk, Table
091	Transmission and Distri-	156	Doors, Glass
	bution Lines	157	Doors, Solid
		158	Electrical Applaince (Small)
100	Enclosures	159	Extension Cords
101	Fence (metal, wire)	160	Files, Book Case, Safe, etc.
102	Fence (wood)	161	Floor and Floor Covering
103	Guard Rails	162	Safes
104	Opening (gate)	163	Shelves
105	Wall (brick, wood, stone)	164	Windows, Window Coverings
106			
		170	Glass and Ceramics
110	Explosives and Flammables	171	Crucibles
111	Commercial Explosives	172	Dishware
112	Compressed Air, Aerosol Cans	173	Flat Glass
113	Coyote Getters	174	Light Bulbs, Fluorescent
114	Dust, Fine Particles	175	Tubing, Flasks, Beakers
115	Explosion followed by Fire	176	
116	Flash Backs		
117	Fuel Oil	180	Handtools - Portable, not
118	Gasoline, Gas Vapours		powered
119	Hot Ashes	181	
120	Hydrogen, Percholoric Acid	182	
121	L.P. Gas		
122	Lightning followed by Fire	190	Brush Cutting Types
123	Open Flame	191	Brush Hook, Pulaski
124	Paint, Varnish	192	Hedge Clipper
125	Solvents	193	Machete
126	Smoke	194	Sandvik Brush Axe
127	Steam	195	Tree Trimmer
128		196	
130	Firearms	200	Burning Types
131	Flare Lighter	201	Blow Torch
132	Handgun	202	Cutting Equipment
133	Rifle	203	Fire Control Burnouts
134		204	Soldering Iron
		205	Welding Equipment
		206	

14. <u>Specific Component Involved - Continued</u>

<u>Code</u>   <u>Code</u>

Code		Code	
210	<u>Cleaning and Cooking Types</u>	270	<u>Sawing and Striking Types</u>
211	Brooms, Brushes, Mops	271	Hammer, Mallet
212	Buckets, Pails	272	Hand Saws
213	Can Openers	273	Maul, Sledge
214	Cutlery	274	Post Driver
215	Pots, Pans	275	
216			
		280	<u>Turning and Lifting Types</u>
220	<u>Earth Moving Types</u>	281	Jack
221	Earth Scraper (hand)	282	Pry Bar
222	Grubhoe	283	Screw Driver
223	Mattock, Pick, Pick Axe	284	Tire Iron
224	Shovel, Hoe, Rake	285	Vise
225	Wheelbarrow, Carts	286	Wrench
226		287	
230	<u>Hacking and Punching Types</u>	290	<u>Tools - Portable & Powered</u>
231	Adze	291	Buffing Machine
232	Awl, Prost Pin	292	Circular Saw, Electric
233	Axe (double blade)	293	Chain Saw, Gas or Electric
234	Chisel	294	Drain Cleaner
235	Crowbar	295	Drill, Electric
236	Drills	296	Edger, powered
237	Hatchet, Axe (single blade)	297	Fireline Flail
238		298	Floor Waxer, Polisher
		299	Grinder
240	<u>Knives and Scythe Types</u>	300	Hedge Trimmer, Electric
241	Drawing Knive	301	Hole Auger
242	Paint Scraper	302	Jackhammer, Air
243	Paper Cutter (hand)	303	Lawnmower, powered - hand
244	Pocket Knife	304	Lawnmower, powered - tractor
245	Scythe, Weed Cutter	305	Router
246		306	Sander, Hand Electrical
		307	Steam Cleaner
250	<u>Office Types</u>	308	
251	Pencil, Pen, Drafting Tools		
252	Scissors	310	<u>Heating Equipment</u>
253	Stapler	311	Blanket, Pad Heaters
254	Thumb Tacks, Clips	312	Fireplace
255		313	Heating Tapes
		314	Stove, electric
260	<u>Pulling and Scraping Types</u>	315	Stove, gas-natural, butane
261	Bolt Clipper	316	Stove, portable type
262	File Rasp	317	Stove, wood, coal, oil
263	Hand Planer	318	
264	Nail Puller		
265	Pliers, Pinchers	320	<u>Hoisting Apparatus</u>
266		321	Air Hoist

14. Specific Component Involved - Continued

Code		Code	
322	Chain Hoist	390	Ice Making Machinery
323	Escalators	391	
324	Hydraulic Lift		
325	Hydraulic Tailgate Hoist	400	Laundry/Dry Cleaning
326	Life Lines	401	Extractor - Washer
327		402	Ironer
		403	Tumbler
330	Kitchen Equipment	404	Dryer
331	Cook Stove, Oven	405	
332	Deep Fat Fryer		
333	Dishwasher	410	Loading Facilities
334	Food Locker, Freezer	411	Cat Walks & Steps
335		412	Loading Arms
		413	Loading Valves
340	Laboratory Equipment	414	Meters
341	Glassware	415	Spouts
342	Kiln	416	
343	Photo Equipment, Cameras		
344	Refrigerators	420	Mechanical, Power Transmission
345		421	Belts, Chains, Ropes, Cables
		422	Drums, Pulleys, Sheaves
350	Ladder Types	423	Fan Blades, Spokes
351	Extension	424	Gears, Sprockets
352	Fixed (permanent)	425	
353	Single		
354	Step	430	Machine Shop Types
355		431	Drillpress
		432	Grinder (stationary)
360	Machines	433	Lathe
361		434	Milling Machine
362		435	Pipe Threader
		436	Power Press
370	Construction Type Machinery	437	
371	Batching Plant		
372	Cable Ways	440	Metal Forming Machines (large)
373	Concrete Vibrators	441	
374	Drilling Equipment		
375	Rock Crusher	450	Office Machinery
376	Screening Plant	451	ADP Machinery
377	Tampers	452	Calculator
378		453	Duplicating Machine
		454	Teletype
380	Food Machinery	455	Terminal
381	Bread Cutter	456	Typewriter
382	Bottling Machine	457	Xerox Machine
383	Dough Brake	458	
384	Dough Mixer	459	
385	Meat Grinder		
386			

14. Specific Component Involved - Continued

Code

460	Printing Machine	518	Wind Storm
461	Corner Cutters	519	
462	Guillotine Papercutter		
463	Platen Press	530	Paper and Paper Products
464	Rotary Slitters	531	Bags, Paper Containers
465		532	Cardboard
		533	Stationery
470	Woodworking Shop Types	534	
471	Jointer		
472	Planer	540	Petroleum Products
473	Sander	541	Alcohol
474	Shaper	542	Asphalt, Tar
475	Table Saws (all types)	543	Class A material (solid
476			combustible)
		544	Class C material (electrical)
480	Metal Items	545	Coal
481	Chain	546	Coke
482	Flanes, elbows, tees &	547	Crude Oil 40° API and lighter
	fittings	548	Crude Oil less than 20° API
483	Heavy Metal over 100 lbs.	549	Crude Oil 20° to 39.9^{o} API
484	Heavy Metal under 100 lbs.	550	Cutback Asphalt
485	Metal Particles (slivers,	551	Flash Point less than 0° C
	chips)		(e.g. Gasoline)
486	Molten Metal	552	Flash Point 0° to 24° C
487	Nails (nails in boards)		(e.g. Toluene)
488	Nuts, Bolts	553	Flash Point 25° to 39° C
489	Pipe		(e.g. Xylene)
490	Plates, Sheet	554	Flash Point 40° to 59° C
491	Reinforcing Steel		(e.g. Kerosene, Domestic
492	Scrap Iron		Heating Oils, Solvents)
493	Structural Steel	555	Flash Point 60° to 99° C
494	Wire, Cable		(e.g. Other Fuel Oils)
495		556	Flash Point 100° C and above
			(e.g. Lubes, Transformer
500	Mineral Items		Oils)
501	Concrete Products	557	Gas, predominantly hydrocarbon
502	Dirt	558	Gas, predominantly not hydro-
503	Lime and Cement Sacks		carbon
504	Masonry Products	559	LNG (liquified Natural Gas)
505	Sand, Gravel, Stone	560	LPG (liquified Petroleum Gas)
506		561	Metal
		562	Sulphur
510	Natural Phenomena	563	Unknown
511	Avalanche, Falling Rock	564	Water
512	Earthquake	565	Wax, Grease
513	Electrical Storm	566	Well Brine
514	Flood, Washout	567	Other
515	Ice, Snow, Sleet	568	Other Class B material
516	Rain		(flammable liquid and vapour)
517	Unreasonable Dry	569	Other Petrochemical, Flash
			Point above 100° C
		600	

14. <u>Specific Component Involved - Continued</u>

<u>Code</u>   <u>Code</u>

571	Regular Gasoline	653	Climbing (general)
572	Premium Gasoline	654	Fishing
573	No-Lead Gasoline	655	Gymnastics
574	Stove Oil	656	Hiking
575	Bunker Fuels	657	Hunting
576	Diesel 20X	658	Mountain Climbing
577	Diesel 40	659	Parachuting
578	Diesel 25	660	Playground Equipment
579	Diesel 0	661	Prospecting
580	Diesel 50	662	Self-Defense
581	Furnace 20X	663	Skating
582	Furnace 25	664	Snow Skiing, Sledding
583	Furnace 16	665	Spelunking (cave exploring)
584	Furnace 40	666	Swimming
585	Furnace 50	667	Team Sports
		668	Under Water activity
610	Plants, Trees, Vegetation	669	Water Skiing
611	Branches, Twigs	670	
612	Grass, Weeds	671	
613	Nettles		
614	Poison Ivy, Sumac, Oak	680	Shipping Vessels Structural Component
615	Poisonous Plants		
616	Thorns, Briers	681	Bulkheads
617	Tree Trunk, Large Limbs	682	Deck
618		683	Foundation
		684	Hull
620	Plastic and Rubber Items	685	Superstructure
621	Pipe and Conduit	686	
622	Plastic Container	687	
623	Rubber Tires		
624		690	Compartments and Tankage
		691	Ballast Tank
630	Pumps and Prime Movers	692	Bilge
631	Prime Movers	693	Bridge
632	Pumps, portable	694	Bunker Tank
633	Pumps - service station	695	Centre Tank
634	Pumps, stationary	696	Engine Room
635	Pumps, vacuum	697	Galley
		698	Living Quarters
640	Radio and T. V. Service	699	Pump Room
641	Antennas	700	Service Tank
642	Radio Sets	701	Slop Tank
643	T. V. Sets	702	Wing Tank
644		703	
		704	
650	Recreational Activities		
651	Boxing, Wrestling	710	Ship Systems
652	Camping, Picnicking	711	Air Conditioning System

14. <u>Specific Component Involved - Continued</u>

<u>Code</u>   <u>Code</u>

712	Bilge and Drainage System		770	Natural Surface Types
713	Cargo Heating System		771	Cliff, Ravine
714	Cargo Pumping System		772	Glacier Areas
715	Cargo Tank Cleaning		773	Hills, Slopes
716	Cargo Tank Gauging		774	Mine and Cave Floor
717	Compressed Air System		775	Stream Bed
718	Condensate System		776	Thermal Areas
719	Distribution System		777	Trails, Pathways
720	Electrical Power Supply		778	
721	Feed Water System			
722	Fresh Water System		780	Vehicles (all types)
723	Fuel Oil System		781	
724	Hydraulic System		782	
725	Lube Oil System			
726	Main Propulsion Drive System		790	Aircraft Types
727	Remote Cargo Handling		791	Aircraft Service Equipment
728	Salt Water Circulating		792	Fixed Wing Aircraft
729	Sanitary System		793	Helicopter (rotary wing)
730	Segregated Ballast System		794	
731	Ship Service Refrigeration		795	
732	Steam Distribution System			
733	Steam Generation System		800	Construction Types (heavy duty)
734	Tank Venting System		801	"A" Frame Trucks
735	Ventilation System		802	Back Hoe, Shovel Loader
736			803	Bulldozer, Tractors
737			804	Capacitor/Stabilizer
			805	Cranes
740	Flat Surface Types		806	Draglines
741	Decks/Vessel, Boat		807	Drill Rigs
742	Floors (buildings)		808	Dumper
743	Lawn Areas		809	Earth Movers
744	Platforms, Loading Docks		810	Graders
745	Raceways, Catwalks, Ramps		811	Line Equipment
746	Roads (composite type)		812	Loaders
747	Roads (dirt)		813	Rollers
748	Roof Surface		814	Rough Terrain Vehicles
749	Sidewalks (composite)		815	Shovels
750	Sidewalks (dirt)		816	
751			817	
752				
			820	Highway Types
760	Temporary Surface Types		821	Ambulance
761	Boatswain's Chair		822	Bicycle
762	Scaffolds		823	Buses, all types
763	Staging		824	Carry-all (panel truck)
764			825	Fire Truck
765			826	Jeep, 500 kg
			827	Motorcycle, Scooters

14. Specific Component Involved - Continued

Code

Code		Code	
828	Pick-Up Truck	878	Vessels under 65'
829	Radar Van	879	
830	Road Brusher, Street Sweeper		
831	Sedan, Coupe, Sport	880	Miscellaneous Types
832	Snow Plow	881	Boat Trailer
834	Stationwagon	882	Brush Chipper
835	Truck, up to 2500 kg gross	883	Horse Trailer
836	Truck, 2501 to 15,000 kg gross	884	House Trailer
837	Truck, 15,001 to 30,000 kg gross	885	Tractor/Trailer Unit
		886	Trailer, Other
838	Truck, 30,001 and over, gross	887	
839		888	
840	Industrial Types (Plant)	890	Water
841	Dollies	891	Canals and Laterals
842	Forklifts	892	Ice
843	Front-End Loaders	893	Icicle
844	Hand Truck	894	Lake, Pond
845	Lumber Carriers	895	Ocean
846	Skids	896	Reservoir, Pool
847	Stackers	897	River, Stream
848		898	Steam
849		899	
850	Railroad Types	900	Wood and Wood Products
851	Engines	901	Chips, Splinters
852	Freight, Passenger	902	Lumber
853	Hand Car (motorized)	903	Plywood Products
854		904	Poles, Logs (utility)
		905	Shaners
860	Stairway Types	906	
861	Fire Escapes, Chutes		
862	Fire Escapes, Step	910	Gulf Agents
863	Steps and Landings	911	
864		912	
865		913	
		914	
870	Water Borne Types	915	
871	Airboats	916	
872	Marsh Buggies	917	
873	Motor Boats	918	
874	Row Boats, Canoes	919	
875	Sail Boats	920	
876	Submersible Boats	921	
877	Vessels over 65'	922	

15. Contaminants

Code	Substance	Code	Substance
0005	Abate	0335	Benzoyl peroxide
0010	Acetaldehyde	0340	Benzyl chloride
0020	Acetic acid	0360	Beryllium and compounds
0030	Acetic anhydride	0477	BGE (n-butyl glycidyl ether)
0040	Acetone	1015	Biphenyl (diphenyl)
0060	Acetonitrile	2053	Biphenyl mixture (phenyl ether)
0065	2-Acetylaminofluorene	2630	Bis (chloromethyl) ether
0870	Acetylene dichloride	0370	Bismuth and compounds
	(1.2-dichloroethylene)	0380	Boron oxide
0070	Acetylene	0381	Boron tribromide
0080	Acetylene tetrabromide	0382	Boron trifluoride
0050	Acids, miscellaneous	0390	Bromine
0090	Acridine	0391	Bromine pentafluoride
0110	Acrolein	0400	Bromoform
0115	Acrylamide	0410	Butadiene (1.3-butadiene)
0120	Acrylonitrile (vinylcyanide)	0420	Butane
0125	Aldrin	0480	Butanethoil (butyl mercaptan)
0130	Allyl alcohol	0430	2-Butanone (MEK) (methyl ethyl
0140	Allyl chloride		ketone)
0145	Allyl glycidyl ether	0435	2-Butoxy ethanol (butyl cello-
0150	Allyl propyl disulfide		solve)
0235	Alpha naphthyl thiourea (ANTU)	0440	n-Butyl acetate
0160	Aluminum and compounds	0441	sec-Butyl acetate
0162	4-Aminodiphenyl	0442	tert-Butyl acetate
1030	2-Aminoethanol (ethanolamine)	0460	Butyl alcohol
0165	2-Aminopyridine	0461	sec-Butyl alcohol
0185	Ammate (ammonium sulfamate)	0462	tert-Butyl alcohol
0170	Ammonia	0470	n-Butylamine
0175	Ammonium chloride	0435	Butyl cellosolve (2-butoxy
0180	Ammonium nitrate		ethanol)
0185	Ammonium sulfamate (ammate)	0473	tert-Butyl chromate
0190	n-Amyl acetate	0477	n-Butyl glycidyl ether (BGE)
0191	sec-Amyl acetate	0480	Butyl mercaptan (butanethiol)
0210	Amyl alcohol	0485	p-tert-Butyl toluene
0220	Aniline	0490	Cadmium (metal dust and
0225	Anisidine (o.p-isomers)		soluble salts)
0230	Antimony and compounds	0491	Cadmium oxide (fume)
0235	ANTU (alpha naphthyl thiourea)	0500	Calcium arsenate
0240	Argon	0505	Calcium carbonate
0260	Arsenic and compounds	0510	Calcium cyanamide
0270	Arsine	0520	Calcium oxide
0290	Asphalt (Petroleum) fumes	0522	Camphor
0300	Azinphos, methyl (Guthion)	0525	Carbaryl (Seven)
2690	Azodrin (monocrotophos)	0527	Carbon black
0310	Barium and compounds	0530	Carbon dioxide
0320	Benzene (benzol)	0540	Carbon disulfide
0330	Benzidine	0560	Carbon monoxide
0320	Benzol (benzene)	0570	Carbon tetrachloride
2222	p-Benzoquinone (quinone)		(tetrachloromethane)

15. Contaminants - Continued

Code	Substance	Code	Substance
2070	Carbonyl chloride (phosgene)	1016	Corundum (A1203)(emery)
2670	Carbophenothion (Trithion)	0735	Cotton dust (raw)
1037	Cellosolve acetate (ethoxyethyl acetate)	0737	Crag
		0750	Creosote
0575	Cellulose	0760	Cresol
0577	Cement, Portland	0740	Cresylic acid
0610	Cerium	0770	Crotonaldehyde
0611	Chlordane	0780	Cumene
0612	Chlorinated camphene (toxaphene)	0790	Cyanides
		0800	Cyanogen
0613	Chlorinated diphenyl oxide	0810	Cycloheane
0640	Chlorine	0820	Cyclohexanol
0614	Chlorine dioxide	0830	Cyclohexanone
0615	Chlorine trifluoride	0840	Cyclohexene
0617	Chloroacetaldehyde	0845	Cyclopentadiene
0618	a-Chloroacetophenone (phenacylchloride)	0846	2,4-D(2,4-dichlorophenoxyaceti acid)
0620	Chlorobenzene (monochlorobenzene)	0847	DDT
		0850	DDVP
0623	o-Chlorobenzylidene malonitrile (OCBM)	0853	Decaborane
		0857	Demeton (Systox)
0627	Chlorobromomethane	0923	DGE (diglycidyl ether)
0680	2-Chloro-1,3-butadiene (chloroprene)	0860	Diacetone alcohol (4-methyl-2-pentanone)
0630	Chlorodiphenyl or Chloronaphthalenes	1130	1,2-Diaminoethane (ethylenediamine)
0645	1-Chloro-2,3-epoxypropane (epichlorohydrin)	0861	Diazomethane
1120	2-Chloroethanol (ethylene chlorohydrin)	0932	Dibrome (dimethyl-1,2-dibromo-2,2-dichloroethyl phosphate)
2580	Chloroethylene (vinyl chloride)	1140	1,2-Dibromoethane (ethylene dibromide)
0670	Chloroform (trichloromethane)	0863	Dibutyl phosphate
		0864	Dibutylphthalate
2640	Chloromethyl ether (methyl chloromethyl ether)	0865	Dichloracetylene
		0867	o-Dichlorobenzene
0660	1-Chloro-1-nitropropane	0868	p-Dichlorobenzene
0675	Chloropierin (nitrotrichloromethane)	0869	Dichlorobenzidine
		0871	Dichlorodifluoromethane
0680	Chloroprene (2-chloro-1,3-butadiene)	0872	1,3-Dichloro-5,5-dimethyl hydantoin
0685	Chromic acid and chromates	1160	1,1-Dichloroethane (ethylidine chloride)
0690	Chromium, sol chromic, chromous salts	0874	1,2-Dichloroethane (ethylene dichloride)
0710	Coal tar naphtha	0870	1,2-Dichloroethylene (acetylene dichloride)
0700	Coal tar pitch volatiles		
0720	Cobalt and compounds	0880	Dichloroethyl ether
0730	Copper (fume, dust, or mist)	1730	Dichloromethane (methylene chloride)

15. <u>Contaminants - Continued</u>

Code	Substance	Code	Substance
0887	Dichloromonofluoromethane	0926	Diphenyl amine
0890	1,1-Dichloro-1-nitroethane	1011	Diphenyl (biphenyl)
0846	2,4-dichlorophenoxyacetic acid (2,4-D)	1013	Diphenylmethane diisocyanate (methylene bis(phenylene phenylene isocyanate)(MDI)
2190	1,2-Dichloropropane (propylene dichloride)	1014	Dipropylene glycol methyl ether
0900	Dichlorotetrafluoroethane	1015	Di-sec-octylphthalate (di(2-ethylhexyl)phthalate)
0905	Dieldrin		
0910	Diethylamine	2680	Di-Syston (disulfoton)
0920	2-Diethylaminoethanol	2680	Disulfoton (Di-Syston)
1010	Diethylene dioxide (dioxane)	1000	Dyes, misc.
0921	Diethylene triamine	1016	Emery (corundum)(A1203)
1210	Diethyl ether (ethyl ether) (ethyl oxide)	2425	Endosulfan (thiodan)
1015	Di(2-ethylhexyl) phthalate (di-sec-octylphthalate)	1017	Endrin
		0645	Epichlorohydrin (1-chloro-2,3-epoxypropane)
0922	Difluorodibromomethane		
0923	Diglycidyl ether (DGE)	1019	EPN
1490	Dihydroxybenzene (hydroquinone)	2215	1,2-Epoxypropane (propylene oxide)
0924	Diisobutyl ketone (2,6-dimethylheptanone)	1365	2,3-Epoxy-1-propanol (glycidol)
0925	Diisopropylamine	1020	Epoxy systems, general
2710	Dimecron (phosphamidon)	1025	Ethane
1655	Dimethoxymethane (methylal)	1220	Ethanethiol (ethyl mercaptan)
0927	Dimethyl acetamide	1030	Ethanolamine (2-aminoethanol)
0928	Dimethylamine	1033	2-Ethoxyethanol (ethylene glycol monoethyl ether)
0929	4-Dimethylaminoazobenzene		
2600	Dimethylaminobenzene (xylidene)	1037	Ethoxyethyl-acetate (cellosolve acetate)
0931	Dimethylaniline	1040	Ethyl acetate
2590	Dimethylbenzene (xylene)	1050	Ethyl acrylate
0932	Dimethyl-1, 2-dibromo-2, 2-dichloroethyl phosphate (dibrome)	1060	Ethyl alcohol
		1070	Ethyl amine
		1075	Ethyl sec-amyl ketone (5-methyl-3-heptanone)
0930	Dimethylformamide	1080	Ethyl benzene
0924	2,6-Dimethylheptanone (diisobutyl ketone)	1090	Ethyl bromide
0940	1,1-Dimethylhydrazine	1100	Ethyl butyl ketone (3-heptanone)
1942	Dimethylnitrosoamine (n-nitrosodimethylamine)	1110	Ethyl chloride
0950	Dimethylphthalate	1120	Ethylene chlorohydrin (2-chloroethanol)
0960	Dimethyl sulfate	1033	Ethylene glycol monoethyl ether (2-ethoxyethanol)
0970	Dinitrobenzene		
0975	Dinitro-o-cresol	1130	Ethylenediamine (1,2-diaminoethane)
0980	Dinitrophenol		
0990	Dinitrotoluene	1140	Ethylene dibromide (1,2-dibromoethane)
1010	Dioxane (diethylene dioxide)		

15. <u>Contaminants - Continued</u>

Code	Substance	Code	Substance
1160	Ethylene dichloride (1,2-dichloroethane)	1675	2-Heptanone (methyl (n-amyl) ketone)
1910	Ethylene glycol dinitrate	1100	3-Heptanone (ethyl butyl ketone)
1170	Ethylene glycol monomethyl ether acetate (methyl cellosolve acetate)	1372	Hexochloroethane
		1373	Hexachloronaphthalene
1175	Ethylene imine	1370	Hexamylenetetramine
1190	Ethylene oxide	1380	Hexane (n-hexane)
1210	Ethyl ether (ether oxide) (diethyl ether)	1690	2-Hexanone
1155	Ethyl formate	1385	Hexone (methyl isobutyle ketone)
1160	Ethylidene chloride 1,1-dichloroethane)	1387	sec-Hexyl acetate
		1390	Hydrazine
1210	Ethyl oxide (ethyl ether) (diethyl ether)	1405	Hydrochloric acid (muriatic acid)
1220	Ethyl mercaptan (ethanethiol)	1410	Hydrogen
1984	Ethyl parathion (parathion)	1420	Hydrogen bromide
1225	Ethylmorpholine	1430	Hydrogen chloride
1230	Ethyl silicate	1440	Hydrogen cyanide
1240	Exhaust of diesel combustion engine (no breakdown)	1460	Hydrogen fluoride
		1470	Hydrogen peroxide
1260	Exhaust of gasoline combustion engine (no breakdown)	1475	Hydrogen selenide
		1480	Hydrogen sulfide
		1567	IGE (Isopropyl glycidyl ether)
1250	Fabrics, misc.	1500	Indene
1263	Ferbam	1510	Indium and compounds
1267	Ferrovanadium dust	1515	Iodine
1300	Fibrous glass	1520	Iron and compounds
1280	Fluoride	1530	Isoamyl acetate
1270	Fluorine	1532	Isoamyl alcohol
1285	Fluorotrichloromethane	1534	Isobutyl acetate
1290	Formaldehyde	1536	Isobutyl alcohol
1310	Formic acid	1538	Isophorone
1320	Freon, unspecified	1540	Isopropyl acetate
1330	Furfuryl alcohol	1560	Isopropyl alcohol
1340	Gasoline	1562	Isopropylamine
1360	Germanium compd.	1565	Isopropylether
1363	Glycerine mist	1567	Isopropyl glycidyl ether (IGE)
1365	Glycidol (2,3-epoxy-1-propanol)	1570	Kerosine (Kerosene)
		1575	Ketene
1366	Graphite	1590	Lead and compounds
0300	Guthion (azinphos, methyl)	1593	Limestone
1367	Gypsum	1595	Lindane
1368	Hafnium	1503	Lithium hydride
1400	Helium	1830	L.P.G. (liquified petroleum gas)(natural gas)
1369	Heptachlor		
1371	Heptane (n-heptane)	1615	Magnesite

15. Contaminants - Continued

Code	Substance	Code	Substance
1610	Magnesium and compounds	1765	o-methylcyclohexanone
1616	Malathion	1013	Methylene bis (phenylene isocyanate)(MDI)(diphenylme-thane diisocyanate)
1618	Maleic anhydride		
1620	Manganese and compounds		
1652	MAPP (methyl acetylene-propadiene mixture)	2650	4,4-Methylene bis (2-chloroaniline)
1073	MDI (methylene bis(phenylene isocyanate)(diphenylmethane diisocyanate)	1730	Methylene chloride (dichloromethane)
0430	MEK (2-butanone)(methyl ethyl ketone)	0430	Methyl ethyl ketone (MEK) (2-butanone)
1630	Mercury and compounds	1770	Methyl formate
1635	Mesityl oxide	1075	5-Methyl-3-heptanone (ethyl sec-amyl ketone)
2700	Meta-Systox R (oxydemeton methyl)	1772	Methyl iodide
1640	Methane	1670	Methyl isobutyl carbinol (methyl amyl alcohol)
1643	Methanethiol (methyl mercaptan)	1385	Methyl isobutyl ketone (hexone)
1660	Methanol (methyl alcohol)	1773	Methyl isocyanate
1646	Methoxychlor	1643	Methyl mercaptan (methanethiol)
0590	2-Methoxyethanol (methyl cellosolve)	1774	Methyl methacrylate
1650	Methyl acetate	1775	Methyl parathion
1651	Methyl acetylene (propyne)	2010	Methyl propyl ketone (2-pentanone)
1652	Methyl acetylene-propadiene mixture (MAPP)	1777	Methyl silicate
1653	Methyl acrylate	1778	a-Methyl styrene
1655	Methylal (dimethoxymethane)	2065	Mevinphos (phosdrin)
1660	Methyl alcohol (methanol)	1780	Mineral Spirits
1665	Methylamine	1790	Molybdenum and compounds
1670	Methyl amyl alcohol (methyl-isobutyl carbinol)	2690	Monocrotophos (Azodrin)
1675	Methyl (n-amyl) ketone (2-heptanone)	1792	Monomethyl aniline
1680	Methyl bromide	1794	Monomethyl hydrazine
1690	Methyl butyl ketone (2-hexanone)	1797	Morpholine
0590	Methyl cellosolve (2-methoxyethanol)	1405	Muriatic Acid (hydrochloric acid)
1170	Methyl cellosolve acetate (ethylene glycol monomethyl ether acetate)	0710	Naphtha (coal tar)
1710	Methyl chloride	2037	Naphtha (petroleum distillates)
1720	Methyl chloroform (1,1,1-trichloroethane)	1810	Naphthalene
2640	Methyl chloromethyl ether (chloromethyl ether)	1815	a-Naphthylamine
1735	Methyl 2-cyanoacrylate	1820	b-Naphthylamine
1740	Methylcyclohexane	1830	Natural gas (L.P.G.)(liquified Petroleum gas)
1760	Methylcyclohexanol	1850	Neon
		1840	Nickel metal & soluble comp.
		1841	Nickel carbonyl
		1855	Nicotine
		1860	Nitric acid
		1890	Nitric oxide
		1865	p-Nitroaniline
		1870	Nitrobenzene

15. Contaminants - Continued

Code	Substance	Code	Substance
1872	p-Nitrochlorobenzene	2053	Phenyl ether - biphenyl mixture
1877	N-Nitrodimethylamine	2280	Phenylethylene (styrene monomer)
1875	4-Nitrodiphenyl		
1880	Nitroethane	2057	Phenyl glycidyl ether (PGE)
1900	Nitrogen	2060	Phenylhydrazine
1903	Nitrogen dioxide	2065	Phosdrin (mevinphos)
1907	Nitrogen trifluoride	2070	Phosgene (carbonyl chloride)
1912	Nitroglycerine	2710	Phosphamidon (Dimecron)
1920	Nitromethane	2080	Phosphine
1930	Nitrophenols	2085	Phosphoric acid
1940	Nitropropane	2090	Phosphorous
1942	N-Nitrosodimethylamine (dimethyl nitrosoamine)	2091	Phosphorous pentachloride
1945	Nitrotoluene	2092	Phosphorous pentasulfide
0675	Nitrotrichloromethane (chloropicrin)	2093	Phosphorous trichloride
		2110	Phthalic anhydride
1953	Nitrous oxide	2120	Pieric acid (2,4,6-trinitrophenol)
0623	OCBM (o-chlorobenzylidene malonitrile)	2125	Pival (2-Pivalyl-1,3-indandione)
1955	Octachloronaphthalene		
1957	Octane	2127	Plaster of Paris
5010	Oil mist, general	2100	Plastic decomposition prod. (unspecified)
1960	Osmium tetroxide		
1970	Oxalic acid	2130	Platinum
2700	Oxydemetonmethyl (Meta-Systox R)	2135	Polytetrafluoroethylene decomposition products
1975	Oxygen difluoride	2137	Potassium bisulfite
1980	Ozone	2140	Potassium hydroxide
2000	Paraffin	2150	Propane
1982	Paraquat	2167	Propargyl alcohol
1984	Parathion (ethyl parathion)	2163	B-Propiolactone
1986	Pentaborane	2180	n-Propyl acetate
1987	Pentaerythritol	2170	Propyl alcohol
1988	Pentachloronaphthalene	2190	Propylene dichloride (1,2-dichloropropane)
1989	Pentachlorophenol		
1990	Pentane	2210	Propylene glycol
2010	2-Pentanone (methyl propyl ketone)	2213	Propylene imine
2020	Perchloroethylene (tetrachloroethylene)	2215	Propylene oxide (1,2-epoxypropane)
2030	Perchloromethyl mercaptan	2185	n-Propyl nitrate
2033	Perchloryl fluoride	1651	Propyne (methyl acetylene)
2037	Petroleum distillates (naphtha)	2215	Pyrethrum
0618	Phenacylchloride (A-chloroacetophenone)	2220	Pyridine
		2222	Quinone (p-benzoquinone)
2057	PGE (phenyl glycidyl ether)	2224	RDX
2040	Phenol	2200	Resins, miscellaneous, unless classified elsewhere
2042	p-Phenylene diamine	2225	Rhodium
2047	Phenyl ether	2226	Ronnel

15. Contaminants - Continued

Code	Substance	Code	Substance
2228	Rotenone	2427	Thiram
2229	Rouge	2430	Tin (inorganic - except tin oxide)
2230	Selenium compounds		
2231	Selenium hexafluoride	2431	Tin (Organic compounds)
9525	Seven (carbaryl)	2432	Tin oxide
2236	Silicon carbide	2440	Titanium dioxide and other titanium compounds
2240	Silver and compounds		
2245	Sodium bicarbonate	2460	Toluene (toluol)
2250	Sodium fluoroacetate	2470	Toluene-2,4-diisocyanate
2260	Sodium hydroxide	2475	o-Toluidine
2263	Starch	2460	Toluol (Toluene)
2267	Stibine	0612	Toxaphene (chlorinated camphene)
2270	Stoddard solvent		
2275	Strychnine	2477	Tributyl phosphate
2280	Styrene monomer (phenylethylene)	1720	1,1,1-Trichloroethane (methyl chloroform)
2290	Sulfur dioxide	2495	1,1,2-Trichloroethane
2300	Sulfur hexafluoride	2490	Trichloroethylene
2310	Sulfuric Acid	0670	Trichloromethane (chloroform)
2320	Sulfur monochloride	2483	Trichloronaphthalene
2321	Sulfur pentafluoride	2324	2,4,5-Trichlorophenoxyacetic acid (2,4,5-T)
2323	Sulfuryl fluoride		
0857	Systox (Semeton)	2510	1,2,3-Trichloropropane
2324	2,4,5-T(2,4,5-trichloro-phenoxyacetic acid)	2485	1,1,2-Trichloro 1,2,2-trifluoroethane
2325	Tantalum	2480	Triethylamine
2327	TEDP	2500	Trifluoromonobromomethane
2329	Teflon decomposition products	2120	2,4,6-Trinitrophenol (pieric acid)
2330	Tellurium		
2332	Tellurium hexafluoride	2410	2,4,6-Trinitrophenylmethylni-tramine (tetryl)
2334	TEPP		
2335	Terphenyls	2530	Trinitrotoluene
2337	1,1,1,2-Tetrachloro-2,2-difluoroethane	2532	Triorthocresyl phosphate
		2535	Triphenyl phosphate
2339	1,1,2,2-Tetrachloro-1,2-difluoroethane	2670	Trithion (carbophenothion)
		2537	Tungsten and compounds
2340	1,1,2,2-Tetrachloroethane	2540	Turpentine
2020	Tetrachloroethylene (perchloroethylene)	2560	Uranium and compounds
		2570	Vanadium and compounds
0570	Tetrachloroniethane (carbon tetrachloride)	2575	Vinyl venzene
		2580	Vinyl chloride (chloroethylene)
2350	Tetrachloronaphthalene	0120	Vinylcyanide (acrylonitrile)
2360	Tetraethyl lead	2582	Vinyl toluene
2390	Tetrahydrofuran	2586	Warfarin
2370	Tetramethyl lead	2590	Xylene (dimethylbenzene)
2380	Tetramethyl succinonitrile	2600	Xylidine (dimethylaminobenzene)
2395	Tetranitromethane	2602	Yttrium
2410	Tetryl (2,4,6-trinitrophenyl-methinitramine)	2606	Zinc and compounds
		2611	Zinc chloride
2420	Thallium	2610	Zinc oxide (fume)
2425	Thiodan (endosulfan)	2620	Zirconium compounds

15. <u>Contaminants - Continued</u>

II. Unidentified Chemicals

Code	Substance	Code	Substance
5440	Bleach	5370	Microbiological pathogens
5340	Brake fluid mist	5009	Oil
5130	Carburetor cleaner	5010	Oil Mist
5250	Caustic cleaner mist	5390	Paint and Varnish thermal decomposition products
5240	Cosmetic solvent vapor		
5140	Degreaser	5180	Paint mist
5400	Detergent	5110	Paint thinner
5330	Drilling fluid mist	5345	Pesticide, unspecified
5360	Duplicator vapors	5260	Photographic developer vapor
5210	Dye	5280	Plant and food products
5220	Dye solvent vapor	5270	Rubber accelerators and anti-oxidants
5450	Enzyme		
5170	Explosives	5410	Soap
5150	Fabric plastizisers, modifiers and finishers	5460	Stainless Steel (composition unknown or unspecified)
5160	Flavoring agents	5430	Tar
5320	Flux	5120	Type cleaner
5100	General solvent vapor	5420	Wax
5348	Glue	5380	Welding gases, unspec. (ozone, NOx, CO, CO2)
5350	Glue vapors		
5015	Grease	5930	Other alcohols
5029	Ink	5950	Other solvents
5030	Ink mist	5999	Other and unidentified chemicals
5230	Ink solvent vapor		
5310	Metal fume		

III. Physical Hazards

Code	Substance	Code	Substance
8110	Noise, continuous	8390	Improper Illumination
8120	Noise, intermittent	8380	Infrared
8130	Noise, impact	8350	Long wave Radio - frequency
8140	Noise, ultrasonic	8340	Microwave
8150	Vibration	8360	Sunlight
		8370	Ultra-violet
8260	Alpha radiation		
8270	Beta radiation	8510	Lasers - infrared
8220	Gamma radiation - sealed source	8505	Lasers - microwave and RF
		8590	Lasers - ultraviolet
8280	Gamma radiation - unsealed	8530	Lasers - visible
8210	X-radiation	8600	Lasers - X-radiation
8290	Other ionizing radiation		
		8410	Low Pressure
8310	Exposure to cold	8420	High pressure
8320	Heat stress - dry		
8330	Heat stress - humid	8870	Electrical shock
		8880	All other physical hazards

Section A Identity   Items 15 to 16   Page 23

15. <u>Contaminants - Continued</u>

IV.   Dust

Code	Substance	Code	Substance
9015	Cristobalite	9090	Graphite
9010	Crystalline free silica	9120	Leather
9013	Fused silica	9075	Mica
9017	Tridymite	9170	Mortar dust
		9100	Organic dust
9050	Amorphous silica, including diatomaceous earth	9180	Plastic dust
		0577	Portland cement
		9190	Rubber dust
9020	Asbestos	9085	Soapstone
9140	Brass dust (composition unknown or unspecified)	9030	Talc
		9087	Tremolite
9150	Brick dust	9210	Wood - sawdust
9160	Clay dust		
9040	Coal	9130	Inert or nuisance dust, not elsewhere classified
9220	Dust containing enzymes		
9200	Flour dust		

16. <u>Phase of Operation at Time of Incident</u>

Code		Code	
01	Normal Operation	20	Unloading Tank, Truck, Tanker, Tank Car, etc.
02	Blending		
03	Blowing out Equipment, Lines Gas Freeing	21	Upset Condition
		22	Well Work
04	Dismantling	23	While Cementing
05	During Drilling	24	Withdrawing Pipe from Hole
06	During Water Draw-Off	25	
07	During Work-Over	26	
08	Excavating		
09	Filling Tank, Truck, Road Tanker, Tank Car, Vehicle Tanker, etc.	30	Ship Location
		31	Anchorage
		32	At Sea
10	Gerry-rigged Operations	33	In Channel - buoyed - with Pilot
11	Idle, Out-of-Service, During Non-Occupancy	34	In Channel - buoyed - without Pilot
12	Installing Minor New Equipment	35	In Channel - not buoyed - with Pilot
13	Maintenance, Repair	36	In Channel - not buoyed - without Pilot
14	Major Additions, Alterations and Expansion	37	In Port
15	New Construction	38	In restricted waters with Pilot
16	Paraffin Control		
17	Regular Operation, Normal Production	39	In restricted waters without Pilot
18	Shutting Down		
19	Start-Up Testing	40	In Shipyard

Section A Identity   Items 16 to 17   Page 24

16. Phase of Operation at Time of Incident - Continued

<u>Code</u>   <u>Code</u>

41   Pier   57   Installing Minor New
42   Sea Berth   Equipment
43   Single Point Mooring   58   Leaving Drydock
44   59   Lightering
45   60   Loading Cargo
46   61   Major Alteration
                                   62   New Construction
50   Ship Operation   63   Pumping Bilge
51   Ballasting   64   Shifting Cargo
52   Berthing   65   Tank Cleaning
53   Bunkering   66   Unberthing
54   De-Ballasting   67
55   Discharging Cargo   68
56   Entering Drydock   69

17. Months Since Last Inspection

Note: 1.   Use separate sheet for each injured employee.

        2.   Include other people involved under Item 71.

21. Last Name

22. Initials

23. Employee Number

24. Address

25. Age in Years - Use 00 if not applicable.

26. Employment Status

        01   Permanent Employee
        02   Temporary Employee
        03   Summer Employee
        04   Contractor
        05   Concessionaire
        06   Employee Family
        07   Public (Visitor)
        08   (Public (Other)

27. Nature of Injury - Illness

Code   Code

01	Abrasion	28	Internal Injuries
02	Alcohol	29	Multiple Injuries
03	Amputation	30	Occupational Skin Disease or Disorder (Dermatitis)
04	Animal, Insect Disease		
05	Asphyxiation (not drowning)	31	Particle in Eye
06	Avulsion (loss of flesh by shearing, tearing)	32	Poisoning (Systematic effects of toxic materials)(include food poisoning)
07	Bites (Trauma)		
08	Blinded (Eye Injury)	33	Puncture
09	Burns (Chemical)	34	Respiratory Conditions due to Toxic Agents
10	Burns (Thermal)		
11	Concussion (Head)	35	Scratches
12	Contagious Disease	36	Shock (electric)
13	Contusion (bruised but surface intact)	37	Smoke Inhalation
14	Crush	38	Sprain
15	Cuts	39	Strain (back)
16	Dislocation	40	Strain (not back)
17	Disorders due to Physical Agents	41	Stress (emotional)
18	Disorders due to Repeated Trauma	42	Stroke
		43	Unknown
19	Diving Illness	44	
20	Drug Abuse	45	
21	Dust Disease of the Lungs	50	Personal Items
22	Fracture	51	Clothing
23	Fungus	52	Dentures
24	Heart Disease	53	Glasses
25	Hernia	54	
26	Immersed in Water	55	
27	Infection	56	

28. Part of Body Injured

Code   Code

00	No part of the body injured	30	Arm-Hand Areas
		31	Elbow
10	Head Area	32	Fingers
11	Ear(s)	33	Hand(s) - not Fingers
12	Eye(s)	34	Lower Arm (below elbow)
13	Face	35	Upper Arm (above elbow)
14	Jaw	36	Wrist
15	Neck		
16	Nose & Throat	40	Trunk Area
17	Skull and Scalp	41	Abdomen
18	Teeth & Gums	42	Back (Spine)
19		43	Buttlocks
20		44	Chest, Rib(s)

28. Part of Body Injured - Continued

Code

			Code	
45	Groin		60	Body System
46	Hip(s)		61	Circulatory
47	Shoulder(s)		62	Digestive
			63	Excretory
50	Lower Extremeties		64	Genito - Urinary
51	Ankle		65	Nervous
52	Foot-Feet-not Toes		66	Reproductive
53	Knee(s)		67	Respiratory
54	Leg(s)			
55	Thigh		70	Not Elsewhere Listed
56	Toe(s)		71	Multiple Area Skin Problem
57				(poison ivy, burns, etc.)
			72	Multiple Body Injuries
			73	

29. Severity of Injury

Code

			Code	
00	No Injury Involved		05	Disabling Injury (permanent partial)
01	No Aid Required (bruises)			
02	First Aid Attention only		06	Disabling Injury (permanent total)
03	Medical attention only			
04	Disabling Injury (temporary)		07	Disabling Injury (fatal)

30. Estimated Days Off Work

This is the Supervisor's best estimate of time loss if an employee is
involved. When actual days off are determined, submit a new A.143 with
Items 1, 2, and 30 only completed (if actual differs from estimate).

Section C
Analysis of Management Problem   Items 41 to 43   Page 28

41. Frequency Potential   1 Frequent   2 Occasional   3 Rare

42. Severity Potential   1 Frequent   2 Serious   3 Minor

43. Human Error

Code		Code	
000	No Human Error	081	Face Protection
		082	Foot/Toe Protection
010	Unsafe Speed (not motor	083	Gloves
	vehicle)	084	Goggles, Glasses
011	Descending unsafely	085	Hard Hat
012	Feeding too fast	086	Hearing Protection
013	Feeding too slowly	087	Life Jacket
014	Misjudging distance	088	Pant/Leg Protection
015	Moving body too fast	089	Respirator - Gas Mask
	(walking, etc.)	090	Skin Cream, Insect Repellent
016	Operating too fast	091	Special Clothing
017	Operating too slowly	092	
018	Throwing not handling	093	
019			
		100	Failed to Dress Properly
020	Violations of Law	101	Glasses worn not safety type
021	Aircraft safety	102	Shoes not right for job done
022	Motor Vehicle safety	103	
023	Watercraft safety	104	
024			
		110	Failed to get First Aid
030	Failure to Follow Regulations		(usually resulting in
031	Disobeyed Rules		infection
032	Violated a Policy	111	Failed to get First Aid
033	Went ahead without checking	112	Small wound left unattended
034		113	
040	Failure to Get Help	120	Body Hygiene Problem (sanitary
041	Failed to get help		conditions a problem
042		121	Clothing Irritants
		122	
050	Failure to Shut Down	123	
051	Machine not shut off		
052	Power Circuit not secured	130	Failure to Warn
053		131	Failed to report bad condition
		132	Failure to Warn
060	Did not use right Tool	133	Inadequate instruction
061	Wrong Tool for job done		(employee to employee - not
062			supervision)
		134	No Time to Warn
070	Did not use right Equipment	135	Time but didn't think to do it
071	Wrong equipment for job done	136	
072			
		140	Distracting, Abusing, Teasing
080	Failed to use Protective	141	Distracted by outside source
	Clothing	142	Fighting - harm
		143	Horseplay - harm
		144	

43. Human Error

Code		Code	
150	Improper Use of Body	213	Unsafe Placing, Loading, or Hanging
151	Insecure grip (hands)		
152	Lifted with back, not legs	214	
153	Over-exertion		
154	Should have used mechanical lift	220	Using Equipment Unsafely
155	Used hands, not tool	221	Using equipment
156		222	
		230	Vehicle Operator Error (applies to any kind of vehicle)
160	Improper Use of Equipment (not motor vehicle)		
161	Failure to check defect	231	Collision pulling out
162	Improper use	232	Did not use helper to guide backing
163	Overloaded		
164	Unauthorized use	233	Disregard for traffic control
165	Unsafe Equipment	234	Distracted
166	Unsecured	235	Driving on Road Shoulders
167		236	Driving too fast, too slow
		237	Driving under influence (alcohol, drugs, etc.)
170	Unexpected - Without Warning		
171	Unpredicted contact	238	Engineer Error
172		239	Failed to check before backing
		240	Failed to clean glass (vision problem)
180	Making Safety Device Inoperative		
		241	Failed to communicate (radio, telephone, signals)
181	Blocked or plugged safeguard		
182	Disconnected protective device	242	Failed to give Right-of-way
183		243	Failed to maintain flying speed
		244	Failed to signal intention
190	Unsafe Position or Posture		
191	Cramped position	245	Failed to use flares, lights, etc.
192	Entering small openings (clearance space)		
		246	Failed to use horn, whistle, etc.
193	Exposed under suspended loads		
194	Insecure footing/seating	247	Failed to use seat/life belts
195	Lengthy work in same position	248	Fell asleep at wheel
196	Riding unsafely (not motor vehicle)(cran hock, buckets)	249	Filed no Flight Plan
		250	Following too closely
197	Unable to swim	251	Getting into or out of vehicle
198	Unnecessary exposure		
199	Working too close	252	Illegal/Improper Turn
200		253	Improper entrance or exit from traffic lane
201			
		254	Improper parked (illegally)
210	Unsafe Mixing/Combining/ Placing	255	Intersectional collision
		256	Keys left in vehicle (unattended, stolen)
211	Unsafe Combining		
212	Unsafe Mixing	257	Lost control, skidding

Section C
Analysis of Management Problem   Items 43 to 44   Page 30

43. Human Error

Code   Code

258   Misinformed/Nor Forewarned   280   Pedestrian Type Errors
259   Misjudged clearance   281   Against traffic light
260   Not driving defensively   282   Coming from behind
261   Not safely secured (brakes)   between
262   Off-road driving incident   283   Lying in street
263   Operating over too long   284   Not standing in Safety Zone
        a period   285   Playing in street
264   Over/Undershoot landing   286   Unsafe crossing of road
        field   287   Walking into path of
265   Passengers without seats   vehicle
266   Passing on Hill/Curve   288   Walking with traffic
267   Pilot Error - Aircraft   289
268   Pilot Error - Vessel   290
269   Started too early/too late
270   Towing or Pushing Other   300   Human Error followed by
271   Unable to stop for object on   Fire
        the road (not another   301   Failed to clean property
        vehicle or animal)   302   Failed to control burning
272   Unauthorized use, misuse   303   Fire not fully
        of vehicles   extinguished
273   Vehicle unattended   304   Improper disposal
274   Vehicle overloaded   305   Improper firing
275   Watch not posted on vessel   306   Improper thawing of Water
        (signal man)   pipe
276   Wrong lane, Position   307   Using matches
277   308
278   309
279   310

44. Condition Defect

Code   Code

000   012   Incorrect length of sleeves
001   Defective Equipment/Tools   013   Ornament worn is hazardous
002   Environmental Hazard   014   Pants not properly secured
003   Fire from Condition Defect   015   Poor bedding facilities
004   Hazardous Placement   (protection from elements)
005   Hazardous Method   016   Shoes not suitable to work in
006   Inadequate or Hazardous Dress   017   Should wear glasses or have
007   Inadequate Protection   them changed (if they are
008   Other   needed)
009   018   Wearing ordinary glasses
                                               (should wear safety type)
010   Dress or Apparrel Defects   019
011   Clothes not right for
        weather

Section C
Analysis of Management Problem   Item 44   Page 31

44. Condition Defect

Code		Code	
020	Hazards in General	064	Guard inadequate
021	Communication difficulty	065	Shoring inadequate
022	Disorderly crowd (riot)	066	Substandard firefighting program
023	Environmental problem		
024	Disorderly person	067	Uncovered/Unsheathed
025	Excessive heat/cold	068	Ungrounded/Poorly grounded
026	Excessive noise	069	Unlabeled/Mis-labeled
027	Excessive smoke	070	Unshielded/Badly shielded
028	Failure of personal protection	071	
029	Hazardous placement	072	
030	Hazards on Play Field Surface	080	Vehicle Condition Defects
031	Inadequate Lighting	081	Accelerator Assembly
032	Inadequate or no labelling	082	Air-Conditioning
033	Lacking bathing facilities (sanitation problem)	083	Auto Jacks (bumper)
034	Lifting facilities inadequate	084	Blade, Plow, Scraper
035	Narrow space problem (overcrowding)	085	Body/Hull/Wing Structure
036	Piled improperly	086	Bumper
037	Placed poorly	087	Brakes, Braking System
038	Poor Housekeeping	088	Cable Controls, Guides, Pulleys
039	Pressure control failure	089	Cargo-Moving/Sliding
040	Road blocked (due to storms, landslides)	090	Chain Drive Assembly
041	Rough Ground	091	Clutch Assembly
042	Secured poorly	092	Cooling System - Radiator, Fan
043	Slippery surface	093	Differential/Bearing
044	Steep hill/incline/slope	094	Doors
045	Supply shortage	095	Exhaust/Muffler System
046	Too bulky for manual handling	096	Foot Pad/Handhold
047	Too heavy for manual handling	097	Foot Pedals
048	Tool/Equipment missing	098	Front-End Stability
049	Traffic control problem	099	Front/Rear Suspension System
050	Ventilation/Wing Current	100	Fuel Line (not Tank)
051	Vertical Space Problem	101	Fuel Tank
052	Visibility Problem	102	Gauges and Instruments
053	Weather Condition	103	Gear Shift
054		104	Glass
055		105	Guard missing
056		106	Heating System
		107	Hood
		108	Lighting System
		109	Mirror
060	Inadequate Protection	110	Motor/Engine
061	Dangerous Work Area (general)	111	Rotor Blade, Propeller
062	First Aid unavailable	112	Seat Belts
063	Guard (covering) needed	113	Seats (front and back)

44. Condition Defect - Continued

Code   Code

114   Shocks/Springs   153   Fires from Dust, Gases, Fumes
115   Signal Devices   154   Flammable Liquids
116   Sprayer Heads, Burner Nozzles 155   Flashbacks
117   Steering Assembly (tie rods)   156   Flying Sparks
118   Tires   157   Heat/Smoke Conductors
119   Undercovering   158   Heating Equipment Failure
120   Visor   159   Open Flame
121   Wheel, Axle, Spindle   160   Overheated Grease, Wax,
122   Windshield Wipers   Tar
123   161   Rubbish/Trash/Litter
                                      162   Spark Arrester Missing
150   Fires from Condition Defects   163   Spontaneous Ignition
151   Exposure fire
152   Electrical short

52. Design Problem

Code   Code

01   Deficiency of design or   13   Lack of adequate use
        layout location   14   Lack of proper instrumenta-
02   Erosion or corrosion not   tion
        anticipated   15   Lack of valve where required
03   Freezing   16   Mechanical Lifting Device
04   Heating, Ventilation Problem   needed
05   Improper Tool/Equipment   17   Poor storage facilities
        design   18   To be determined
06   Inaccessible location   19   Visibility problem
07   Inadequate drainage   20   Without safeguards
08   Inadequate Exhaust   21
09   Inadequate fire escapes   22
10   Inadequate parking facilities   23
11   Inadequate protection   24
        against fire spread   25
12   Inadequate spacing   26
                                      27
                                      28

53. Construction Problem

Code   Code

01   Contractor is not meeting his   04   Welding defect
        safety responsibilities   05
02   Equipment installed/Placed   06
        Wrongly   07
03   Improper or deficient con-   08
        struction   09

Section C
Analysis of Management Problem   Items 54 to 55   Page 33

54. Maintenance Problem

Code		Code	
01	Badly Positioned	23	No warning signs, not barricaded
02	Beyond effective maintenance	24	Part broken or bent
03	Deferred inspection	25	Part missing
04	Deferred or inadequate maintenance	26	Pipe or welding defect
05	Disposal problem	27	Polluted atmosphere
06	Electrical short circuit	28	Poor communication facilities
07	Equipment control problem	29	Poor electrical maintenance
08	Equipment installed or placed Wrong	30	Poor instrument maintenance
09	Equipment operating instructions	31	Poor mechanical maintenance
10	Erosion or corrosion	32	Poor or incomplete operating instructions
11	Exhaust inadequate	33	Poor road conditions
12	Heating, Ventilating problem	34	Substandard personal protective equipment
13	Housekeeping problem	35	Unsanded, unsalted, not cleaned
14	Improper cleaning method (floor, machines etc.)	36	Unsuitable location
15	Improperly cleaned	37	Visibility problem
16	Inadequate control of pesticides, chemicals, etc.	38	Weather a main factor in this incident
17	Inadequate Crew	39	Without safeguards
18	Inadequate Fire	40	
19	Inadequate fire protection	41	
20	Inadequate or faulty inspection	42	
21	Mechanical short circuit	43	
22	Mechanical lifting equipment needed	44	
		45	

55. Operating Problem

Code		Code	
01	Authority imposed requirement	11	Make-Do situation
02	Available procedure inadequate	12	Operating beyond capacity
03	Defective Equipment	13	Polluted atmosphere
04	Failure to follow proper operating standards	14	Poor operating practice
05	Force Majeure	15	Poor storage practice
06	Hazard posting inadequate	16	Procedure not available
07	Hazardous work site	17	Rescue and life saving systems inadequate
08	Inadequate Organization	18	Storage handling problem
09	Lack of frequent patrolled procedure	19	Unit shutdown
10	Lack of periodic test and inspections procedure	20	Unsuitable product used
		21	Utility failure - Air

Section C
Analysis of Management Problem   Items 55 to 57   Page 34

55. Operating Problem - Continued

Code		Code	
22	Utility failure - Cooling water	26	Water pollution problem
23	Utility failure - Electricity	27	
24	Utility failure - Inert Gas	28	
25	Utility failure - Steam	29	
		30	
		31	

56. Supervision Problem

Code		Code	
00	Planning - insufficient or non existent	20	Directing - problems
01	Planning - area of responsibility is unclear	21	Communication problems
02	Planning - no emergency procedure	22	Poor communication facilities
03	Planning - no standard procedure	23	Unauthorized work activity
04	Inadequate control of pesticides	24	Verbal orders given but not followed
05	Inadequate Crew - a utilization of manpower problem	25	Written orders exist but not followed
06	Sanitary facilities inadequate	26	
07	Weather instruction procedure lacking	30	Controlling - problems
08		31	Close supervision not possible
09		32	Inadequate performance by contractor
10	Organizing - inadequate	33	
11		34	
12		35	
		36	

57. Training Problem

Code		Code	
01	A better Trained person needed, but not available	06	Heavy equipment operators course lacking
02	Defensive driving course lacking	07	Inadequate fire training
03	Driver training and licensing lacking	08	Inadequate indoctrination of employee
04	Emergency task with insufficient time to train properly	09	Job orientation training lacking
05	Fork lift truck operators course lacking	10	Rescue and lifesaving systems inadequate

Section C
Analysis of Management Problem   Items 57 to 59   Page 35

57. Training Problem - Continued

Code		Code	
11	Services of Loss Management Co-ordinator lacking	16	Work pressures did not allow time to train properly
12	Skill proficiency needs up-grading due to technological changes	17	
13	Supervisory training lacking	18	
14	Technical training lacking	19	
15	Training insufficient. Has received training but did not follow routine.	20	
		21	
		22	

58. Personnel Problem

Code		Code	
01	Allergy problems	11	Unsuitable work clothing
02	Emotional problem	12	Unsuited for this type of work
03	Employee conduct	13	Work fatigue due to excessive overtime
04	Illness	14	Work fatigue due to manual exertion
05	Mentally handicapped	15	Work fatigue due to moon-lighting
06	Physical fitness questioned	16	
07	Physically handicapped person - deserves special treatment	17	
08	Recurrence of old physical problem	18	
09	Suspected influence of alcohol/drugs	19	
10	Under or overweight	20	

59. Finance/Budget Problem

Code		Code	
01	An evaluation of the "risk" (cost to benefit) is needed to determine the economics of the situation to be controlled	06	Funds were budgeted and alloted but the funds were used for other purposes
02	Excess property problem	07	Funds were budgeted and alloted but the problem developed before they were used
03	Facilities poor	08	Funds were budgeted but the full amount, as budgeted, was not received
04	Funds insufficient to purchase suitable product	09	Funds were not budgeted to take care of this situation
05	Funds not provided for mechanical lifting equipment		

59. Finance/Budget Problem - Continued

Code		Code	
10	Improper building/facility design	17	Protection against fire spread
11	Improper tool/equipment design	18	Storage/handling problem
12	Inadequate fire escape funds	19	Substandard personal protection
13	Inadequate fire protection facilities	20	Too expensive to replace
14	Obsolete building, office, yard storage facilities, etc.	21	
		22	
		23	
15	Overage, obsolete	24	
16	Property disposal problem	25	

60. Purchasing Problem

Code		Code	
01	Awaiting material previously ordered	03	Other
		04	Poorly designed
02	Equipment supplied does not meet specifications	05	Substitute material supplied
		06	

61. Industrial Relations/Legal Problem
(Because of policies, regulations or practices imposed)

Code		Code	
01	Classification/qualification problem	09	Product liability
02	Grievance issue	10	Recruitment problem
03	Labour management issue	11	Retirement problem
04	Local staffing problem	12	Student handling problem
05	Medical problem	13	Training to be provided
06	Pay management problem	14	Workers Compensation issue
07	Physical examination problem	15	
08	Physically handicapped	16	
		17	

62. Medical Problem

Code		Code	
01	Abdominal pain	07	Fever
02	Addiction	08	Flu
03	Allergies	09	Gastro Intestinal
04	Blood pressure	10	Gynecology and obstetrics
05	Cyanosis	11	Headache
06	Fatigue	12	Indigestion
		13	Infection

Section C
Analysis of Management Problem   Items 62 to 64   Page 37

62. Medical Problem - Continued

Code

		Code	
14	Inflamation	20	Tumour and growths
15	Nervous/mental disorder	21	Vertigo - faint, dizzy
16	Pain - chest, muscular, etc.	22	Tests - blood
17	Pallor	23	Tests - general
18	Respiratory - colds, coughs, etc.	24	Tests - urine
		25	Tests - other
19	Swelling		

63. Security Problem

Code

		Code
01	Reported burglary	10
02	Reported theft	11
03		12
04		13
05		14
06		15
07		16
08		17
09		18

64. Problems not Related to Employee or Company Operation

Code

		Code	
01	Attacks	08	Suicide
02	Display of warning signs inadequate	09	Theft by others
		10	Vagrancy
03	Evidence of arson, incendiarism	11	Vandalism by others
		12	
04	Evidence of trespassing	13	
05	Fraud or Embezzlement	14	
06	Murder	15	
07	Other party did not use reasonable prudence and care	16	
		17	

GULF OIL CANADA LIMITED
LOSS MANAGEMENT INCIDENT REPORT (INJURY, DAMAGE , LOSS) — LOMIS

( ) INDICATES PAGE IN CODE BOOK   DATE: ___________________

A — IDENTITY

1. Organization Code (5)
2. Report Number (5)
3. Location of Incident (5)

Service Station Code No.
4. Date of Incident   Year   Mo.   Day
5. Time of Incident (24 Hour Clock)
6. Total Cost of Incident (Estimate) (5) $
7. Liability Claim Expected (5) $
8. Amount Awarded (5) $

9. Type of Incident (6)   Prim.   Secon.   Other
10. Type of Property (6)
11. Ownership of Property (7)
12. Specific Unit Involved (7)
13. Model or Construction Year (9)
14. Specific Components Involved (10)
15. Contaminants (14)
16. Phase of Operation at Time of Incident (18)
17. Months Since Last Inspection (18)

B — PEOPLE

Use Separate Sheet for Each Injured Employee

21. Last Name
22. Initials
23. Employee Number
24. Address

25. Age in Years

26. Employment Status (19)
27. Nature of Injury (20)
28. Part of Body Injured (20)
29. Severity of Injury (20)
30. Estimated Days Off Work (20)

C — MANAGEMENT PROBLEM

40. IMMEDIATE CAUSE
41. Frequency Potential — 1 Freq., 2 Occas., 3 Rare (21)
42. Severity Potential — 1 Major, 2 Serious, 3 Minor (21)

43. Human Error (21)
44. Condition Defect (22)

UNDERLYING CAUSE   50. Primary   51. Secondary

52. Design Problem (24)
53. Construction Problem (24)
54. Maintenance Problem (24)
55. Operating Problem (24)
56. Supervision Problem (25)

57. Training Problem (25)
58. Personnel Problem (25)
59. Finance/Budget Problem (25)
60. Purchasing Problem (26)
61. Ind'l Rel./Legal Problem (26)

62. Medical Problem (26)
63. Security Problem (26)
64. Problems Not Related to Employee or Company Operation (26)

D — DESCRIPTION

71. STORY OF INCIDENT — Attach a diagram if necessary. First aid treatment and disposition; describe clearly how the incident occurred; nature of damage; loss; liability, type and amount of product; name and address of claimant. Particulars if reported to authorities. Names of non-injured people involved and witnesses. Use an extra sheet if required.

WCB Form Completed? Yes   No

ANALYSIS

72. ANALYSIS — State precisely the direct cause. Then determine the underlying cause or system breakdown as per Section 50.

PREVENTION

73. CORRECTIVE ACTION — Changes in system taken or planned to eliminate the basic and underlying causes.

ORIGINATOR: _________________________   INVESTIGATED BY: _______________________   DATE: _______________________

Distribution: White — Insurance/Data Centre   Canary — Management   Pink — Co-ord. Loss Mgt.   Golden Rod — Oringinator

INSTRUCTIONS — A143

This form is designed for recording downgrading incidents such as:

Industrial Injury and Disease
Property Loss and Damage
Fires and Explosions
Burglary, Theft and Breaches of Security
Public and Product Liability

Pollution — air, water, ground, noise
Off the job Safety and Health Incidents
and other incidents resulting in, or
with the potential for injury, damage,
loss or Company liability.

The underlying causes should be determined and preventive action taken.
The form shall be completed by the immediate supervisor at the earliest possible time after such an occurrence. Additionally in the case of a serious incident involving death, injury or major damage, immediate notice shall be given by telephone or wire to Dept. H.Q., to the Regional Employee Services Unit and to the H.O. Manager — Insurance. If the incident involves a Company-owned or operated vehicle Form A 144 shall be used.

DEFINITIONS

Personal Injury — any injury or industrial disease arising out of and in the course of employment. In addition to form A 143, the applicable Workmen's Compensation forms shall be completed.

First Aid — treatment requiring first aid or services of a plant nurse only.

Medical Aid — treatment required in a hospital or by qualified medical practitioner as defined by applicable Workmen's Compensation Legislation.

Lost Time — requiring time away from work beyond the day of injury.

Other Compensable — any other incident covered by Workmen's Compensation Legislation.

Fire — any damage or loss resulting from fire.

Explosion — any damage or loss resulting from explosion.

Mix — any loss or downgrading of Company products resulting from unplanned mixing of products or any other materials.

Spill — any loss or damage resulting from spillage of products or any other materials.

Theft/Burglary — any theft or loss or damage resulting from forced entry into Company vehicles or property or loss of Company assets as a result of real or implied threats.

Other — any other loss or liability to the Company except that resulting from normal "wear and tear". "Wear and tear" can only be determined by the supervisor most directly involved.

PL — Public and Product Liability.

PD — Damage to the property of others.

ESTIMATED COSTS

Include only *additional* costs to the operation incurred as a result of the incident. Provide your best estimates of all costs to clean up, repair and/or replace loss or damaged property.

Make-up pay — the difference between regular pay and Workmen's Compensation for time lost by injured employee.

Other costs — additional wages payable for overtime, training or replacement (where necessary), transportation costs, meal allowances, accommodation, etc., as required.

SEPARATE SET BEFORE COMPLETING THIS SIDE

ADDITIONAL INFORMATION

FIELD COMMITTEE

☐ Endorses action taken by supervisor.

☐ Makes new or additional recommendations.

ACTION RECOMMENDED ______________________

Date ________________________ Signature ________________________

CENTRAL COMMITTEE

☐ Endorses action indicated above.

☐ Makes new or additional recommendations.

ACTION RECOMMENDED ______________________

Date ________________________ Signature ________________________

APPENDIX E

A CLIENT-BASED INSURANCE COMPANY SYSTEM

The following client-based system, developed by David Hansen and
Alan Josefsek of the Fred S. James Company of New York, Inc.,
uses a unique concept in obtaining code information for entry into
the data base. The accident report form is a two-sided document.
On the front upper half it merely serves as an accident report form.
However, on the lower half of the front and on the back it contains
the source codes. These codes are entered on the accident report
form in the appropriate sections. This makes the accident report
form the input form as well. Thus, the system has an all-purpose
document that is simple for the client company to use and is easy for
the insurance company to enter. Rechecking proper codes is quick,
because the "code book," so to speak, is right on the form. The
only limitation of the system is the size of the form, which limits the
number of available codes. This, however, is not that restrictive.
The system can generate excellent computer graphics reports
and regular printed reports for each client. The graphics reports
can highlight such conditions as cost analysis, cause analysis, numbers of cases, and recordable cases. The printed reports list multiple files for analysis and provide the system user with easy-to-under-
stand information.
Three samples of accident report/input and two samples of regular printed report forms follow.

Client-Based Insurance System

WC/LIABILITY ACCIDENT/INCIDENT INVESTIGATION REPORT

INSURANCE CLAIM #

PROPERTY	REGION	PROP. TYPE	DISTRICT MANAGER	NAME OF INJURED (Last, First, Mid. Init.)	#

CODE IF EMPLOYEE: POSITION TITLE	TYPE EMPLOYEE	YRS. EMP.	CODE AGE	SEX	MARRIAGE STATUS	DATE OF ACCIDENT	TIME (AM/PM)
Q-	R-		Y-			/ /	

INJURED — ADDRESS & PHONE #

CODE   LOCATION OF ACCIDENT (Code & Explain)
G-

CODE JOB ATTEMPTED	CODE ACTUAL BODY ACTIVITY	CODE TYPE OF ACCIDENT
S-	T-	U-

CODE AGENT INFLICTING THE INJURY OR DAMAGE	CODE PART OF BODY INJURED	CODE TYPE OF INJURY
V-	W-	X-

SELF INFL.	NON INFL.	FOR OFFICE USE ONLY ➤	INJURY EFFECT	DAYS LOST	

CODE   BASIC CAUSE (Code & Explain)   COST OF CLAIM
M-

CODE   CONTRIBUTORY CAUSES OR FACTORS (Code & Explain)   ENVIRONMENTAL COND.
M-   O-   P-

CORRECTIVE ACTION TAKEN AND OTHER RELEVANT COMMENTS (Use back of form as necessary)	ACCIDENT PREVENTABLE (Yes/No)
	WITNESSES INTERVIEWED (Yes/No/None)

WITNESSES — NAME & ADDRESS & PHONE #

DOES TENANT HAVE INSURANCE? _____ Yes _____ No   IF YES, PLEASE GIVE COMPANY, NUMBER, ETC.

LITIGATION STATUS:

SUIT 1ST NOTICE	REPRESENTATION LETTER REC'D	OTHER ATTORNEY NAME & ADDRESS
Yes or No	Yes or No	

HEARING SCHEDULED/DATE	SUMMONS REC'D	
Yes or No / /	Yes or No	

DEMAND	ATTORNEY ASSIGNED:	NAME AND ADDRESS:
$	_____ Yes _____ No	

DATE OF THIS REPORT	SIGNATURE

CODE Q

Q1   Resident Manager
Q2   Clerk
Q3   Maintenance Supervisor
Q4   Maintenance
Q5   Security Guard
Q6   Groundsman
Q7   Officer of Corporation
Q8   Regional Managers
Q9   District Managers
Q10 Vice President of Subsidiary
Q11 Department Managers
Q12 Assistant Controllers
Q13 Attorney
Q14 Accountant
Q15 Secretary
Q16 Word Processor
Q17 Accounts Payable Clerk
Q18 Clerk Typist
Q19 Data Control Clerk
Q20 Elevator Operator
Q21 Inside Sales
Q22 Outside Sales
Q23 Carpenter
Q24 Laborer
Q26 Third Party Inj.
Q99 Other (Specify)

TYPE EMPLOYEE CODE R

R1   Temporary
R2   Seasonal
R3   Probationary
R4   Permanent
R5   Resident–Elderly
R6   Resident–Handicap
R7   Resident–HUD
R8   Resident–Market Rate
R9   Non-Resident
R99 Other (Specify)

ACCIDENT LOCATION G

G1   Elevator
G2   Maintenance Room
G3   Penthouse
G4   Laundry Room
G5   Corridor/Hall
G6   Lobby
G7   Roof
G8   Common Spaces
G9   Ceiling Spaces
G10 Boiler Room
G11 Office
G12 Grounds
G13 Steps
G14 Living Room
G15 Dining Room
G16 Kitchen
G17 Bathroom
G18 Bedroom
G19 Vehicle
G20 Loading Area
G21 Trash Area
G22 Stairway
G23 Parking Lot
G24 Swimming Pool
G25 Playground
G26 Mail Room
G27 Rest Room
G28 Cafeteria
G29 Community Room
G30 Storage
G31 Xerox Area
G32 Coffee Area
G33 Garage
G34 Balcony
G35 Patio
G36 Curb
G37 Models
G38 Wall Spaces
G39 Trash Containers/Chutes
G78 More Than One Room
G99 Other (Specify)

ENVIRONMENTAL COND. O

O1   Daylight
O2   Dusk
O3   Dark
O4   Dark & Artificial Light
O99 Other (Specify)

CODE P

P1   Clear
P2   Fog
P3   Ice
P4   Rain
P5   Snow/Sleet
P99 Other

JOB ATTEMPTED S

S1   Driving Vehicle
S2   Handling Materials
S3   Maintenance–Building
S4   Maintenance–Electric
S5   Maintenance–Machinery
S6   Maintenance–Grounds
S7   Maintenance–Other
S8   Office Work
S9   Operating Equipment
S10 Painting
S11 Repair–Electric
S12 Repair–Machinery
S13 Repair–Plumbing
S14 Repair–Structural
S15 Repair–Other
S16 Riding Vehicle
S17 Mowing
S18 Hand Task
S19 Inspecting
S20 Assisting Resident
S21 Cleaning

S22	Hand Tool Use	U3	Animal Contact	U39	Fire—Friendly	**PART OF BODY W**		X9	Crushing
S23	Open/Close Door	U4	Bitten by People	U40	Fire—Hostile	W1	Ankles	X10	Cut
S24	Carrying	U5	Bitten by Insects	U99	Other (Specify)	W2	Arm	X11	Dislocation
S25	Emptying Trash	U6	Bitten by Animals			W3	Back	X12	Electric Shock
S26	Patrolling	U7	Body Reaction			W4	Eyes	X13	Faint
S27	Moving to/from apts.	U8	Burglary/Theft			W5	Face	X14	Fracture
S28	Domestic Task	U9	Caught By/Betw.	**INJURY AGENT V**		W6	Feet	X15	Hernia
S29	Playing	U10	Contact—Chem.	V1	Auto	W7	Fingers	X16	Infection
S30	Swimming	U11	Contact—Cold	V2	Chemical	W8	Hand	X17	Internal
S31	Enter/Leave Prem.	U12	Contact—Hot	V3	Dust	W9	Head	X18	Irritation
S99	Other (Specify)	U13	Contact—Object Falling	V4	Electrical Apparatus	W10	Internal	X19	Multiple
		U14	Corpse	V5	Equipment	W11	Legs	X20	Nausea
BODY ACTIVITY T		U15	Electrical Contact	V6	Hand Tools	W12	Neck	X21	Occup. Ill.
		U16	Explosion	V7	Machine	W13	Toes	X22	Puncture
T1	Riding	U17	Fall—Elevation	V8	Tractor	W14	Trunk	X23	Sprain
T2	Sitting	U18	Fall—Level	V9	Insect	W15	Wrist	X24	Sting
T3	Standing	U19	Fall—Ladder	V10	Other Person	W16	Multiple	X25	Strain
T4	Walking	U20	Fall—Stairs/Steps	V11	Animal	W17	Lungs	X26	Unknown
T5	Hand Task	U21	Fall—Vehicle	V13	Stairs	W18	Teeth	X27	Emotional
T6	Pulling	U22	Fall—Water	V14	Hammer	W19	Ears	X28	Heart Attack
T7	Pushing	U23	Fall—Other	V15	Work Material	W20	Groin	X29	Death
T8	Bending	U24	Hit and Run	V16	Fire	W21	Mouth	X99	Other
T9	Lifting	U25	Inhalation	V17	Door	W22	Elbow		
T10	Driving	U26	Ingestion	V18	Ice	W23	Chest	**AGE Y**	
T11	Kneeling	U27	Overexertion	V19	Smoke/Fume	W24	Nervous System	Y1	under 21
T12	Reaching	U28	Particle in Eye	V20	Truck	W99	Other (Specify)	Y2	21-25
T13	Carrying	U29	Poison	V21	Auto			Y3	26-30
T14	Lying Down	U30	Rape	V22	Golf Cart	**TYPE OF INJURY X**		Y4	31-35
T15	Climbing	U31	Slipping	V23	Work Surfaces	X1	Abrasion	Y5	36-40
T16	Running	U32	Solicitation	V24	Foreign Object	X2	Amputation	Y6	41-50
T17	Swinging	U33	Struck Against/By	V25	Falling Object	X3	Asphyxiation	Y7	51-60
T99	Other (Specify)	U34	Tripping	V26	Sharp Object	X4	Bite	Y8	61-70
		U35	Vandalism	V27	Glass	X5	Blister	Y9	over 70
ACCIDENT TYPE U		U36	Verbal Abuse	V28	Grass	X6	Bruise		
U1	Automotive	U37	Water Damage—Property (Patron)	V29	Water	X7	Burn		
U2	Burns	U38	Water Damage—Property (Owner)	V30	Defective Equipment	X8	Concussion		
				V99	Other (Specify)				

BASIC CAUSE AND CONTRIBUTING CAUSE (M)

M1	Unsafe Personal Factor — Bodily Effects	M24	Unsafe Act — Failure to Warn
M2	Unsafe Personal Factor — Disregard of Safety Rule	M25	Unsafe Act — Failure to Use Protective Equipment
M3	Unsafe Personal Factor — Fail to Report Accident	M26	Unsafe Act — Horseplay
M4	Unsafe Personal Factor — Improper Attitude	M27	Unsafe Act — Improper Method
M5	Unsafe Personal Factor — Inattention Fellow Workers	M28	Unsafe Act — Operating without Authority
M6	Unsafe Personal Factor — Inattention — Injured	M29	Unsafe Act — Safety Device Made Inoperative
M7	Unsafe Personal Factor — Inexperience	M30	Unsafe Act — Unsafe Loading
M8	Unsafe Personal Factor — Lack of Instructions	M31	Unsafe Act — Unsafe Posture or Position
M9	Unsafe Personal Factor — Lack of Knowledge or Skill	M32	Unsafe Act — Unsafe Use of Agent
M10	Personal Factor — Other (Explain)	M33	Unsafe Act — Using Unsafe Equipment
		M34	Unsafe Act — Using Unsafe Speed
		M35	Unsafe Act — Working on Moving Equipment
M13	Hazardous Condition — Congested Area	M37	Unsafe Act — Other (Explain)
M14	Hazardous Condition — Defect of Agent		
M15	Hazardous Condition — Defective Equipment		
M16	Hazardous Condition — Housekeeping	M38	Patron Action — Vehicular
M17	Hazardous Condition — Improper Design	M39	Patron Action — Horseplay
M18	Hazardous Condition — Improper Dress	M40	Patron Action — Other (Explain)
M19	Hazardous Condition — Improper Illumination		
M20	Hazardous Condition — Slick Surface		
M21	Hazardous Condition — Improperly Guarded	M41	Patron Action — Carelessness
M22	Hazardous Condition — Inclement Weather	M42	Patron Action — Improper Housekeeping
M23	Hazardous Condition — Other (Explain)	M99	Other (Specify)

If the above codes do not apply, a detailed explanation of the suspected causes
must be included in the space provided. The *"other relevant comments"* section
and the space below should be used when additional writing space is needed.

____

____

____

____

____

____

____

Please complete and mail as follows: 1. White orig. to: Insurance Department

                                                                   Attn: Claims Administrator
                                              2. Yellow copy to: Regional Office
                                              3. Green copy to: Site File

AUTOMOBILE ACCIDENT/INCIDENT INVESTIGATION REPORT

INSURANCE CLAIM #

DRIVER LIC. #

PROPERTY	REGION	PROP. TYPE	DISTRICT MANAGER	NAME OF DRIVER (Last, First, Mid. Init.)

CODE	IF EMPLOYEE; TITLE	CODE	TYPE EMPLOYEE	YRS. EMP.	CODE	AGE	SEX	SOCIAL SEC. #	DATE OF ACCIDENT	TIME (AM/PM)
Q-		R-			Y-					

CODE	LOCATION OF ACCIDENT (Code & Explain)	CODE	CODE	ENVIRONMENTAL CONDITIONS	ADDRESS OF DRIVER & PHONE #
G-		O-	P-		

POLICE NOTIFIED   Yes   No   VEHICLE TYPE   SERIAL #   LIC. PLATE #   CODE TYPE OF ACCIDENT   U-

CODE	AGENT INFLICTING THE INJURY OR DAMAGE	CODE	IF BODILY INJURY, PART OF BODY INJURED	CODE	TYPE OF INJURY
V-		W-		X-	

OWNER CAUSED   3RD PARTY CAUSED   FOR OFFICE USE ONLY ➤   INJURY EFFECT

CODE	DESCRIBE INCIDENT / BASIC CAUSE (Code & Explain)	CODE	TRAFFIC CONTROL
M-		AA-	

CODE	CONTRIBUTORY CAUSES OR FACTORS (Code & Explain)	COST OF CLAIM	CODE	ACTIVITY INVOLVED
N-			BB-	

CORRECTION ACTION TAKEN AND OTHER RELEVANT COMMENTS	TYPE OF DAMAGE	ACCIDENT PREVENTABLE (Yes/No)
	CODE Z-	WITNESSES INTERVIEWED (Yes/No/None)

WITNESSES NAME & ADDRESS & PHONE #

OTHER DRIVER NAME, ADDRESS, PHONE #, LIC. #   DATE OF THIS REPORT   SIGNATURE

OWNER OF OTHER VEH. IF DIFFERENT FROM DRIVER

OTHER VEH. TYPE	SERIAL #	LIC. PLATE #

DESCRIBE DAMAGE TO VEH. NO. 1	CHECK ONE OF THE 8 DIAGRAMS BELOW IF IT ADEQUATELY DESCRIBES THE ACCIDENT OR DRAW YOUR OWN DIAGRAM IN THE SPACE TO THE RIGHT ➤	ACCIDENT DIAGRAM	NUMBER THE VEHICLES OUR VEHICLE IS NO. 1	DESCRIBE DAMAGE TO VEH. NO. 2

Indicate north

REAR END   1   RIGHT TURN   5
OVERTAKING   2   RIGHT TURN   6
LEFT TURN   3   HEAD ON   7
INTERSECTION   4   SIDE SWIPE   8   9

OTHER INSURANCE: IF YES, GIVE COMPANY, NUMBER, ETC.   Please complete and mail as follows:

Yes   No

INJURED PARTIES NAME(S), ADDRESSES, PHONE #'S:

1.
2.
3.
4.

1. White orig. to: Insurance Department

       Attn: Claims Administrator
2. Yellow copy to: Regional Office
3. Green copy to: Site File

ACCIDENT LOCATION G

Other Area

G40 Driveway
G41 Parking Lot/Property
G42 Parking Lot/Public
G43 Garage
G44 Property Grounds
G45 Other (Specify)

Highway

G46 Bridge/Overpass
G47 Curve
G48 Entrance/Exit
G49 Hill
G50 Intersection
G51 Off Roadway
G52 Straight & Level
G53 Other (Specify)

Metropolitan Road

G54 Bridge/Overpass
G55 Curve
G56 Entrance/Exit
G57 Hill
G58 Intersection
G59 Off Roadway
G60 Straight & Level
G61 Other (Specify)

Suburban Road

G62 Bridge/Overpass
G63 Curve
G64 Entrance/Exit
G65 Hill
G66 Intersection
G67 Off Roadway
G68 Straight & Level
G69 Other (Specify)

Rural Road

G70 Bridge/Overpass
G71 Curve
G72 Entrance/Exit
G73 Hill
G74 Intersection
G75 Off Roadway
G76 Straight & Level
G77 Other (Specify)
G99 Other

TRAFFIC CONTROL AA

AA1   None
AA2   Traffic Signal
AA3   Stop Sign
AA4   Flashing Light
AA5   Yield Sign
AA6   Police Officer
AA7   Flagman
AA8   No Passing Zone
AA9   School Zone
AA10 RR Crossing
AA99 Other (Specify)

BASIC CAUSE & CONTRIBUTING CAUSE M

M43 Vehicle Action—Vehicle
M44 Vehicle Action—Golf Cart
M45 Vehicle Action—Truck
M46 Vehicle Action—Motorcycle
M47 Vehicle Action—Other
M48 Vehicle Action—Wreckless Driving
M49 Vehicle Action—Driving While Under Influence
M50 Vehicle Action—Authorized Use
M51 Vehicle Action—Unauthorized Use
M52 Vehicle Action—Other Vehicle at Fault
M53 Vehicle Action—Owner Vehicle at Fault
M54 Vehicle Action—Theft
M55 Vehicle Action—Fire
M56 Vehicle Action—Vandalism
M57 Vehicle Action—Property Damage
M99 Other (Specify)

ENVIRONMENTAL CONDITIONS P

P1   Clear
P2   Fog
P3   Ice
P4   Rain
P5   Snow/Sleet
P99 Other (Specify)

ENVIRONMENTAL CONDITIONS O

O1   Daylight
O2   Dusk
O3   Dark
O4   Dark & Artificial Light
O99 Other (Specify)

POSITION TITLE Q

Q1   Resident Manager
Q2   Clerk
Q3   Maintenance Supervisor
Q4   Maintenance
Q5   Security Guard
Q6   Groundsman
Q7   Officer of Corporation
Q8   Regional Managers
Q9   District Managers
Q10 Vice President of Subsidiary
Q11 Department Managers
Q12 Assistant Controllers
Q13 Attorney
Q14 Accountant
Q15 Secretary
Q16 Word Processor
Q17 Accounts Payable Clerk
Q18 Clerk Typist
Q19 Data Control Clerk
Q20 Elevator Operator
Q21 Inside Sales
Q22 Outside Sales
Q23 Carpenter
Q24 Laborer
Q25 Other Aut. Driver
Q99 Other (Specify)

TYPE EMPLOYEE R

R1   Temporary
R2   Seasonal
R3   Probationary
R4   Permanent
R99 Other (Specify)

ACCIDENT TYPE U

U1   Collision
U2   Collision—Property Damage
U3   Collision—Bodily Injury
U4   Collision—BI/PD
U5   Collision—Multiple
U6   Struck by Animal
U7   Struck by Others
U8   Hit and Run
U9   Glass
U10 Non-owned/Hired
U11 Theft
U12 Fire
U13 Vandalism
U14 Flood
U99 Other (Specify)

INJURY AGENT V

V1   Wet Surface
V2   Drugs
V3   Alcohol
V4   Other People
V5   Fire
V6   Vandalism
V7   Theft
V8   Inclement Weather
V9   Other Vehicle
V10 Wreckless Driving
V11 Unauthorized Driver
V12 Animal
V13 Mechanical Failure
V14 Negligence
V15 Obstruction
V00 Other (Specify)

PART OF BODY W

W1   Ankles
W2   Arm
W3   Back
W4   Eyes
W5   Face
W6   Feet
W7   Fingers
W8   Hand
W9   Head
W10 Internal
W11 Legs
W12 Neck
W13 Toes
W14 Trunk
W15 Wrist
W16 Multiple
W17 Lungs
W18 Teeth
W19 Ears
W20 Groin
W21 Mouth
W22 Nervous System
W23 Elbow
W24 Chest
W99 Other (Specify)

TYPE OF INJURY X

X1   Abrasion
X2   Amputation
X3   Asphyxiation
X4   Blister
X5   Bruise
X6   Burn
X7   Concussion
X8   Crushing
X9   Cut
X10 Dislocation
X11 Electric Shock
X12 Faint
X13 Fracture
X14 Hernia
X15 Infection
X16 Internal
X17 Irritation
X18 Multiple
X19 Nausea
X20 Puncture
X21 Sprain
X22 Sting
X23 Strain
X24 Unknown
X99 Other

AGE Y

Y1   Under 21
Y2   21-25
Y3   26-30
Y4   31-35
Y5   36-40
Y6   41-50
Y7   51-60
Y8   61-70
Y9   Over 70

PROPERTY DAMAGE Z

Z1   Left Fender
Z2   Left Rear Door
Z3   Left Front Door
Z4   Left Tires
Z5   Panels
Z6   Right Fender
Z7   Right Rear Door
Z8   Right Front Door
Z9   Right Tires
Z10 Right Panels
Z11 Accessories
Z12 Interior
Z13 Front Bumpers
Z14 Front Grill
Z15 Front End
Z16 Rear Bumpers
Z17 Rear End
Z18 Total Loss
Z19 Trunk
Z20 Hood
Z21 Roof
Z22 Another Vehicle
Z23 Buildings
Z24 Trees/Shrubs
Z25 Fence
Z26 Signs
Z99 Other

ACTIVITY INVOLVED BB

BB1   Delivery
BB2   Installation
BB3   Personal Use
BB4   Sales Call
BB5   Service Call
BB6   Business Errand
BB99 Other (Specify)

PROPERTY LOSS INVESTIGATIVE REPORT

JAMES MANAGEMENT ACCIDENT PROFILING SYSTEM

INSURANCE CLAIM #

NUMBER

APT #

PROPERTY NAME | REGION | SITE MANAGER | DISTRICT MANAGER | REGIONAL MANAGER

CODE PRIMARY LOSS **A-** | CODE ADD'L LOSS—PROPERTY **B-** | CODE ADD'L LOSS—CASUALTY **C-** | CODE CONSEQUENTIAL LOSS **D-**

DATE OF ACCIDENT | TIME (AM/PM) | CODE TYPE OCCUPANCY **E-** | CODE CONTRIBUTING FACTORS **F-** | CODE LOCATION OF OCCURRENCE **G-** | CODE PROPERTY TYPE **H-**

CODE TYPE OF PROTECTION **K-** | CODE PERIL **L-** | CODE ORIGIN/CAUSE **M-** | CODE CONSTRUCTION **I-** | CODE LOCATION DESCRIPTION **J-**

SUBROGATION | OTHER INSURANCE | INSURANCE CO. & POLICY # | CODE **O-** | CODE **P-** ENVIRONMENTAL CONDITIONS | MONTHLY RENT $ | EST. COST, REPAIRS $

WITNESS NAME, ADDRESS, PHONE | WITNESSES INTERVIEWED (Yes/No/None) | ACCIDENT AVOIDABLE (Yes/No)

ACTION TAKEN/COMMENTS | DATE OF THIS REPORT | REVISION DATE

INVESTIGATING SUPERVISORS SIGNATURE

PRIMARY LOSS A

A1   Boiler Machinery
A2   Builders Risk Materials.
A3   Builders Risk—Real Property
A4   Business Personal Property
A5   Loss of Rents
A6   Real Property
A7   Third Party Pd
A99   Other (Specify)

ADDITIONAL LOSS (PROPERTY) B

B1   Boiler Machinery
B2   Builders Risk Materials
B3   Builders Risk—Real Property
B4   Business Personal Property
B5   Loss of Rents
B6   Real Property
B7   Third Party Pd
B99   Other (Specify)

ADDITIONAL LOSS (Casualty) C

C1   Bodily Injury
C2   Property Damage
C3   Personal Injury

CONSEQUENTIAL LOSS D

D1   Boiler Machinery
D2   Builders Risk Materials
D3   Builders Risk—Real Property
D4   Business Personal Property
D5   Loss of Rents
D6   Real Property
D7   Third Party Pd
D99   Other (Specify)

TYPE OCCUPANCY E

E1   Elderly
E2   Handicapped
E3   HUD
E4   Market Rate
E5   Mixed Family
E99   Other

CONTRIBUTING FACTORS F

F1   Faulty Electrical Equipment
F2   Lightning
F3   Arson
F4   Friction
F5   Matches/smoking
F6   Spontaneous Combustion
F7   Failed Pressure Vessel or Valve
F8   Other Mechanical or Material Failure
F9   Faulty Installation
F10   Windstorm
F11   Sand or Dust Storm
F12   Hurricane
F13   Tornado
F14   Rain Storm
F15   Wave Wash

F16   Flood
F17   Snow Storm
F18   Vandalism
F19   Riot/Insurrection
F20   Insect/vermin
F21   Faulty Maintenance
F22   Unprotected access
F23   Combustible Construction
F24   Storage Conditions
F25   Flammable or Combustible Liquids
F26   Flammable or Combustible Interior Finish or Trim
F27   Failure of Protective Systems or Equipment
F28   Inadequate Fire Divisions
F99   Other (Specify)

LOCATION OF OCCURRENCE G

G1   Elevator
G2   Maintenance Room
G3   Penthouse
G4   Laundry Room
G5   Corridor/Hall
G6   Lobby
G7   Roof
G8   Common Spaces
G9   Ceiling Spaces
G10   Boiler Room
G11   Office
G12   Grounds
G13   Steps
G14   Living Room
G15   Dining Room
G16   Kitchen
G17   Bathroom
G18   Bedroom
G19   Vehicle
G20   Loading Area
G21   Trash Area
G22   Stairway
G23   Parking Lot
G24   Swimming Pool
G25   Playground
G26   Mail Room
G27   Rest Room
G28   Cafeteria
G29   Community Room
G30   Storage
G31   Xerox Area
G32   Coffee Area
G33   Garage
G34   Balcony
G35   Patio
G36   Curb
G37   Models
G38   Wall Spaces
G39   Trash Containers/Chutes
G78   More Than One Room
G99   Other (Specify)

PROPERTY TYPE H

H1   Garden Apartments
H2   High Rise
H3   Low Rise
H4   Mid Rise
H5   Townhouse
H6   Detached House
H7   Office Building
H99   Other (Specify)

CONSTRUCTION I

I1   Frame 100%
I2   Frame with Masonry Veneer
I3   Masonry with Wood Floors and Roof
I4   Non-combustible Masonry or Concrete with Exposed Steel
I5   Fire Resistive Concrete/Masonry. No Exposed Steel
I6   Mixed
I99   Other (Specify)

LOCATION DESCRIPTION J

J1   Urban
J2   Suburban
J3   Rural
J99   Other (Specify)

TYPE OF PROTECTION K

K1   Burglar Alarm
K2   Evacuation Alarm (Local)
K3   Fire Alarm Central Station
K4   Fire Alarm (Local)
K5   Individual Smoke Detectors
K6   Smoke Detectors and Alarm
K7   Sprinklers—Wet System
K8   Extinguishers—Water
K9   Other Extinguishing System
K10   Attended Front Desk
K11   Security Guard—Unarmed
K12   Standpipe and Hose
K13   None
K14   Fire Alarm—Manual
K15   Heat Detection
K16   Extinguishers—Chemical
K17   Security Guard—Armed
K18   Unattended Front Desk Sign In/Out Sheet
K19   Sprinklers—Dry System
K20   Outside Hydrant
K99   Other (Specify)

PERIL L

- L1 Fire—Friendly
- L2 Smoke
- L3 Fire and Smoke
- L4 Wind
- L5 Ice
- L6 Water
- L7 Third Party
- L8 Subsidence
- L9 Land/Mud Slide
- L10 Aircraft/Vehicle Collision
- L11 Explosion
- L12 Collapse
- L13 Implosion
- L14 Act of God
- L15 Flood
- L16 Theft
- L17 Robbery
- L18 Mysterious Disappearance
- L20 Fire—Hostile
- L21 Earthquake
- L22 Lightning
- L23 Freezing
- L24 Breakage
- L25 Civil Commotion
- L26 War/Riot
- L27 Vandalism/Malicious Mischief
- L28 Falling Objects
- L29 Spillage
- L30 Wear & Tear
- L99 Other (Specify)

ENVIRONMENTAL COND. O

- O1 Daylight
- O2 Dusk
- O3 Dark
- O4 Dark & Artificial Light
- O99 Other (Specify)

CODE P

- P1 Clear
- P2 Fog
- P3 Ice
- P4 Rain
- P5 Snow/Sleet
- P99 Other

ORIGIN/CAUSE M

- M57 Faulty Electrical Equipment
- M58 Lightning
- M59 Arson
- M60 Friction
- M61 Matches/Smoking
- M62 Spontaneous Combustion
- M63 Failed Pressure Vessel or Valve
- M64 Other Mechanical or Material Failure
- M65 Faulty Installation
- M66 Windstorm
- M67 Sand or Dust Storm
- M68 Hurricane
- M69 Tornado
- M70 Rain Storm
- M71 Wave Wash
- M72 Flood
- M73 Snow Storm
- M74 Vandalism
- M75 Riot/Insurrection

- M76 Insect/Vermin
- M77 Faulty Maintenance
- M78 Unprotected Access
- M79 Smoking in Bed
- M80 Attractive Nuisance
- M81 Hail
- M82 Sleet
- M83 Larceny
- M84 Burglary
- M85 Sink Hole
- M86 Erosion
- M87 Unknown
- M88 Vehicle
- M89 Wear & Tear
- M90 Extreme Hot
- M91 Extreme Cold
- M92 Housekeeping
- M99 Other (Specify)

If the above codes do not apply, a detailed explanation of the suspected causes must be included in the space provided. The *"other relevant comments"* section and the space below should be used when additional writing space is needed.

Please complete and mail as follows:

1. White orig. to:

2. Yellow copy to: Regional Office
3. Green copy to: Site File

JAMES - MANAGEMENT ACCIDENT PROFILING SYSTEM
ABC COMPANY
SAMPLE REPORT
BASIC CAUSE/JOB ATTEMPTED/INJURY AGENT
PERIOD ANALYZED: 1/1/81-12/31/81   VALUED: 12/31/81

BASIC CAUSE	JOB ATTEMPTED	INJURY AGENT	NUMBER OF ACCIDENTS	PERCENT OF TOTAL NUMBER
HAZARDOUS CONDITION-IMPROPER DESIGN-HAZARD	OTHER (SPECIFY)	EQUIPMENT	1	0.44
	TOLL COLLECTION-OTHER	EQUIPMENT	1	0.44
*TOTAL HAZARDOUS CONDITION-IMPROPER DESIGN-HAZARD			5	2.20
HAZARDOUS CONDITION-IMPROPER ILLUMINATION	INSPECTING/POLICING	OTHER (SPECIFY)	1	0.44
HAZARDOUS CONDITION-INCLEMENT WEATHER	MAINT.-ROAD	OTHER (SPECIFY)	1	0.44
	OTHER (SPECIFY)	EQUIPMENT	1	0.44
*TOTAL HAZARDOUS CONDITION-INCLEMENT WEATHER			2	0.86
HAZARDOUS CONDITION-NO UNSAFE CONDITION	CLEANING	TRUCK	1	0.44
	HANDLING MATERIALS	EQUIPMENT	1	0.44
	MOVING TO/FROM LOC.	WORK SURFACES	1	0.44
	OFFICE WORK	OTHER (SPECIFY)	1	0.44
	OPERATING TOLL-ENTRY	OTHER (SPECIFY)	1	0.44
		WORK SURFACES	1	0.44
*TOTAL OPERATING TOLL-ENTRY			2	0.88
	OTHER (SPECIFY)	OTHER (SPECIFY)	1	0.44
	REPAIR-ELECTRIC	TRUCK	1	0.44
	REPAIR-MACHINERY	WORKPIECE	1	0.44
	REPAIR-VEHICLE	OTHER (SPECIFY)	1	0.44
*TOTAL HAZARDOUS CONDITION-NO UNSAFE CONDITION			10	4.41
HAZARDOUS CONDITION-OTHER (EXPLAIN)	HANDLING MATERIALS	TRUCK	1	0.44
	LANE CLOSING/OPEN	WORK SURFACES	1	0.44
	OPERATING EQUIP.	OTHER (SPECIFY)	1	0.44
	OPERATING TOLL-EXIT	OTHER (SPECIFY)	1	0.44
*TOTAL HAZARDOUS CONDITION-OTHER (EXPLAIN)			4	1.78

```
              JAMES - MANAGEMENT ACCIDENT PROFILING SYSTEM
                            ABC COMPANY
                            SAMPLE REPORT
              JOB ATTEMPTED/BODY ACTIVITY/INJURY AGENT ANALYSIS
              PERIOD ANALYZED: 1/1/81-12/31/81    VALUED: 12/31/81
```

JOB ATTEMPTED	ACTIVITY	INJURY AGENT	NUMBER OF ACCIDENTS	PERCENT OF TOTAL NUMBER
HANDLING MATERIALS	LIFTING	EQUIPMENT	3	1.32
		FIRE HAND TOOLS	1	0.44
		OTHER (SPECIFY)	5	2.20
		TRUCK	1	0.44
		WORKPIECE	5	2.20
*TOTAL LIFTING			18	7.93
	OTHER (SPECIFY)	DUST	1	0.44
	REACHING	EQUIPMENT	1	0.44
		OTHER (SPECIFY)	1	0.44
		TRUCK	1	0.44
		WORKPIECE	1	0.44
*TOTAL REACHING			4	1.76
	STANDING	FIRE HAND TOOLS	1	0.44
	WALKING	DOOR	1	0.44
		OTHER (SPECIFY)	2	0.88
		WORK SURFACES	3	1.32
*TOTAL WALKING			6	2.64
*TOTAL HANDLING MATERIALS			32	14.10
HANDTOOL USE	LIFTING	FIRE HAND TOOLS	1	0.44
	REACHING	WORK SURFACES	1	0.44
	SWINGING	HAMMER	1	0.44
*TOTAL HANDTOOL USE			3	1.32
INSPECTING/POLICING	LIFTING	DUST	1	0.44
		OTHER (SPECIFY)	3	1.32
*TOTAL LIFTING			4	1.76
	OTHER (SPECIFY)	OTHER (SPECIFY)	1	0.44
		WORK SURFACES	1	0.44
*TOTAL OTHER (SPECIFY)			2	0.88
	STANDING	WORK SURFACES	1	0.44

APPENDIX F

A PAPER COMPANY'S SAFETY DATA SYSTEM

The following system was designed by Charles F. Vejvoda, Safety
Director, Container Division, Westvaco Corp., for use on a Wang
expanded word-processor machine. It uses an investigation report
form called an Unintended Event (U.E.) Flash Report. The system
is symbolized by U.E., the character adorning the form. There are
posters, depicting U.E. having various injuries, that are to be used
as reminders to report non-accident events and to identify places
where such events have occurred. A sample U.E. character is shown
on the following page.

This system can produce the OSHA Form 200 and reports that
indicate types of injuries, locations, and whether or not prompt corrective action was taken by plant management. For such a small system it has proved quite effective in pinpointing areas for more training
and in measuring management's responsiveness. A copy of the system's
Unintended Event (U.E.) Flash Report, an input document, and some
U.E. poster samples follow.

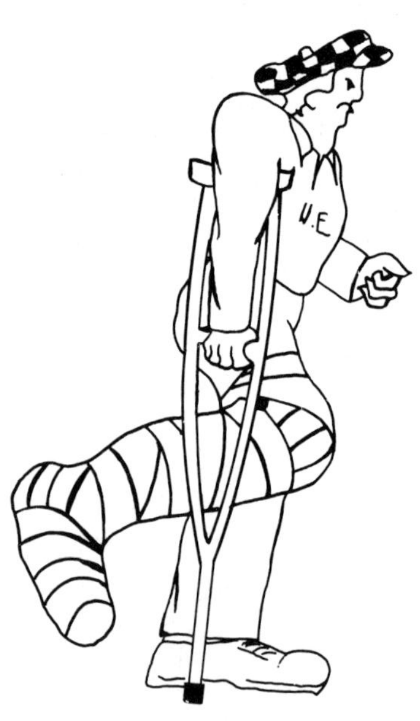

Report #_____________

UNINTENDED EVENT (U.E.)
FLASH REPORT

LOCATION_____________________________PLANT#__________

1. When did Event Occur?____/____/____________AM/PM
2. Where did Event Occur?_____________________________

3. Type of Unintended Event (U.E.) Check appropriate box(es)
     A. [ ] Injury - [see reverse side]
   (OSHA definitions) 1.[ ] Lost Workday
                       2.[ ] No Lost Workday   B. [ ] Near Miss
                         a. [ ] Medical Treatment
                         b. [ ] Light duty
                         c. [ ] First Aid
     C. [ ] Damage - [ ] Machinery   D. [ ] An Unsafe Condition(s)
                        [ ] Building   or Act, which caused
                        [ ] Product   or could cause:
                        [ ] Other   [ ] Injury [ ] Damage
                                                [ ] A slowdown or
                                                    production stoppage

COMPLETE AS MUCH AS POSSIBLE:

4. Who got hurt?______________________or Who could get hurt?______________
5. Who treated the injury?_______________ Part of body injured___________
    Describe treatment ____
    ____

6. What got damaged?____________________or What could get damaged?_________
7. Who fixed the damage?________________ What did they do?________________
    ____

8. How did it happen?___________________or what could have happened?_______
    ____
    ____*
    ____

9. Has this event occurred before?   Why did it happen (again)?________
    [ ] Yes, when?____/____/____ [ ] No
                                          ____
    ____

    SAFETY SUGGESTION:_____________________ How can we keep it from happening?______
    ____
    ____Put into effect by: / /

    SUBMITTED BY________________________   DEPARTMENT:_______________________
              [Signature is optional if B. C. or D. is checked above]

    DEPARTMENT SUPERVISOR   FUNCTIONAL MANAGER   GENERAL MANAGER

    Date / /   Date / /   Date / /

    Copies To: Division Manager   Regional Manager
                Division Safety Manager
                Employee Personnel file**
    [within 24 hrs. if 3A1, 3A2a or 3A2b, is checked   Date / /
    if 3A2c, B, C or D is checked, local plant
    policy on distribution applies.]

*   Space available for additional information on reverse side.
** This form can be used to document supervisor safety talks with employees.

11-80

If the following information is filled in, this document will comply with the
information requirements of the OSHA 101 form.

About the injured employee:

Social security #_________________, Age_____, Sex_______
Home address____

Occupation____

Name and address of Physician, Clinic or Hospital where
injured employee received treatment:

____
____
____
____

The following information will help in the administration of Worker Compensation.

When was the injury reported?____/____/____   Time_____:____AM/PM
Who reported the injury?_________________________________
Who was the injury reported to?__________________________

Additional Information

____
____
____
____
____
____
____

Document # 1587A
Page # 1
Go to last line of page (up to 10 screen loads)

While pressing the shift key, press command and then the underline
key

Press insert   (INPUT DOCUMENT - U.E. Flash Report # 11-80
Entering the following in:
Column #

1. _________Plant #
2. _________Log #
3. _________Date XX/XX (no year)
4. _________Time (24hr. clock - add 12 to PM hours)
5. _______________Name of injured
6. _________Occupation
7. _________Department
8. _________Supervisor
9. _______________Description of injury
10. ____Nature of injury
11. ____Part of the body injured
12. ____Source of injury
13. ____Accident type
14. ____Hazardous condition
15. ____Unsafe act
16. ____enter an X
17. ____enter an X
18. ____enter # ____
19. ____enter # ____
20. ____enter an X
21. ____enter a ____
22. ____enter an R

APPENDIX G

INVESTIGATION FORMS

The investigation forms contained in this Appendix are examples of different approaches and concepts in form design. Some are used with data systems; others are from manual systems. One combines injury information with fleet accident data, while another stresses personal injury and property damage equally. Each of these forms offers ideas and exhibits subtle differences that can contribute to the custom design of a particular company's investigation form.

The following forms have been reproduced with the permission of (listing of companies in order).

Form Idnt.	Company
NER -4	Mobil Oil Corporation, New York, NY
W9-2309	Westvāco Corporation, New York, NY
ABC-2361	American Broadcasting Company, New York, NY
RF 2	Construction Safety Association of Ontario, Toronto, Canada
Form #70	Westvāco Corporation, New York, NY
916-090-03	Schering-Plough Corporation, Kenilworth, NJ
Form 8526-4	Pfizer Corporation, New York, NY
No number	Schering-Plough Corporation, Kenilworth, NJ
CLV-1359-C	Westvāco Corporation, New York, NY

NORTHEAST REGION
SUPERVISORS INVESTIGATION REPORT

Todays Date _______________

Submit within 72 hrs. of
incident

M/V Collision (CO 735) ☐	Contamination of Spill (CO 735) ☐	Industrial Incident (CO 889) ☐	Other ☐
Minor 1st Aid Injury (Comp. Rpt.) ☐	Personal Injury (Dr. Treatment) (Comp. Rpt., CO 8915, Log on OSH 100) ☐	Lost Time Injury Comp. Rpt., CO 3915, Log on OSH100 ☐	

1. Terminal Plant or District _____________________ Location of Incident _______________________________

2. Employee ______________________ Age ______ Date Employed ________________ Job Title ________________
    Time on present job ___________________ Name of Immediate Supervisor _______________________________

3. Date of Collision/Injury/Incident ___________Time ________ # of hrs. employee worked prior to incident ___________. If injury, was
    employee doing regularly assigned work? ________ if not explain _______________________ Days off ____________

INJURY EXPERIENCE PREVIOUS 24 MONTHS

4. Date of Injury	NLT	LT	Days Lost	Type of Incident	Disciplinary Action
______________	______	______	__________	_________________	__________________
______________	______	______	__________	_________________	__________________
______________	______	______	__________	_________________	__________________

5. If an injury —
     (a)   What safeguards were provided (Guards, hard hats, goggles, etc.)? _________________________________
     (b)   Was employee using them? _______ If not explain ____
     (c)   What equipment, machine, tool or object was closely associated with the injury? ___________________
     (d)   What was lacking, defective or unsafe about (c) or the work area or equipment which caused the injury? ___________________
     (e)   Was anyone's negligence a contributing factor in this injury? ______ if so please explain _______________
     (f)   How could the injury have been prevented? ____
     (g)   How can the area, operation or equipment be made safer? _______________________________________
     (h)   What could the employees do to avoid a similar injury? ____
     (i)   What could supervision do to prevent another similar injury? ____________________________________
     (j)   Has action been taken already?________ if so what? ____
     (k)   If lost time injury estimate costs involved $ ______________

DRIVERS M/V RECORD PREVIOUS 24 MONTHS

6. Date of Collision	Avoidable/Unavoidable	Type of Collision	Disciplinary Action
______________	_______________________	________________	________________
______________	_______________________	________________	________________
______________	_______________________	________________	________________

7. If M/V Collision (a) Investigation revealed the following facts ____

     (b)   Based on the above facts the incident is judged (1) Avoidable ☐   (2) Unavoidable ☐
     (c)   Could the collision have been prevented?_____ if so explain ____________________________________
     (d)   Was the tacograph information analysed? ______________

8. Complete for all reports.
     (a)   Did the supervisor visit the scene? ______ if not why? ____
     (b)   Are photos available? ________________
     (c)   Did employee report the incident at his earliest opportunity? ___________________________________
     (d)   Was disciplinary action taken? ________ if so what? ____
     (e)   Additional comments: ____

        Date _______________ Supervisor _______________________________
     (f)   Comments of Facility Manager ____

        Date _______________ Manager _________________________________

    This report is to accompany all incident reports as noted above.
Distribution: (White) Field Op/Plant Mgr.   (Blue) Safety   (Pink) Claims   (Green) M/V Super   (Yellow) File

NER-4 April 73

SAFETY INVESTIGATION REPORT

(MUST BE COMPLETED WITHIN 2 DAYS OF OCCURRENCE)

	CASE #

OFFICE USE ONLY

LT ☐ DL ______
MT ☐ DR ______
FA ☐ ILL ☐
PD ☐
NM ☐
HC ☐
AG ______
LS ______
OCC ______

Report at Least: 1. All accidents which cause medical treatment injuries and/or $500.00 property damage.
2. All near misses and/or hazardous conditions that we should think about.

Accident: An event in an activity that results in injury and/or property/equipment damage.

Near Miss: An event that could, under different circumstances result in injury and/or property/ equipment damage.

Hazardous Condition: Conditions, circumstances or situations that have the potential of causing injury and/or property/ equipment damage.

///// **THIS SECTION TO BE COMPLETED BY PERSON REPORTING** /////

ORIGINATOR

Type of event/condition check applicable box(s): ☐ Injury ☐ Property/Equipment damage ☐ Near Miss ☐ Hazardous Condition ☐ Other (Specify) ________________

Name of Person: Reporting ☐ Injured ☐ | Date of Occurrence S M T W T F S | Time | Location of Occurrence | Supervisor's Name

Describe what happened. As appropriate, list type of injury, property damage and story of accident or event.

Immediate action taken to control the hazard. What would prevent it from happening again?

SIGNATURE

///// **THIS SECTION TO BE COMPLETED BY SUPERVISOR** /////

SUPERVISOR

What would prevent it from happening again? Identify, Severity and Frequency Potential. (See Instructions on Reverse Side)

FOR NEXT SIMILAR EVENT

SEVERITY POTENTIAL	FREQUENCY POTENTIAL
☐ Major	☐ Frequent
☐ Serious	☐ Occasional
☐ Minor	☐ Rare

SIGNATURE | DATE

For Injury Treatment: Doctor's Name/Hospital

///// **THIS SECTION TO BE COMPLETED BY SUPERINTENDENT** /////

SUPERINTENDENT

CAUSES: (Check all that apply)   (See Reverse Side For Definitions)

DEFICIENCY OF: ☐ Execution ☐ Knowledge ☐ Design ☐ Procedure ☐ Equipment ☐ Materials ☐ Facilities
☐ Contractors ☐ Other (Specify) ________________

LOSS ANALYSIS: Estimated Actual Loss $ ____________ Estimated Potential Loss $ ____________

Explanation of Estimate

What additional action is planned or should be done to minimize/eliminate this or similar events? Include action steps, responsibilities, completion dates. (See Reverse For Instructions)

SIGNATURE | DATE

DISTRIBUTE COPIES TO: (Note — Originator Keeps Bottom Copy)

TOP — Originator ——→ Supervisor ——→ Superintendent ——→ Production Manager
Prepares 5 —————→ Location Safety
or more Copies ——→ Supervisor ——→ Originator
—————→ File
—————→ Other Addressees as Appropriate

BOTTOM — Originator (Keeps)

W9-2309 Rev. 2/81

FORM COMPLETION INSTRUCTIONS

SUPERVISOR:

Following the investigation of the event, the supervisor should consider what changes in methods, procedures, facilities or equipment will minimize or eliminate the causes of the event. In addition, consideration could be given to similar operations and practices which may be subject to the same problem and provide suggestions as practical.

To complete the boxes marked Frequency Potential and Severity Potential the following definitions apply.

Severity Potential — The most probable result of the next similar event or hazardous condition.

Major — An amputation, fatality or damage greater than $75,000.

Serious — A medical treatment or lost time injury or damage from $1,000 to $75,000.

Minor — First aid or damage less than $1,000.

Frequency Potential — The expected number of times this event will occur.

Frequent — Once per week to twice per month.

Occasional — Once per month to twice per year.

Rare — Has been known to occur.

SUPERINTENDENT:

1. **Deficiencies:** When reviewing the report, consider the causes of how this event occurred and identify the appropriate deficiency(ies) which apply. As an aid the following definitions apply:

   Execution — The operator knew what to do but didn't do it.

   Knowledge — The operator didn't know the proper way to perform the job task, activity, etc.

   Procedure — The procedure or instructions were less than adequate.

   Design — The manner in which facilities or equipment are made or constructed was less than adequate.

   Equipment — Includes any type of wheeled, or portable devices, machinery or vehicles not functioning.

   Contractors — The performance on the part of an outside contractor was less than adequate.

   Materials — Includes all types of raw materials, process materials, finished goods, and maintenance and cleaning materials.

   Facilities — Pertains to deficiencies due to determination such as rough road, poor drainage, poor ventilation.

   Other (Specify) — This includes a specific problem which cannot accurately be identified through the use of the six categories above. When using this category, a written explanation must be included.

2. **Loss Analysis:** So that cost analysis can be made part of the overall safety analysis of accidents, near misses and hazardous conditions two estimates are needed. Estimated actual loss can range from no dollars for a near miss to many dollars for an injury. A rough figure is all that is necessary. If the accident could have been more severe or the near miss could have caused serious injury, this is where the Estimated Potential Loss is important. An explanation on how these estimates were arrived at will complete this portion.

3. **Additional Actions:** Simply state what additional actions are planned, who is responsible for the work and when it will be completed. This can be an estimate depending on the circumstances.

⊙abc EMPLOYEE ACCIDENT & INVESTIGATION REPORT

INSTRUCTIONS:

ANSWER ALL QUESTIONS

– PRINT ALL INFORMATION

– USE BALL POINT PEN OR TYPE

DISTRIBUTION: YELLOW COPY – ABC MEDICAL DEPT.   WHITE COPY – DEPT. HEAD (AFTER   BLUE COPY
GREEN COPY – ABC INSURANCE DEPT.   COMPLETION FORWARD TO ABC   CHAIRMAN
PINK COPY – DEPT. HEAD   INSURANCE DEPT.)   APPP COMMITTEE
                                                                               (DIRECTOR OF INSURANCE)

TO BE COMPLETED BY ABC NURSE OR SUPERVISOR

IS EMPLOYEE WORKING? ☐ YES ☐ NO | DEPARTMENT HEAD | SUPERVISOR

NAME OF EMPLOYEE | SOCIAL SECURITY # | ABC EMPLOYEE # | DEPARTMENT

ADDRESS OF EMPLOYEE | ☐ MALE ☐ FEMALE | AGE | JOB TITLE | LENGTH OF SERVICE ______ YRS. ______ MOS.

ADDRESS WHERE ACCIDENT OCCURRED (INCLUDE COUNTY) | WAS THIS ABC PREMISES ☐ YES ☐ NO

DATE OF ACCIDENT: ________ 19___ DAY OF WEEK ________ HOUR OF DAY ________ A.M. ________ P.M.   WAS INJURED PAID IN FULL FOR THIS DAY? ________

WHAT HAPPENED? BE SPECIFIC

STATE NATURE OF INJURY AND PART OR PARTS OF BODY AFFECTED (AS CUT RIGHT INDEX FINGER)

WAS MEDICAL CARE PROVIDED?   IF SO, WHEN?

NAME AND ADDRESS OF PHYSICIAN:

NAME AND ADDRESS OF HOSPITAL:

WITNESSES | DATE OF REPORT

PREPARED BY (NAME & TITLE)

IMPORTANT: THIS SECTION TO BE COMPLETED BY DEPT. HEAD. RETAIN PINK COPY AND FORWARD WHITE COPY TO ABC INSURANCE DEPT. PROMPTLY. BLUE COPY TO DIRECTOR OF INSURANCE

TO BE COMPLETED BY DEPT. HEAD

HAS EMPLOYEE RETURNED TO WORK? ☐ YES ☐ NO | IF SO, GIVE DATE | IS DESCRIPTION OF ACCIDENT ACCURATE? IF NOT, DESCRIBE FURTHER

WHY DID IT HAPPEN?
      HUMAN ELEMENT

      EQUIPMENT/ENVIRONMENT

WHAT HAVE YOU DONE TO PREVENT A RECURRENCE?
      EDUCATION/TRAINING

      EQUIPMENT/ENVIRONMENT

      PLANNED FOLLOW-UP

WILL EMPLOYEE BE PAID WHILE OUT? | WHEN WILL EMPLOYEE RETURN TO WORK?

EXPLAIN IF YOU HAVE TAKEN NO ACTION

DATE OF INVESTIGATION | INVESTIGATED BY (NAME & TITLE)

MEDICAL DEPARTMENT

Construction Safety Association of Ontario

74 Victoria Street, Toronto, Ontario M5C 2A5 • (416) 366-1501

ACCIDENT CAUSAL DATA

CLAIM NO. _______________________________

EMPLOYEE'S NAME____

DATE & TIME OF ACCIDENT____

BACKGROUND INFORMATION

1. Type of project where accident occurred (Examples: bridge, industrial, high-rise apartment) _______________

2. What type of construction was your company doing on this project? (Examples: carpentry, painting, home building) _______

ACCIDENT DETAILS

1. Employee activity at time of accident (Examples: grinding, walking, climbing, drilling)_______________

2. Tools, equipment, machinery involved (Examples: cement truck, hammer, rolling scaffold) _______________

3. What personal protective equipment was being worn? (Examples: safety glasses, goggles) _______________

4. (a) If worker or object fell, how far?_______________ft.

   (b) If object was being lifted or fell, how heavy was it?_______________lb.

5. Working surface at time of accident (Examples: ladder, ground, staging, trench)_______________

6. Condition of working surface (Examples: littered; wet, confined and muddy; icy and unprotected) _______________

7. Severity of injury or occupational disease (Examples: cut, bruise, sprain, hearing loss) _______________

ACCIDENT ANALYSIS

What were the apparent causes of the accident?

1. Environmental (Examples: wind, rain, mud, ice) _______________

2. Equipment (Examples: broken shovel, truck gears jammed) _______________

3. Human (Examples: fatigue, lack of training) _______________

4. Material (Examples: lumber piled loosely, cement spilled) _______________

5. Other

RF2   (See Over)

ACCIDENT PREVENTION

What action can or will be taken to prevent recurrence? (please indicate in appropriate box)

1. ☐ Worker Training   2. ☐ Equipment Inspection Procedure

3. ☐ Closer Supervision   4. ☐ Written Task Procedures

5. ☐ Better Jobsite Housekeeping   6. ☐ Provision & Use of Proper Equipment for the
                                                         Task
7. Other ____

OPTIONAL

APPROXIMATE INDIRECT COSTS	FOR REFERENCE ONLY
1. Salary to injured man while not working	$_____________________
2. Cost of cleanup time and equipment	$_____________________
3. Equipment loss or damage	$_____________________
4. Materials loss or damage	$_____________________
5. Costs for lost production time	$_____________________
6. Salary & training of temporary personnel	$_____________________
7. Insurance increases (Projected)	$_____________________
8. Delay in project completion (Projected)	$_____________________
9. Management time lost in handling 1 to 8 above	$_____________________

COMMENTS

State any further information about this accident which may be helpful.

____
____
____
____
____
____
____
____
____
____
____
____
____

SUPERVISOR'S ACCIDENT INVESTIGATION REPORT

COMPANY OR BRANCH		DEPARTMENT		
EXACT LOCATION		DATE OF OCCURRENCE	TIME ☐ A.M. ☐ P.M.	DATE REPORTED

PERSONAL INJURY		PROPERTY DAMAGE	
INJURED'S NAME		PROPERTY DAMAGED	
OCCUPATION	INJURED PART OF BODY	ESTIMATED COSTS $	ACTUAL COSTS
NATURE OF INJURY		NATURE OF DAMAGE	
OBJECT/EQUIPMENT/SUBSTANCE/INFLICTING INJURY		OBJECT/EQUIPMENT/SUBSTANCE/INFLICTING DAMAGE	
PERSON WITH MOST CONTROL OF OBJECT/EQUIPMENT/SUBSTANCE		PERSON WITH MOST CONTROL OF OBJECT/EQUIPMENT/SUBSTANCE	

DESCRIPTION

DESCRIBE CLEARLY HOW THE ACCIDENT OCCURED: ATTACH ACCIDENT DIAGRAM FOR ALL MOTOR VEHICLE ACCIDENTS.

ANALYSIS

WHAT ACTS, FAILURES TO ACT AND/OR CONDITIONS CONTRIBUTED MOST DIRECTLY TO THIS ACCIDENT?

WHAT ARE THE BASIC OR FUNDAMENTAL REASONS FOR THE EXISTENCE OF THESE ACTS AND/OR CONDITIONS?

LOSS SEVERITY POTENTIAL			PROBABLE RECURRENCE RATE		
☐ MAJOR	☐ SERIOUS	☐ MINOR	☐ FREQUENT	☐ OCCASIONAL	☐ RARE

PREVENTION

WHAT ACTION HAS OR WILL BE TAKEN TO PREVENT RECURRENCE? PLACE X BY ITEMS COMPLETED.

INVESTIGATED BY	DATE	REVIEWED BY	DATE

• THIS REPORT MUST BE
FILED WITHIN 2 WORK DAYS.

SCHERING-PLOUGH CORPORATION

INJURY REPORT

DATE OF REPORT

LOCATION OF OCCURRENCE

DATE AND TIME

☐ FIRST AID   ☐ MEDICAL TREATMENT   ☐ LOST TIME

NAME OF EMPLOYEE

DEPARTMENT/LOCATION

SUPERVISOR

TYPE OF INJURY		PART OF BODY INJURED		REMARKS BY MEDICAL PROFESSIONAL:
LACERATION	STRAIN, SPRAIN	FINGER	EYE	
ABRASION	PUNCTURE	HAND	HEAD	
BRUISE CONTUSION	DERMATITIS ALLERGY	ARM, WRIST	FACE	
THERMAL BURN	FRACTURE	TOE	SHOULDER	
CHEMICAL BURN	OTHER OCCUP. ILL. OR INJURY	FOOT	BACK, SPINE	
EYE IRRITATION	INFECTION	ANKLE	TRUNK, GROIN	
INHALATION	ANIMAL BITE	LEG	CHEST	
OTHER		OTHER		SIGNATURE

TO BE COMPLETED BY IMMEDIATE SUPERVISOR

HOW DID INJURY OCCUR?

IN YOUR OPINION, WHAT MIGHT HAVE BEEN DONE TO PREVENT THIS INJURY?

SUPERVISOR'S SIGNATURE

TITLE

TO BE COMPLETED BY SUPERVISOR AND/OR MANAGER

INJURY REVIEW

SIGNATURE

TITLE

TO BE COMPLETED BY SAFETY

SAFETY REVIEW

COPY DISTRIBUTION:
WHITE ⟶ HEALTH SERVICES ⟶ DIVISION SAFETY
CANARY ⟶ HEALTH SERVICES ⟶ EMPLOYEE'S SUPERVISOR
PINK ⟶ HEALTH SERVICES ⟶ EMPLOYEE'S SUPERVISOR ⟶ DIVISION SAFETY
GOLDENROD ⟶ HEALTH SERVICES ⟶ EMPLOYEE'S SUPERVISOR ⟶ LOCAL SAFETY UNIT

916-090-03
(REV. 3/76)

OCCUPATIONAL INCIDENT REPORT
PART III CORPORATE COPY
(COMPLETE AND TRANSMIT WITHIN 10 DAYS OF INCIDENT)

Pfizer

INCIDENT NO.

Division	Facility	Department Assigned	OSHA/MSHA Recordable ☐ Yes ☐ No	Date of Incident/Death

Employee's Name	Social Security No.	Age	Sex	Job Title

Time of Incident (24 hour clock) example: 10PM=2200	During overtime? ☐ Yes ☐ No	Years experience in the job with the company.

Did the incident result in a disability beyond the day of occurrence? ☐ Yes ☐ No If yes, complete boxes below:

LAST DAY WORKED	DAYS AWAY FROM WORK		SUB TOTAL (Days)	RESTRICTED WORK ACTIVITY		SUB TOTAL (Days)	TOTAL LOST WORKDAYS
	DATE BEGAN	DATE RETURNED TO WORK		DATE BEGAN	DATE RETURNED TO WORK		
		PENDING	NA		PENDING	NA	NA

Briefly describe the nature of the incident and the extent of injury/illness.

Would the confidential sharing of information regarding this incident be beneficial to prevent recurrence in other plants? ☐ Yes ☐ No

INJURY CODES
- ☐ 10 Occupational Injury
- ☐ 11 First Aid Treatment
- ☐ 12 Non-Employee Injury or Illness
- ☐ 21 Skin Disease or Disorder
- ☐ 22 Lung Disease from Dust
- ☐ 23 Respiratory Illness from Toxic Agents
- ☐ 24 Systemic Poisoning from Toxic Materials
- ☐ 25 Disorders from Physical Agents Other than Toxic
- ☐ 26 Disorders Associated with Repeated Trauma
- ☐ 29 All Other Occupational Illnesses

NATURE OF INJURY CODES
- ☐ 100 Amputation
- ☐ 110 Asphyxia
- ☐ 120 Burn or Scald (Heat)
- ☐ 130 Chemical Burn
- ☐ 140 Concussion
- ☐ 150 Contagious Disease
- ☐ 160 Contusion, Crushing, Bruise Intact Skin Surface
- ☐ 170 Cut, Laceration, Puncture
- ☐ 180 Dermatitis, Rash, Tissue Inflammation
- ☐ 190 Dislocation
- ☐ 200 Electric Shock, Electrocution
- ☐ 210 Fracture
- ☐ 220 Freezing, Frostbite
- ☐ 230 Hearing Loss or Impairment
- ☐ 240 Heatstroke, Sunstroke
- ☐ 250 Hernia, Rupture
- ☐ 260 Inflammation or Irritation of Joints
- ☐ 270 Systemic Poisoning
- ☐ 280 Pneumoconiosis
- ☐ 290 Effects of Radiation, Sunburn
- ☐ 300 Scratches, Abrasions
- ☐ 310 Sprains, Strains
- ☐ 400 Multiple Injuries
- ☐ 990 Occupational Disease Not Otherwise Classified
- ☐ 999 Occupational Injury Not Otherwise Classified

PART OF BODY CODES
- ☐ 100 Head
- ☐ 120 Ears
- ☐ 130 Eye(s), Optic Nerves and Vision
- ☐ 140 Face
- ☐ 144 Mouth (Lips, Teeth, Tongue, Throat, and Sense of Taste)
- ☐ 146 Nose (Nasal Passages, Sinus, and Sense of Smell)
- ☐ 200 Neck
- ☐ 300 Upper Extremities
- ☐ 310 Arm(s) (Above Wrist)
- ☐ 313 Elbow
- ☐ 320 Wrist
- ☐ 330 Hand (Not Wrist or Fingers)
- ☐ 340 Finger(s)
- ☐ 400 Trunk
- ☐ 410 Abdomen (Includes Internal Organs)
- ☐ 420 Back (Includes Back Muscles, Spine)

PART OF BODY CODES (Cont'd)
- ☐ 430 Chest (Includes Ribs, Breastbone, and Internal Chest Organs)
- ☐ 440 Hips (Includes Pelvis, Buttocks)
- ☐ 450 Shoulder(s)
- ☐ 500 Lower Extremities
- ☐ 510 Leg(s) (Above Ankle)
- ☐ 511 Thigh
- ☐ 513 Knee
- ☐ 515 Lower Leg
- ☐ 520 Ankle
- ☐ 530 Foot (Not Ankle or Toes)
- ☐ 540 Toe(s)
- ☐ 700 Multiple Parts (More than one part of body)
- ☐ 800 Entire Body System
- ☐ 850 Respiratory System (Lungs, etc.)
- ☐ Not Elsewhere Classified-[ ][ ][ ]*

SOURCE OF INJURY CODES
- ☐ 0200 Animals (Insects, Birds, Reptiles)
- ☐ 0500 Boilers, Pressure Vessels
- ☐ 0540 Pipes/Valves
- ☐ 0600 Boxes, Barrels, Containers, Packages
- ☐ 0700 Buildings, Structures
- ☐ 0900 Chemicals
- ☐ 0901 Acids
- ☐ 0910 Alkalies
- ☐ 1000 Clothing, Apparel, Shoes
- ☐ 1100 Coal and Petroleum Products
- ☐ 1200 Cold (Atmospheric)
- ☐ 1300 Conveyors
- ☐ 1400 Drugs, Medicines
- ☐ 1401 Biologic Products (Sera, Plasma, etc.)
- ☐ 1490 Other Medicinals
- ☐ 1500 Electric Apparatus
- ☐ 1520 Switchboard Bus Struct., Switches, Fuses
- ☐ 1600 Excavations, Trenches and Tunnels
- ☐ 1700 Flame, Fire, Smoke
- ☐ 1800 Food Products
- ☐ 1900 Furniture, Fixtures, Furnishings
- ☐ 2000 Glass Items (Not Bottles, Jars or Flasks)
- ☐ 2200 Hand Tools, Not Powered
- ☐ 2300 Hand Tools, Powered
- ☐ 2400 Heat (Atmospheric or Environmental)
- ☐ 2500 Heating Equipment
- ☐ 2600 Hoisting Apparatus
- ☐ 2700 Infectious or Bacterial Agents
- ☐ 2800 Ladders
- ☐ 3001 Agitators, Mixers, Tumblers
- ☐ 3400 Office Machines
- ☐ 3450 Packaging and Wrapping Machines
- ☐ 3600 Presses (Not Printing Presses)
- ☐ 3750 Saws
- ☐ 3850 Shears, Slitters, Slicers
- ☐ 3999 Machines, Not Elsewhere Classified
- ☐ 4000 Mechanical Power Trans. Apparatus
- ☐ 4100 Metal Items, Not Elsewhere Classified
- ☐ 4300 Mineral Items, Nonmetallic
- ☐ 4400 Noise
- ☐ 4500 Paper and Pulp Items
- ☐ 4600 Particles, Dust
- ☐ 4700 Plants, Trees, Vegetation

SOURCE OF INJURY CODES (Cont'd)
- ☐ 4800 Plastic Items
- ☐ 4900 Pumps and Prime Movers
- ☐ 5050 Sun
- ☐ 5060 Ultra Violet Equipment
- ☐ 5070 Arc Welding Equipment
- ☐ 5080 X-Ray and Fluoroscope Equipment
- ☐ 5099 Radiation, Not Elsewhere Classified
- ☐ 5100 Soaps, Detergents, Cleaning Compounds
- ☐ 5300 Scrap Debris, Waste Material
- ☐ 5400 Steam
- ☐ 5620 Highway Vehicles, Powered
- ☐ 5631 Handtrucks, Dollies, Non-powered Trucks
- ☐ 5635 Forklift and Other Powered Trucks
- ☐ 5638 Tractors and Other Towing Vehicles
- ☐ 5801 Floor of a Building
- ☐ 5810 Ground (Out of Doors)
- ☐ 5815 Ramps
- ☐ 5825 Runways, Platforms
- ☐ 5830 Sidewalks, Paths, Walkways
- ☐ 5840 Stairs, Steps
- ☐ 5845 Street or Road
- ☐ 5899 Working Surfaces, Not Elsewhere Classified
- ☐ 8801 Athletic Event
- ☐ 9800 Unknown, Unidentified
- ☐ Not Elsewhere Classified-[ ][ ][ ][ ]*

UNSAFE ACT CODES
- ☐ 050 Lack of Maintenance of Electrically Energized/Pressurized Equipment
- ☐ 100 Failure to Use Available Personal Protective Equipment
- ☐ 200 Failure to Secure or Warn
- ☐ 250 Horseplay
- ☐ 300 Improper Use of Equipment
- ☐ 305 Overloading
- ☐ 350 Improper Use of Hands or Body Parts
- ☐ 400 Inattention to Footing or Surroundings
- ☐ 450 Making Safety Devices Inoperative
- ☐ 500 Operating or Working at Unsafe Speed
- ☐ 550 Taking Unsafe Position or Posture
- ☐ 552 Entering Confined Spaces w/o Supervision
- ☐ 556 Exposed Under Suspended Loads
- ☐ 558 Exposure to Moving Materials or Equipment
- ☐ 600 Driving Errors
- ☐ 650 Unsafe Placing, Mixing, Combining
- ☐ 750 Using Unsafe Equipment (Not Defective)
- ☐ 998 No Unsafe Act
- ☐ Not Elsewhere Classified-[ ][ ][ ]*

ACCIDENT TYPE CODES
- ☐ 010 Struck Against
- ☐ 020 Struck By
- ☐ 030 Fall From Elevation
- ☐ 034 Fall from Vehicle
- ☐ 035 Fall on Stairs
- ☐ 036 Fall into Shafts, Floor Openings, etc.
- ☐ 050 Fall on Same Level
- ☐ 060 Caught In, Under or Between
- ☐ 080 Rubbed or Abraided
- ☐ 084 Foreign Matter in Eyes

ACCIDENT TYPE CODES (Cont'd)
- ☐ 085 Repetition of Pressure
- ☐ 121 Overexertion in Lifting Objects
- ☐ 122 Overexertion in Pulling or Pushing Objects
- ☐ 123 Overexertion in Wielding Objects
- ☐ 129 Overexertion, Not Elsewhere Classified
- ☐ 130 Contact with Electric Current
- ☐ 150 Contact with Temperature Extremes
- ☐ 151 General Heat
- ☐ 152 General Cold
- ☐ 153 Hot Objects or Substances
- ☐ 181 Inhalation of Toxic, Noxious or Radioactive Substances
- ☐ 182 Ingestion of Toxic, Noxious or Radioactive Substances
- ☐ 183 Absorption of Toxic, Noxious or Radioactive Substances
- ☐ 200 Public Transportation Accident (In which injured was Passenger)
- ☐ 300 Motor Vehicle Accident (In which injured was Occupant)
- ☐ 999 Unclassified, Insufficient Data
- ☐ Not Elsewhere Classified-[ ][ ][ ]*

HAZARDOUS CONDITION CODES
- ☐ 010 Improperly Compounded, Constructed or Assembled
- ☐ 015 Improperly Designed
- ☐ 025 Sharp
- ☐ 030 Slippery
- ☐ 035 Worn, Cracked, Frayed, Broken, etc.
- ☐ 061 Caught in Inrunning or Meshing Objects
- ☐ 062 Caught Between a Moving and a Stationary Object
- ☐ 100 Dress or Apparel Hazards
- ☐ 110 Lack of Necessary Personal Prot. Equip.
- ☐ 205 Excessive Noise
- ☐ 210 Inadequate Aisle Space, Exits, etc.
- ☐ 220 Inadequate Clearance (For Moving Objects or Persons)
- ☐ 230 Inadequate Traffic Control (On Premises)
- ☐ 240 Inadequate Ventilation (General)
- ☐ 250 Insufficient Workspace
- ☐ 260 Improper Illumination (Glare, etc.)
- ☐ 299 Environmental Hazards, Not Elsewhere Classified
- ☐ 300 Hazardous Methods or Procedures
- ☐ 350 Improper Assignment of Personnel
- ☐ 400 Placement Hazards (Material, Equip., etc.)
- ☐ 410 Improperly Riled (Piling)
- ☐ 430 Inadequately Secured Against Undesired Motion
- ☐ 500 Inadequately Guarded
- ☐ 530 Lack of or Inadequate Shoring
- ☐ 540 Underground (Electrical)
- ☐ 550 Uninsulated (Electrical)
- ☐ 580 Inadequately Shielded (Radiation)
- ☐ 590 Inadequately Labeled Materials
- ☐ 600 Hazards of Outside Work Environment
- ☐ 990 Undetermined—Insufficient Information
- ☐ 999 No Hazardous Condition
- ☐ Not Elsewhere Classified-[ ][ ][ ]*

*Fill in 3- or 4-number code found in Code Listing

SIGNATURE FACILITY SAFETY AND HEALTH COORDINATOR	DATE

| **Pfizer** | **OCCUPATIONAL INCIDENT REPORT**
PART I FACILITY COPY | **INCIDENT NO.** |

Division	Facility	Department Assigned	OSHA/MSHA Recordable ☐ Yes ☐ No	Date of Incident/Death

Employee's Name	Social Security No.	Age	Sex	Job Title

Time of Incident (24 hour clock) ________________ example: 10PM=2200	During overtime? ☐ Yes ☐ No	Years experience in the job with the company.

Did the incident result in a disability beyond the day of occurence? ☐ Yes   ☐ No **If yes, complete boxes below.**

LAST DAY WORKED	DAYS AWAY FROM WORK		SUB TOTAL (Days)	RESTRICTED WORK ACTIVITY		SUB TOTAL (Days)	TOTAL LOST WORKDAYS
	DATE BEGAN	DATE RETURNED TO WORK		DATE BEGAN	DATE RETURNED TO WORK		

Briefly describe the nature of the incident and the extent of injury/illness.

Would the confidential sharing of information regarding this incident be beneficial to prevent recurrence in other plants?   ☐ Yes ☐ No

What made the incident happen?

What action is being taken to prevent a recurrence?

Other information (for local use)

SIGNATURE	DATE

Form 8526-4   (3/84)

SAFETY & HEALTH INVESTIGATION REPORT
(MUST BE COMPLETED WITHIN 2 WORK DAYS OF OCCURRENCE)

REPORT DATE

SECTION ONE

CASE LOCATION
BLDG FLOOR PARKING LOT ETC

TYPE OF CASE
☐ FIRST AID   ☐ MEDICAL TREATMENT   ☐ LOST TIME   ☐ FATALITY
☐ INCIDENT (NEAR MISS)   ☐ PROPERTY DAMAGE   ☐ OCCUPATIONAL ILLNESS

CASE DATE	TIME	EMPLOYEE NAME	EMPLOYEE NO	WORK LOCATION	SUPERVISOR'S NAME

Description by Medical Professional: Include nature of injury, occupational illness, part of body, treatment and recommended restrictions of the patient

SIGNATURE	TITLE	APPROX TREATMENT TIME

TO BE COMPLETED BY IMMEDIATE SUPERVISOR

SECTION TWO

Describe what happened and list several probable causes. (See reverse of goldenrod copy for instructions)

In your opinion, based on the probable causes, what actions do you recommend?

SUPERVISOR'S SIGNATURE	EMPLOYEE NO	NAME (PRINT)	TITLE	APPROX INVEST TIME

TO BE COMPLETED BY THE NEXT LEVEL OF SUPERVISOR

SECTION THREE

What changes in methods, procedures or equipment will minimize or eliminate the probable causes identified?
Are there other causes? If so, what should be done about these?

DATE	COMPLETED BY	TITLE	SIGNATURE	APPROX REVIEW TIME

TO BE COMPLETED BY SAFETY

SAFETY REVIEW

DATE	REVIEWED BY	CASE NUMBER

FURTHER REVIEW

DATE	REVIEWED BY

DISTRIBUTION

WHITE	— Originator ⟶ Location Safety
CANARY	— Originator ⟶ Employee's Supervisor ⟶ Next Level Supervision ⟶ Files
PINK	— Originator ⟶ Employee's Supervisor ⟶ Next Level Supervision ⟶ Location Safety
GOLDENROD	— Originator ⟶ Employee's Supervisor ⟶ Next Level Supervision ⟶ Location Safety ⟶ Division Safety

Form Completion Instructions

Medical Professional ▬▬▬▬▬▬▬▬▬▬▬▬▬▬▬▬▬▬▬▬▬▬▬▬▬▬▬▬▬▬▬▬

Complete top third of form. Send white copy to Division Safety Services; send remaining copies to Expense Center Manager.

Injury — Describe fully. If chemicals were involved, list kind of chemicals. If occupational illness, mention possible source, e.g., animal dander, dust, fumes, etc.

Treatment — Describe. Note whether X-rays were taken; if splints, ace bandages, etc. were used. Describe any additional treatment required.

Disposition — Note whether patient was returned to work; sent home; transferred to hospital; given light duty, etc.

Supervisor ▬▬▬▬▬▬▬▬▬▬▬▬▬▬▬▬▬▬▬▬▬▬▬▬▬▬▬▬▬▬▬▬▬▬▬▬

Be complete. List any substandard acts or conditions which may have contributed to incident. Look for basic causes which may have "set up" situation. To assist you, here are some areas to consider in your investigation.

Equipment/Operation — Describe type, placement; correct type; list batch no. if appropriate; current inspection.

Weather — Describe if a factor.

Environment — Adequate lighting; ventilation; work area.

Procedures — Written, understood, followed.

Management — Direction, guidance, support.

Employee — Physical condition, mental condition; job experience, task knowledge; use of personal protective equipment; time spent away from job.

Recommendations

List the items in the methods, operations, equipment or management system which you feel may reduce or eliminate the possibility of the next incident. Note whether or not this situation is economically controllable.

Manager ▬▬▬▬▬▬▬▬▬▬▬▬▬▬▬▬▬▬▬▬▬▬▬▬▬▬▬▬▬▬▬▬▬▬▬▬▬▬

Review incident, investigation and recommendations. Describe what methods, equipment, operations, or management system changes you are making or recommending. If none, so state. If more assistance is required from Safety, request such assistance.

SUPERVISOR'S REPORT OF OCCUPATIONAL INJURY/ILLNESS

ACCIDENT NO.

EMPLOYER

COMPANY | DEPT | DIV., REF'Y., UNIT

ADDRESS | LOCATION CODE

INJURED EMPLOYEE

FIRST NAME | INITIAL | LAST NAME | HOME ADDRESS

AGE | SEX | OCCUPATION | YEARS SERVICE WITH CO | ON THIS JOB | SOC. SEC. NO

INJURY/ILLNESS INFORMATION

DATE OF OCCURRENCE | TIME A.M. P.M. | TIME AND DATE REPORTED TO CO A.M. P.M. | ADDRESS WHERE INJURY/ILLNESS OCCURRED | ON EMPLOYER'S PREMISES ☐ YES ☐ NO

EXACT LOCATION ON PROPERTY

NATURE OF INJURY/ILLNESS (INDICATE SPECIFIC INJURY TO EACH PART OF BODY INVOLVED)

IF INJURY/ILLNESS WAS NOT REPORTED TO COMPANY IMMEDIATELY - STATE REASON WHY

WAS THIS AN "INDEX INJURY" BY COMPANY DEFINITION? (SEE REVERSE SIDE) ☐ YES ☐ NO | IF "YES" CIRCLE CATEGORY NO 1 2 3 4 5 | IF TIME WAS LOST- DATE OF FIRST DAY OFF | WAS THIS A "RECORDABLE" INJURY/ILLNESS AS DEFINED BY FEDERAL STANDARDS ☐ YES ☐ NO | IF "YES" INDICATE CATEGORY IN BLOCKS BELOW

NON-FATAL CASES (SEE REVERSE SIDE FOR DEFINITIONS)

☐ WITH NO LOST WORKDAYS   ☐ WITH LOST WORKDAYS   ☐ OCCUPATIONAL ILLNESS WITH CODE NO. ______   ☐ **FATALITY**

WAS EMPLOYEE TRANSFERRED TO ANOTHER JOB OR TERMINATED BECAUSE OF THIS INJURY OR ILLNESS? ☐ YES ☐ NO

NAME OF DOCTOR | NAME OF HOSPITAL/CLINIC

ADDRESS | ADDRESS

CITY | DATE OF FIRST TREATMENT | CITY | DATE OF FIRST VISIT

INJURY/ILLNESS DESCRIPTION

DESCRIBE FULLY HOW ACCIDENT HAPPENED (INCLUDE WHAT EMPLOYEE WAS DOING WHEN INJURED AND NAME OF OBJECT OR SUBSTANCE WHICH DIRECTLY INJURED EMPLOYEE)

DO NOT WRITE IN THESE SPACES

NATURE

BODY PART

ACTIVITY

TYPE

UNSAFE ACTS

UNSAFE COND

PROT. EQUIP.

WITNESSES TO ACCIDENT:

PREVENTIVE ACTION (SEE REVERSE SIDE)

HOW COULD INJURY/ILLNESS HAVE BEEN PREVENTED?

WHAT ACTION HAS OR WILL BE TAKEN TO PREVENT SIMILAR OCCURRENCES?

ACTION WILL BE TAKEN ON OR BEFORE THIS DATE

DATE OF THIS REPORT | SUPERVISOR'S TITLE AND SIGNATURE | NOTED BY SUP'T. OR MGR.

CLV-1359-C

INDEX INJURY INSTRUCTIONS

If the injury described in this report resulted in any of the disabilities listed below, it is considered an "Index Injury" and should be so reported on the form by category number.

CATEGORY

1. DISABLING (Lost Time)
2. NON-DISABLING EYE INJURY requiring two or more treatments by a physician. (Excluding windblown particles in cases where there was no evident need for eye protection.)
3. NON-DISABLING FRACTURE
4. NON-DISABLING LACERATION requiring sutures.
5. OTHER NON-DISABLING REQUIRING MODIFIED WORK.

A MODIFIED WORK injury is an injury that prevents the injured man from performing all of the essential functions of his regular job. Assignment to less strenuous work, even though within the injured's job classification, shall be considered modified work.

DEFINITIONS FOR FEDERAL STANDARDS
(OCCUPATIONAL SAFETY AND HEALTH ACT)

OCCUPATIONAL INJURY is an injury such as a cut, fracture, sprain, amputation, etc., which results from a work accident or from exposure in the work environment.

RECORDABLE OCCUPATIONAL INJURIES AND ILLNESSES are any occupational injuries or illnesses which result in:

1) FATALITIES, regardless of the time between the injury and death, or the length of the illness; or
2) LOST WORKDAYS CASES, other than fatalities that result in lost workdays; or
3) NONFATAL CASES WITHOUT LOST WORKDAYS, which result in transfer to another job or termination of employment, or require medical treatment (as defined below), or involve loss of consciousness or restriction of work or motion. This category also includes any diagnosed occupational illnesses which are reported to the employer but are not classified as fatalities or lost workday cases.

MEDICAL TREATMENT includes treatment administered by a physician or by registered professional personnel under the standing orders of a physician. Medical treatment does NOT include first aid treatment (one-time treatment and subsequent observation of minor scratches, cuts, burns, splinters, and so forth, which do not ordinarily require medical care) even though provided by a physician or registered professional personnel.

WORK ENVIRONMENT is comprised of the physical location, equipment, materials processed or used, and the kinds of operations performed by an employee in the performance of his work, whether on or off the employer's premises.

OCCUPATIONAL ILLNESS of an employee is any abnormal condition or disorder, other than one resulting from an occupational injury, caused by exposure to environmental factors associated with his employment. It includes acute and chronic illnesses or diseases which may be caused by inhalation, absorption, ingestion, or direct contact.

The following listing gives the categories of occupational illnesses and disorders that will be utilized for the purpose of classifying recordable illnesses.

Illness Codes

21 Occupational skin diseases or disorders
22 Dust diseases of the lungs (pneumoconioses)
23 Respiratory conditions due to toxic agents
24 Poisoning (Systemic effects of toxic materials)

25 Disorders due to physical agents (other than toxic materials)
26 Disorders due to repeated trauma
29 All other occupational illnesses

PREVENTIVE ACTION INSTRUCTIONS

The value of this report depends largely on the time and effort you spend to secure proper answers to the questions in this "Preventive Action" section. To be most effective the following points should be considered in answering each of the questions:

"How Could Accident Have Been Prevented?"

Avoid general statements. State specifically what act should or should not have been performed to prevent the accident. Comments such as, "Be More Careful," "Use Greater Caution," and "Be More Safety Conscious" are of little value and should be avoided. If, in your opinion, the primary preventive measure depends on correction of working conditions or with the use of a specific tool or item of protective equipment, state the specific condition or item involved.

"What Action Has Been or Will be Taken to Prevent Similar Accidents?"

In determining your preventive action consider each of the following six questions, in sequence. Base your preventive action on your answers to these questions.

1. Was the operation necessary? If not, can it be eliminated?
2. Can the operation, tool, or equipment be changed to eliminate the hazard?
3. Can the operation, tool, or equipment be guarded to protect the employee?
4. Can the employee be protected by the use of protective clothing or equipment?
5. Will further instruction and training enable employees to avoid injuries from this cause? Are written procedures advisable?
6. Will stricter enforcement of safety rules and regulations prevent future accidents?

Don't limit preventive action to the injured employee or to the specific condition involved. Correct all other similar conditions and instruct all other employees doing the same type of work. In addition to stating what is to be done, say how you have done it, or plan to do it, such as: "Will hold safety meeting on this subject," "Each employee will be personally contacted and instructed," "Necessary steps being taken to correct condition in this manner—."

"Estimate Date Action Will be Taken"

Estimate the date all proposed preventive action will be completed. Setting this target date will serve to expedite follow-up and completion.

INDUSTRIAL HYGIENE APPLICATIONS

The following articles appeared in the *American Industrial Hygiene Association Journal*. They discuss two approaches to the computerization of industrial hygiene data. The first article, "A Functionally Oriented Computer System for Industrial Hygiene Exposure Data," by Verminski, Protopapas, and Toca, discusses a system that uses intelligent CRT terminals with plain English lead-through and prompting. The other article, "The Computerization of Industrial Hygiene Records," by Snyder, Bell, and Samelson, describes a system that approaches the problem through the use of input forms and a large mainframe data-base file. Both systems can be made compatible with a CSDS and thus are excellent addition modules. Both articles have been reproduced with the permission of the American Industrial Hygiene Association, Akron, Ohio.

A program for computerizing the results of industrial hygiene monitoring data has been developed and implemented. The methodology and approaches taken in data collection, retrieval, and utilization are presented.

The computerization of industrial hygiene records

PHILIP J. SNYDER, M.P.H., ZEB G. BELL, Jr., Sc.D. and RICHARD J. SAMELSON, B.S.
PPG Industries, Inc., Chemicals Group, One Gateway Center, Pittsburgh, PA 15222

introduction

The evaluation of employee exposure has always been a fundamental element in the practice of industrial hygiene. Today, however, a greater degree of reliance is being placed on the evaluation of employee exposure and its significance to occupational health and compliance with applicable regulations. The value of accurate and representative employee exposure data is becoming more widely acknowledged in the evaluation and design of engineering controls; in determining the need for, and type of medical surveillance and respiratory protection to be afforded, and often as an integral component of properly conducted epidemiological studies.

Recognizing the necessity and importance of these monitoring programs and faced with a growing accumulation of sampling data, there have been numerous efforts made by general industry to apply today's available computer technology to the retention and efficient utilization of industrial hygiene records. Often these efforts have been made in conjunction with, or as an integral part of a larger effort to computerize the data generated by the combined functions of the Industrial Hygiene, Medical, and Personnel Departments.[1-4]

Although such an integrated Occupational Health Surveillance System is being developed for PPG, this paper will only present the work successfully completed to date by the Chemicals Group in developing and implementing a computer program for industrial hygiene monitoring records.

PPG's Chemicals Group consists of nine principal manufacturing facilities located in the United States and Canada. These plants are involved in the manufacturing or use of several hundred different chemicals. This list includes, among others: chlorine, benzene, asbestos, chromium, ethylene dibromide, lead, perchloroethylene, trichloroethylene, mercury, and vinyl chloride monomer.

In order to maintain an active surveillance and monitoring program throughout the Chemicals Group, there is an industrial hygiene specialist located at each major manufacturing facility. These plant programs in turn receive direction and guidance from the industrial hygiene staff of the Department of Environmental Affairs, located in Pittsburgh. The necessary industrial hygiene analytical backup is available at each of our facilities and this is augmented by our more elaborate AIHA accreditted laboratory located in Barberton, OH.

program development and implementation

The Area and Personnel Computerized Monitoring Program was developed in 1975 with the technical assistance of in-house Management Information System personnel. The program has been fully operational since July 1, 1976, and has at the present time in excess of 20,000 analytical determinations on computer file.

The computer system has been written in the ANSI-COBOL programming language and

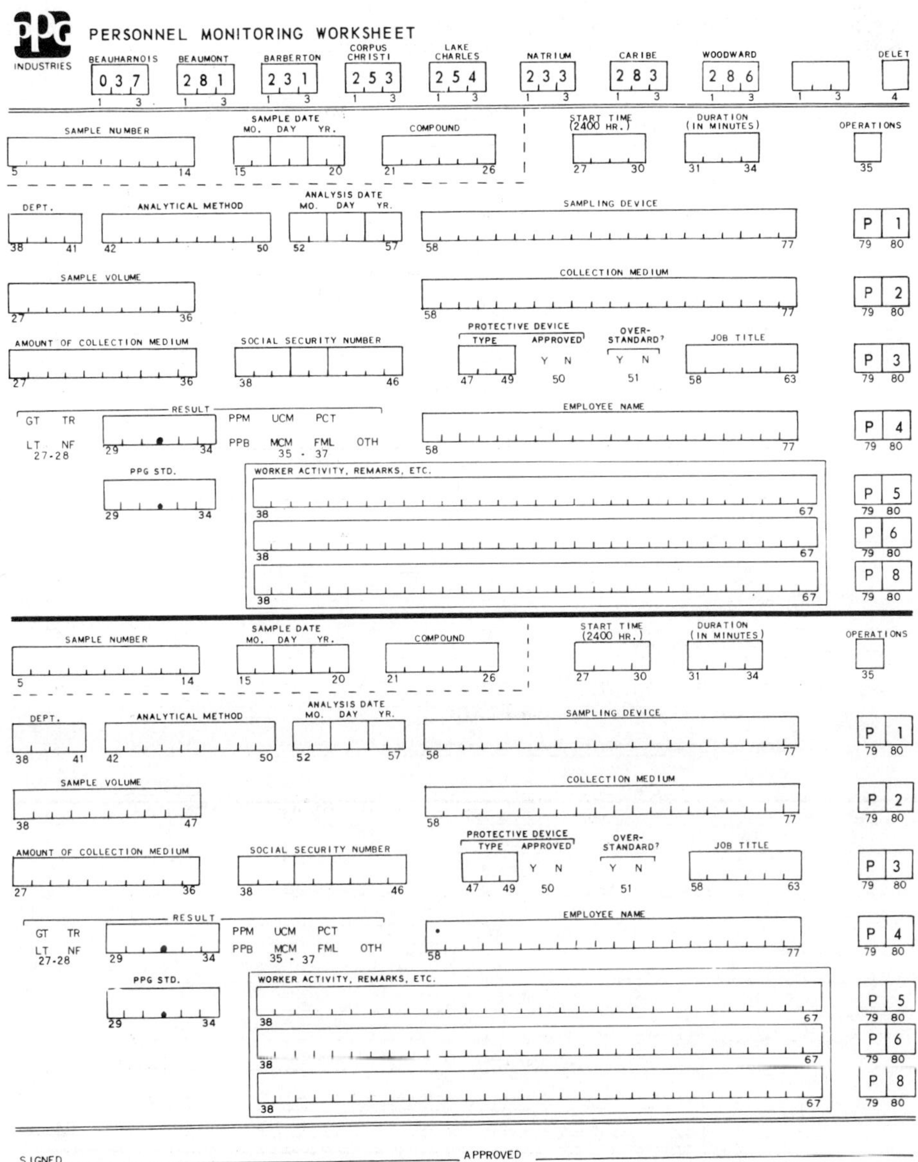

Figure 1 — Worksheet for reporting the results of personnel monitoring.

currently runs under an IBM 303X SVS system configuration. With few exceptions, most jobs run in the standard 256K region size. The maximum number of records is a function of hardware rather than software; however, this can be increased to accommodate system growth as required.

Although many benefits have been derived from the implementation of this system, the primary objective behind this undertaking was

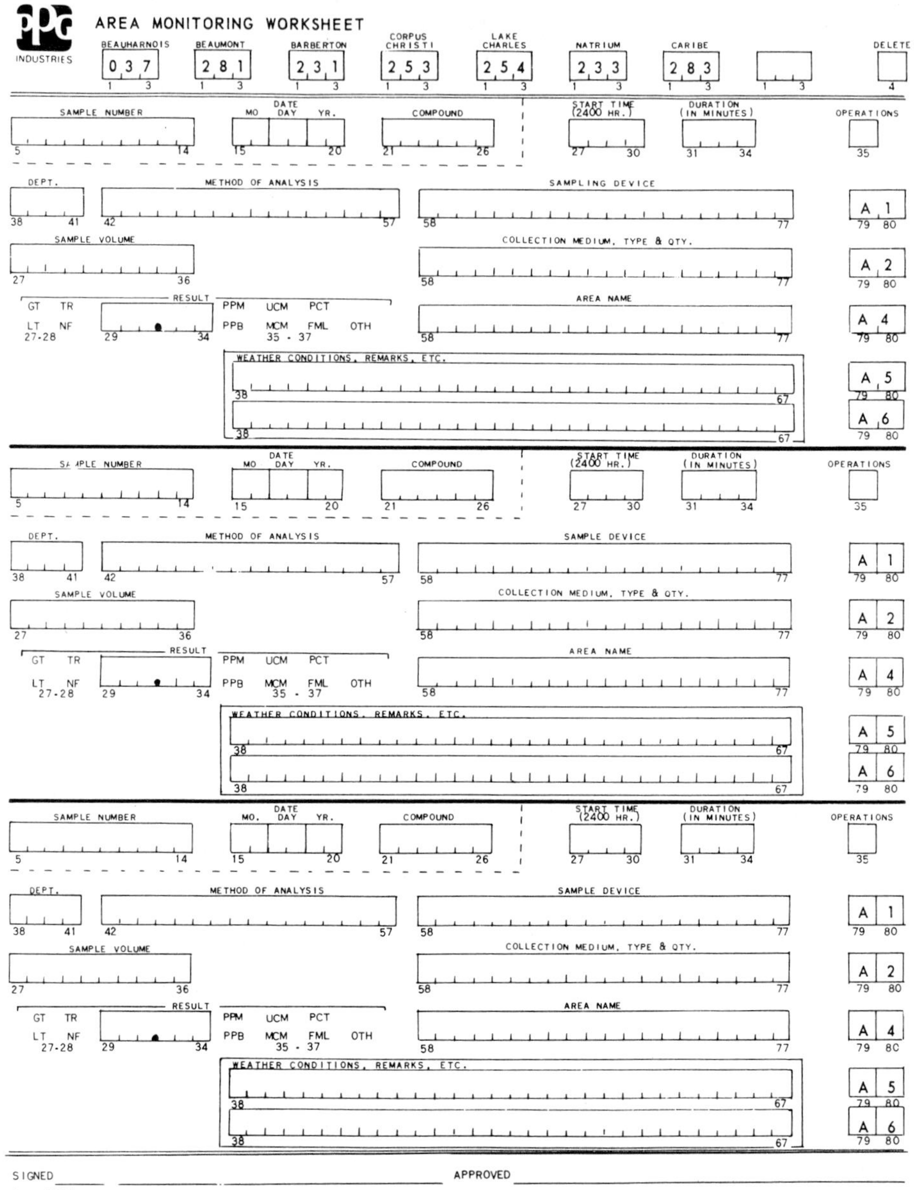

Figure 2 — Worksheet for reporting the results of area monitoring.

to facilitate the day-to-day utilization of the growing quantities of data on workplace exposure, while affording a permanent record-keeping capability.

For reasons of practicality, an effort has not been made at this time to place on the computer file the results of air sampling collected prior to this program's inception. While this does represent in some cases a considerable amount of air sampling, it has been our perhaps not uncommon experience that these older monitoring records are, in general, not as complete or as detailed as required by present recordkeeping standards.

data collection

It has been our experience that some of the most important considerations in developing a useful and viable recordkeeping computer program include: establishing what you want the program to accomplish, identifying what information will be required, and then incorporating this into the initial design of the system.

The worksheet shown as Figure 1 is an example of one of the four types of sampling data worksheets that were developed to select and standardize what information is to be reported when air samples are collected and analyzed. This particular form is one which is used typically for collecting the results of personnel monitoring.

To assure that all the required facts and observations pertaining to the sampling are recorded consistently and are in a coded form standardized for that plant, it was necessary to incorporate quality control checks into the program's system logic. Such control checks include character/numeric validation code verification using previously established tables. Entry codes are maintained by General Office personnel for the fields of "compound," "dept.," "protective device," "job title," "PPG std.," and the associated unit of measure. Data entry requires the proper completion of all appropriate fields and an accurate determination of over/under standard for each sample. When a sample worksheet is incomplete or whenever invalid codes are used, it will, upon being keypunched and processed, be flagged as being in error and requiring correction. Entire transactions must be reentered by either designated plant or General Office personnel.

Although personnel sampling is the predominate method used in assessing actual and representative employee exposure, a second form was developed to address the different types of information needed in evaluating the results of area air measurements. This sampling data worksheet is shown as Figure 2.

To accommodate the reporting of multiple analytical determinations of carbon tube sampling collected in areas of known potential exposure to more than one chemical, an expanded version of these worksheets was developed. Provided as an example is the Expanded Personnel Monitoring Worksheet (Figure 3). While this form requires the same information to be recorded as on the basic worksheet, up to 12 separate analytical determinations can be accommodated. A similar modification was made for the area sampling data worksheet.

After the results of the air sampling are recorded on the appropriate worksheets, the forms are keypunched and processed at the plant terminal. If remote job entry (RJE) is not yet available at the plant location, whereby plants may submit data and retrieve reports, the sampling worksheets are forwarded to the Corporate Computer Center for direct entry into the system. No interactive (i.e., immediate feedback) capabilities exist at the present time.

employee and management notification of monitoring results

When the results of personnel monitoring are recorded, a procedure has been established for simultaneously informing both the employee and his immediate supervisor. If the measurement was found to be in excess of PPG's Internal Standard, an investigation is also initiated to establish the cause of the exposure and what corrective action was taken.

Shown as Figure 4 is the general format of the second page of all personnel monitoring worksheets. By using a blocked carbon design between the pages of the form, this "Environmental Health Notification of Employee Exposure" is automatically completed whenever any sample is reported.

For measurements determined to be in excess of the applicable PPG Standard, there is a "Notice of Investigation and Remedial Action" on the reverse side of the exposure notification form (Figures 5a and 5b). To respond properly to such overstandard measurements and to increase worker awareness, both the employee

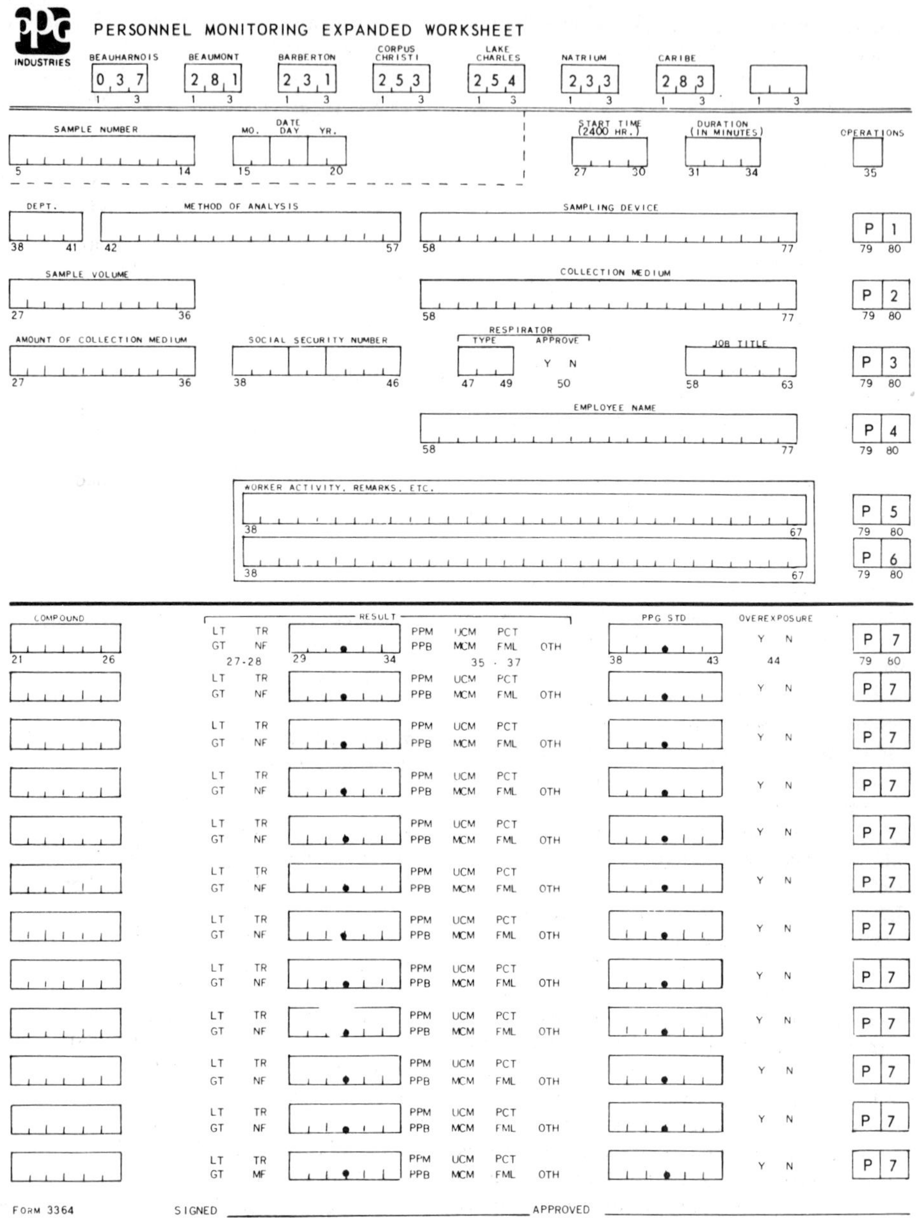

Figure 3 — Personnel monitoring worksheet used for reporting multiple analytical determinations.

and his or her supervisor are involved in completing this investigation. After the form is completed and their signatures obtained, the completed investigation form is kept on file for future reference. A copy of the employee notification is also provided to the Medical

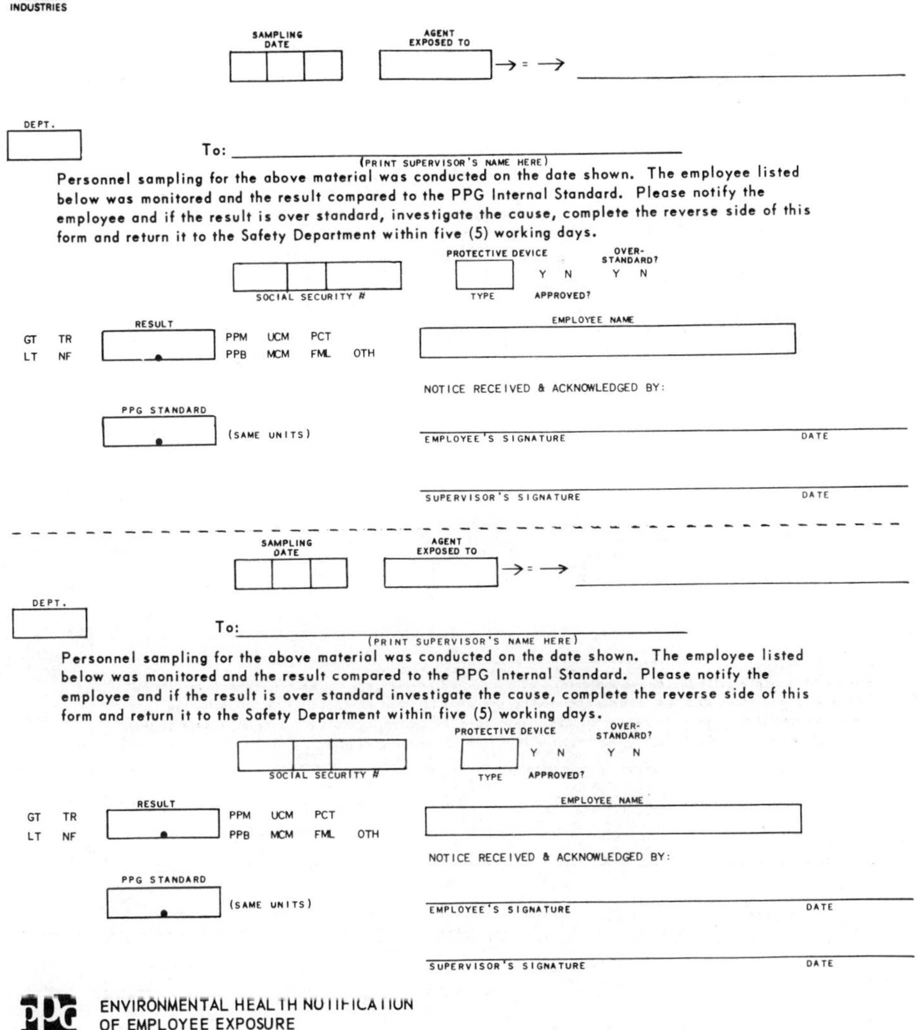

Figure 4 — Notification form for informing both employee and supervisor of monitoring results.

Department for review by the plant physician. Although this is only one way in which the employee is advised of his work-related exposure potential, it has been found to be an effective vehicle for following up on the results of employee monitoring.

retrieval programs

Once entered into the computer record file, all monitoring information is accessible by means of a master file printout which can be listed for any plant, period of time, type of chemical

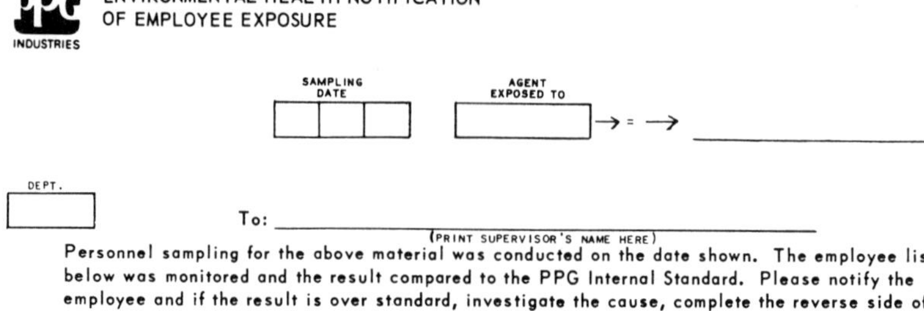

Figure 5a — Notice of investigation and remedial action, front side.

exposure, job category, and for any employee previously monitored.

To supplement this basic listing capability, several subprograms were developed to afford a systematic sorting of the more vital information on file. Through the use of specific retrieval reports, corporate, division, and plant managements are able to review the sampling information in a manner which is compatible with a rapid and efficient evaluation of the data.

ENVIRONMENTAL HEALTH NOTIFICATION OF EXPOSURE FOR:_________________________
(PRINT EMPLOYEE'S NAME)

Job Assignment:____

Was Appropriate Personal Protective Device Used? ☐ Yes ☐ No   If Yes, what Type _________________________

Cause of Overexposure:____

____

____

____

____

____

Remedial Action to be Taken:____

____

____

____

____

____

____

Notice of Investigation & Remedial Action Acknowledged By:_________________________
(EMPLOYEE'S SIGNATURE)

ENVIRONMENTAL HEALTH NOTIFICATION OF EXPOSURE FOR:_________________________
(PRINT EMPLOYEE'S NAME)

Job Assignment:____

Was Appropriate Personal Protective Device Used? ☐ Yes ☐ No   If Yes, what Type _________________________

Cause of Overexposure:____

____

____

____

____

____

Remedial Action to be Taken:____

____

____

____

____

____

____

Notice of Investigation & Remedial Action Acknowledged By:_________________________
(EMPLOYEE'S SIGNATURE)

FORM 3362 REV. 7-77

Figure 5b — Notice of investigation and remedial action, reverse side.

Report No. 1 (Figure 6) provides an opportunity to evaluate the results of large quantities of both employee and area monitoring in terms of the particular stress evaluated. In effect, this subprogram sorts the monitoring data on file by stress or type of

AREA/PERSONNEL MONITORING

COMPOUNDS SAMPLED

REPORT 1

03/01/76 TO 04/01/76   EXAMPLE

PERSONNEL INFORMATION
COMPOUND: NOISE
DURATION: TWA
OVERSTANDARD INDICATION

RESULT	PPG STD.	PROTECTION	DURATION	SAMPLE NO.	SAMPLE DATE	SSAN	JOB TITLE	DEPARTMENT	LOC.
191.000%	100.000	EAR PROTECTION	480	4201	03/23/76	219-90-3848	PIPEFITTER	MAINTENANCE	102
135.000%	100.000	EAR PROTECTION	480	4219	03/14/76	412-80-2140	BLACKSMITH	MAINTENANCE	102

NOT OVERSTANDARD
DURATION: TWA

RESULT	PPG STD.	PROTECTION	DURATION	SAMPLE NO.	SAMPLE DATE	SSAN	JOB TITLE	DEPARTMENT	LOC.
88.000%	100.000	EAR PROTECTION	495	4718	03/02/76	465-12-1019	MACHINIST	MAINTENANCE	102
57.000%	100.000	NO PROTECTION	475	4315	03/29/76	464-13-2119	WELDER	MAINTENANCE	102
35.000%	100.000	NO PROTECTION	480	4441	03/28/76	466-13-2919	PIPEFITTER	MAINTENANCE	102
29.000%	100.000	NO PROTECTION	480	4440	03/14/76	439-42-1515	WELDER	MAINTENANCE	102

NUMBER SAMPLES: 6

Figure 6 — Retrieval of monitoring data by type of exposure.

AREA/PERSONNEL MONITORING

SAMPLE COVERAGE

REPORT 2

03/01/77 TO 05/01/77   EXAMPLE

PERSONNEL INFORMATION ONLY
LOCATION: 102
DEPARTMENT: RESEARCH
JOB TITLE: CHEMIST

COMPOUND	RESULT	PPG STD.	PROTECTION	DURATION	SAMPLE NO.	SAMPLE DATE	SSAN
VINYL CHLORIDE	0.050 ppm	1.000	NO PROTECTION	445	5592	03/01/77	215-89-7629
VINYL CHLORIDE	0.200 ppm	1.000	NO PROTECTION	480	5942	04/20/77	217-42-2190
VINYLIDENE CHLORIDE	1.800 ppm	2.500	CARTRIDGE RESP	475	6111	04/21/77	314-42-2912

JOB TITLE: LAB TECHNICIAN

COMPOUND	RESULT	PPG STD.	PROTECTION	DURATION	SAMPLE NO.	SAMPLE DATE	SSAN
VINYL CHLORIDE	1.500 ppm	5.000	NO PROTECTION	15	5399	03/01/77	260-44-4689
VINYL CHLORIDE	0.100 ppm	1.000	NO PROTECTION	475	5418	04/02/77	260-44-4689

Figure 7 — Retrieval of monitoring data by job category.

<pre>
                         AREA/PERSONNEL MONITORING

                            PERSONNEL REPORT

                               REPORT 3

                         03/01/76/ TO 11/01/76   EXAMPLE

PERSONNEL INFORMATION ONLY
LOCATION: 102
SSAN: 214-89-7629
COMPOUND: PERCHLOROETHYLENE

                                                                SAMPLE   SAMPLE
  RESULT   PPG STD   PROTECTION   DURATION   DEPT.   JOB TITLE NUMBER   DATE

  1.600 ppm   100.000   NO PROTECTION   445   RESEARCH CHEMIST   5592   09/01/76

COMPOUND: TRICHLOROETHYLENE

  LT 0.100 ppm 100.000   NO PROTECTION   465   RESEARCH CHEMIST   5422   08/09/76
  LT 0.100 ppm 100.000   NO PROTECTION   445   RESEARCH CHEMIST   5592   09/01/76

COMPOUNT: VINYLIDENE CHLORIDE

  LT 0.300 ppm   2.500   NO PROTECTION   445   RESEARCH CHEMIST   5592   09/01/76
</pre>

Figure 8 — Retrieval of monitoring data by individual.

exposure evaluated and then again by the duration of the sampling period. The recorded measurements are listed by decreasing quantity value, thereby allowing a quick assessment of compliance with present or proposed regulations. To aid in making this type of determination quickly, the number of samples collected for any one type of exposure is totaled before the next type of exposure is listed. As with the other reports, a listing can be obtained over varying periods of time and within any combination of the major parameters.

To compliment Report No. 1 which presents the sampling data in terms of the *type of exposure,* Report No. 2 (Figure 7) sorts the same quantitative information by *job category monitored.* A review of this type of report:
— *Provides for a rapid comparison of the exposure potential between jobs and departments. This may be in one plant or across a division.*
— *Highlights possible problem areas where additional emphasis is warranted. This may be in the form of workpractices, medical surveillance, engineering, or personal protective equipment.*
— *Can assist in determining whether or not the personnel monitoring frequency is adequate for each job or occupational group. As an example, additional monitoring may be indicated based on representative exposure levels and the specified requirements of an OSHA standard. On the other hand, a reduction in sampling may be suggested based on a solid background of consistently low measurements.*

A third report (Figure 8) was designed to aid in assessing the actual exposure history of the *individual.* While this report alone provides valuable information to the industrial hygienist, plant physician, and management, an additional perspective on employee exposure can often be obtained when the measurements on a single individual are evaluated and compared to those of other employees assigned to the same job category. In some instances such a review may suggest the need for special individual attention.

A fourth report is capable of listing only those samples that may have been collected and analyzed to be *in excess of PPG Standards.* Although this report contains the same information that is retrievable by any one of the subprograms, it can, if necessary, selectively highlight only the apparent problem areas.

AREA/PERSONNEL MONITORING

SUMMARY REPORT

REPORT 5

09/01/76 TO 09/30/76   EXAMPLE

PLANT: 233

DEPARTMENT: PER/TRI

JOB CATEGORY: OPERATOR A

COMPOUND: TRICHLOROETHYLENE

MEAN TWA	CURRENT PPG STD	MAX TWA	NO. TWA SAMPLES	NO. EMPLOYEES ASSIGNED	NO. EMPLOYEES SAMPLED	NO. OVERSTD	NO. OVER-EXPOSURES	NO. EMPLOYEES OVERSTANDARD	DATE LAST SAMPLE
8.20	100.000 ppm	17.00	12	7	7	0	0	0	09/28/76

MEAN STEL	CURRENT PPG STD	MAX STEL	NO. STEL SAMPLES	NO. EMPLOYEES ASSIGNED	NO. EMPLOYEES SAMPLED	NO. OVERSTD	NO. OVER-EXPOSURES	NO. EMPLOYEES OVERSTANDARD	DATE LAST SAMPLE
51.00	200.000 ppm	69.00	8	7	1	0	0	0	09/29/76

MEAN INVEST SAMPLE	CURRENT PPG STD	MAX INVEST	NO. INVEST SAMPLES	NO. EMPLOYEES ASSIGNED	NO. EMPLOYEES SAMPLED	NO. > CT ALLOWED	NO. OVER-EXPOSURES	NO. EMPLOYEES OVERSTANDARD	DATE LAST SAMPLE
40.00	N/A	89.00	5	7	5	0	0	0	09/15/76

Figure 9 — Summary report of monitoring data.

The last report that was developed is, appropriately, a program to summarize all the data collected and recorded on the master file for a specified period of time. Taking into consideration several parameters which have been valuable in assessing the adequacy and coverage of a plant's monitoring program, the report format illustrated in Figure 9 was developed.

In considering the use of such reporting schemes, it is important to keep in mind that the value of the information generated is limited by one's ability to establish standardized and meaningful job categories and consistently apply this reporting logic when the results of monitoring are recorded. Other factors that should be considered in interpreting the significance of the data collected and summarized, include sampling coverage, adequacy, and appropriateness.

conclusion

This computer program has been found to be a vital part of our administrative system to assure employee exposures are being measured at the proper frequency and for the appropriate substances. The computer program helps direct our efforts rather than serving just as a simple method of data storage. Flexibility has been built into the design of the system and enhancements or additional retrieval reports can be developed as the need arises.

The available retrieval capability provides concise, summarized reports on employee exposure and this permits managers to assess if their programs to control exposures are effective, and where more emphasis may be required. As previously indicated, information to be retrieved can be varied selectively to meet the individual needs of those performing monitoring, medical surveillance, and health hazard evaluations. In addition, the impact of proposed regulations can be rapidly determined for any substance in terms of which job categories and departments will be most affected.

The development in the near future of a compatible computer program for handling the results of biological monitoring will further increase the value and utilization potential of this computer program. For example, if an employee's biological test results are abnormal, the program will highlight that employee for incorporation in the next round of personnel monitoring. Likewise, excessively high exposure determinations will, among other things, result in that worker being scheduled for more frequent medical evaluations. Ultimately, the tying together of these two programs will provide for the automatic scheduling of monitoring based on the results of both air and biological measurements.

Naturally, this program also serves as a permanent recordkeeping system, serving to meet the requirements of certain regulations and as an accessible data base for future epidemiological investigations. However, our primary concern is to identify and correct situations on a continuing basis rather than to compile data for retrospective research purposes.

references

1. **Ott, M.G.**: Linking Industrial Hygiene and Health Records. *J. Occup. Med. 19*:388-390 (1977).
2. **Barrett, C.D., et al.**: A Computerized Occupational Medical Surveillance Program. *J. Occup. Med. 19*:732-736 (1977).
3. **Kerr, P.S.**: Recording Occupational Health Data for Future Analysis. *J. Occup. Med. 20*:197-203 (1978).
4. **Jennings, H.R., et al.**: A Computerized Industrial Hygiene Program. *Plant Eng. 30*:149-151 (1976).

Accepted January 12, 1979

A research and development effort was undertaken by the Plastics Business Operations of the General Electric Company to establish a functional computer system for industrial hygiene exposure data using a zone concept for categorizing employee groups. Two basic file structures were established. The first was the factor data file which contains specific information on the potential exposure agents (such as CAS number, agent name, molecular weight, TLV, etc.) and provides the background data necessary to develop output reports. The second was the zone description which contains basic demographic information and exposure data for the zone. The overriding emphasis in the development of this system has been to develop a system of data retrieval which provides useful and timely outputs. The use of intelligent CRT terminals with plain English lead-through and prompting provides user orientation, immediate validation, and minimizes the amount of computer experience necessary to effectively utilize this system.

A functionally oriented computer system for industrial hygiene exposure data

E.R. VERMINSKI, M.S.[A], C. PROTOPAPAS[B] and F.M. TOCA, Ph.D.
[A]Plastics Business Operations, General Electric Company, Lexan Lane, Mt. Vernon, IN 47620; [B]Copeland and Roland, Inc., Dublin, OH 43017; [C]Plastics Business Operations, General Electric Company, Plastics Avenue, Pittsfield, MA 01201

introduction

Presently there is a considerable amount of activity within industry in the development of computer systems for handling industrial hygiene information. The scope, as well as the approaches, to this type of information handling is wide and varied. In order to develop a system that was meaningful for our particular organization, the following criteria were initially established:

> **First**, there must be identifiable functional applications of the system beyond archieval data storage . . . *Functionality*

> **Second**, the system must be user oriented and facilitate data entry and retrieval by personnel not generally familiar with computer operations . . . *Usability*

> **Third**, there must be some quality assurance mechanisms built-in in order to develop a meaningful information system . . . *Self Editing*

The Plastics Business Operations of the General Electric Company is in the process of implementing mechanisms for the computer compilation of industrial hygiene exposure data utilizing an exposure zone concept for categorizing employee groups. "A zone is the characteristic grouping of workers based on their job similarity and the similarity of the environment in which they work."[1] Because the zone concept is not fully applicable in all cases, it is also essential to provide a flexible mechanism for the handling and documentation of such data. For example, one employee in a zone may be assigned a task which only he or she performs on an infrequent basis. Any exposure encountered during this task would not typify exposures for the rest of the zone population. Therefore, a mechanism is needed to accommodate the entry and segregation of any exposure data for this specific task.

The first phase of this project, research and development, began with these initial concepts. It was conducted by Copeland & Roland, Inc. of Dublin, Ohio. Throughout the duration of this first phase, the authors collaborated closely to further define the project. As ideas rapidly evolved, it was realized that in order to provide scientifically valid analyses of these changing data, and at the same time not cause the need to modify the software continually, a universal file handling system would have to be designed. That is, a system needed to be developed which could adapt to the changes in data types and amounts on the data file and still provide access for the analysis of data as this phase progressed.

The result was a system which accessed user-defined displays to edit and display the data kept in the data files. These user-defined displays could be changed at any time by the user, and as old data files were later reviewed, the data were linked to their correct positions on the new display. If the new display contained locations for data not found in the data file, holes were created in the data file for the new information. Conversely, if the new display omitted data that had been previously included, the user was notified of the omission, and that data was not displayed.

The data files themselves consisted of sequential collections of data. In the development stage there were no fields or other limitations imposed upon the field structure; it was purely free-form of data for the various types requested by the displays. Data link words were keys by which the computer recognized these various data types. This provided the link between the data and the display.

Thus, as the project evolved, and the authors discovered the need to collect new types of data, the computer system was readily adapted to the change.

The final results of the research and development phase were the definitions of the types of data files needed and the

FACTOR DATA

① CAS #   75-09-2   ⑤ DATE   : 3-30-79

② FACTOR   : METHYLENE CHLORIDE   ⑥ TLV   : 700 mg/m^3 STEL: 870 mg/m^3

③ PHYSICAL FORM : COLORLESS LIQUID   ⑦ OSHA STD: 1740 mg/m^3 ACTION LEVEL: 870 mg/m^3

④ MOLECULAR WEIGHT: 84.93   ⑧ ZONES   : SM09, PV02, AB04

⑨ ROUTES OF ENTRY : INHALATION, SKIN

    LOCAL EFFECTS   : DRY, SCALY, FISSURED DERMATITIS

        LIQUID AND VAPOR IRRITATE EYES AND UPPER RESPIRATORY TRACT

        LIQUID MAY BE CAUSTIC

    SYSTEMIC EFFECTS: MILD CNS DEPRESSANT

        ELEVATES CO Hb

        LIVER AND KIDNEY DAMAGE, AT HIGH CONCENTRATION IN ANIMALS

    MEDICAL SURREVILLANCE: NERVOUS & RESPIRATORY SYSTEM, LIVER, KIDNEY, SKIN, EYES, & COMPLETE BLOOD COUNT

    CERTIFICATION: RESPIRATOR

    PERSONAL PROTECTIVE MEASURES: GLOVES AND RESPIRATORY PROTECTION

    REFERENCES: 1) NIOSH 77-1B1 209-210   2) PROCTOR: CHEMICAL HAZARDS OF THE WORK PLACE. 342-343 1978

    COMMENTS: (TOXICITY INFORMATION OR ANY SPECIAL PRECAUTIONS)

Figure 1 — Factor Data Sample.

data therein which were necessary to provide data retrieval and results meaningful to the industrial hygienist. Equally as paramount was the necessity for ease of operation of the entire system by noncomputer oriented personnel.

These results are discussed in greater detail in the following sections.

user oriented

One of the major obstacles to computer systems for industrial hygiene exposure data in the past has been the increase in man hours to meet the demands of the system. The use of key punch operations as well as mark sense forms are burdensome and include many opportunities for inputting errors. In order to promote user acceptance (for various plant sites) and to provide quality assurance mechanisms, this system was developed utilizing intelligent Cathode Ray Tube (CRT) terminals with emphasis on man/machine interface functions (such as conversational lead through and prompting for inexperienced operators). Immediate error messages are displayed when data entered is incorrectly formatted or incomplete.

In addition to providing a user input mechanism that provides ease of data entry, it is also essential to provide the user with information retrieval and outputs that can be obtained in a timely fashion. Again the intelligent terminal with a printer suits these needs. Exposure data can be input through a CRT that has quality assurance mechanisms such as data entry editing. Immediately after entry, this data plus all previously entered data can be reported and printed through a flexible menu type selection program.

development of functional data elements

For the purpose of formatting data, two basic files were established. The first was the factor data file which contains specific information on the potential exposure agents and provides the background data necessary to develop reports and functionality in the system. The second was the zone description which contains the background information and exposure data for the zone.

The basic quality control mechanisms are described with the data element descriptions below.

factor data file

Figure 1 illustrates the information format for all potential exposure agents including chemical substances or physical agents and is basically identical to the CRT screen display.

The term factor will herein be used to encompass both chemical substances and physical agents.

① The first item is the Chemical Abstract Service Registry number (CAS) for the factor. This number is used as the basic link between exposure data for a specific factor and information on that factor. It is recognized that the CAS number has some limitations.[2] However, due to TSCA and a basic industry trend, the CAS number appears to have the most universal application.

Where synonyms exist, the name used in specific standards or the TLV list is used. Where there are no standards or TLV's, the name most commonly used in the particular business concerned is used.

In order to cope with physical agents or chemical substances without CAS numbers, a system or number assignment was developed. Because in industrial hygiene exposure monitoring it is necessary to sample for individual constituents of a mixture, each mixture must have its makeup defined, and a separate entry must be made for each in the factor data file.

② The second item is the factor (potential exposure agent). The input here is the actual chemical name followed by an abbreviation with a limited number of characters to be used in the zone description file. For example, tetrahydrafuran would be entered with (THF) beside it.

③ The physical form is entered as background information with respect to the exposure potential.

④ The molecular weight of the factor is entered to provide data consistency for output reports. A conversion from ppm to mg/m^3 can readily be made where data have been input in both units over a period of time. Thus, units can be standardized for data comparisons.

⑤ The fifth item is the date which defines the last time the factor data was updated. The purpose of this entry is to keep the information up to date with changing standards or other information.

⑥ & ⑦ Applicable American Conference of Governmental Industrial Hygienists TLV's® and OSHA standard information is input here. This area has the capability of including any type of limit desired by the user. This could include NIOSH recommended standards as well as "in-house" standards as an arbitrary action level for situations where no OSHA action level has been promulgated. Including this information provides the specific output capacity to compare exposure data to a standard of some sort and to provide better perspective of exposure monitoring results to management.

⑧ All zones in which this factor is listed as having a potential employee exposure are listed in this file. This provides a quick reference to locate where potential exposures to specific agents exist.

⑨ The remainder of the factor data file deals with background information for quick reference. This enables plant physician or nurse to readily call up information for a brief outline on a factor plus specific reference locations for further information.

The input of the factor data information on a CRT has specific program emphasis on conversational lead through, where the CRT tells you what the next step is, and field edits, so that mistakes are either not accepted by the computer or an error message provides a means for immediate correction.

zone description

Figure 2 illustrates the input format for exposure data within a zone. This is also basically identical to the CRT screen display. The key issue here is to provide adequate differentiation of data inputs to provide functional outputs. There is a great volume of records information on exposure data to include equipment calibration information, specific information on analytical equipment, climatic conditions, etc. While it is essential to record and keep this type of information, this task may be accomplished better through manual recordkeeping or microfilming of records. Emphasis has been placed on data elements that are needed for the manipulation of data, quality assurance, and output records that will give a historical view of exposures.

The items above the columns provide basic demographic information for the zone. This provides the basic identification parameters for a particular zone which is flexible enough to accommodate groups of employees as well as particular employee work areas.

① The zone code is displayed which includes some basic information on the business and location (or group of people) of the zone. This code generally identifies the site, location, and process that has been sampled.

② The definition date is very important since it describes when changes in potential exposures (*e.g.*, addition or deletion of a factor) have occurred. It also provides a mechanism to describe the history of a zone. Whenever changes occur, the old zone definition date sample information is kept. This will reflect the changes within a zone when historical retrieval is undertaken in the future.

③ The total number of personnel in the zone is displayed.

④ A basic description of the activities of the zone is displayed. This can be an area title or a description of the activities of a group of employees.

The columns on this display provide the data input format. All previously entered data are either "scrolled-up" or paged so that new data input is placed at the end. Once this information has been input, edited, and verified, a program resequences the data placing a newly added factor value in sequence with like factors.

The following is a description of the column information:

⑤ *Date.* This is the date of the sample in a normal MM-DD-YY sequence.

⑥ *Time.* The time of day (on a 24 hour clock) that the sample was initiated is input to differentiate shifts or just morning versus afternoon samples.

⑦ *CAS#.* The Chemical Abstracts Service number is again the basic link back to factor data information for output information. One of the features of this program is that, after the CAS number is entered, the computer will search the factor data file to insure that this factor is on file and then

ZONE DESCRIPTION

① CODE #   AB04   ③ TOTAL PERSONNEL: 41

② DEFINITION DATE   2-23-79   ④ ACTIVITIES   : Resin Building, 2nd Floor

⑤ DATE	⑥ TIME	⑦ CAS #	⑧ FACTOR	⑨ LEVEL	⑩ METHOD	⑪ DUR.	⑫ EXPLANATION	⑬ EMPLOYEE #
11-10-78	0800	71-43-2	Benzene	.042 mg/m^{3}A	CT-GC-I	420	Centrifuge area	
12-20-78	0820	71-43-2	Benzene	.035 mg/m^3	CT-GC-I	470		1834
12-20-78	0820	71-43-2	Benzene	.043 mg/m^3	CT-GC-I	470		2053
01-04-79	0825	71-43-2	Benzene	.032 mg/m^3	CT-GC-I	455		1674
01-04-79	0830	71-43-2	Benzene	.039 mg/m^3	CT-GC-I	450		2509
01-04-79	0835	71-43-2	Benzene	.036 mg/m^3	CT-GC-I	455		1908
01-11-79	0835	71-43-2	Benzene	.05 mg/m^3	CT-GC-I	455		1632
01-11-79	0836	71-43-2	Benzene	.08 mg/m^3	CT-GC-I	454		1205
03-08-79	0830	71-43-2	Benzene	.18 mg/m^3	CT-GC-I	435		1834
11-10-78	0800	67-66-3	Chloroform	.021 mg/m^{3}A	CT-GC-I	420	Centrifuge area	
12-20-78	0820	67-66-3	Chloroform	.009 mg/m^3	CT-GC-I	470		1834
12-20-78	0820	67-55-3	Chloroform	.013 mg/m^3	CT-GC-I	470		2053
01-04-79	0825	67-66-3	Chloroform	.014 mg/m^3	CT-GC-I	455		1674
01-04-79	0830	67-66-3	Chloroform	.067 mg/m^3	CT-GC-I	450		2509

Figure 2 — Zone Description Sample.

automatically print out the factor name or its appropriate abbreviation if it is on file. If the factor is not on file, the computer will produce a message to that effect. This provides an immediate edit mechanism and quality assurance for the input of data. Where a factor is not on file, the necessary background information must be developed; and the new factor added to the file before data on that factor can be entered.

⑧ *Factor*. The name of the factor or its designated abbreviation in the factor data file is automatically displayed upon accurate entry of the CAS number.

⑨ *Level*. The measured level of the substance is entered with its appropriate units. The numerical level is first checked for numerical units only and preceded by a < or > sign only. Next the units are checked against a table of recognized units. The data element is not accepted until an appropriate unit is entered.

If no code follows the level and units, the sample was a personal sample, and an employee number must be placed in the last column. The codes A for area or B for breathing zone allow the input of these types of sample data. In the case of an area sample, no number will be placed in the last column. A range of specific level values more accurately describes a single sample value depending on the sample and analytical method. The sample and analytical information located in the next column can be used to help establish these ranges. For example, when NIOSH sampling and analytical methods are used as a guide, the sample, and analytical method will have a corresponding Coefficient of Variation documented under precision and accuracy in the NIOSH Manual of Analytical Methods.

⑩ *Method*. The method input is a sequence of three codes: sample method, analytical method, and laboratory. This sequence gives the basic background information on how the sample was taken, analyzed, and the type of laboratory used for analyses. Each part of this method sequence is immediately edited against the codes listed below as the information is entered.

SAMPLE METHOD CODES:

CT = Charcoal Tube
MI = Midget Impinger
FC = Filter, Closed Face
FO = Filter, Open Face
SG = Silica Gel Tube
PD = Passive Dosimeter
DT = Detector Tube
LT = Long Duration Detector Tube
TT = Tenax Tube
SL = Sound Level Meter
ND = Noise Dosimeter
DR = Direct Reading Instrument
PT = Phosgene Tape
O = Other

ANALYTICAL METHOD CODES:

AA = Atomic Absorbtion Spectroscopy
C = Colorimetric
GC = Gas Chromatography
IE = Ion Specific Electrode

UV = Ultra Violet
G = Gravimetric
MC = Microscope Count
O = Other

LABORATORY CODES:

I = Internal
GE = GE Accredited Lab
OA = Outside Lab Accredited
O = Outside Lab

⑪ *Duration*. The length of time of the sample in minutes is entered here, providing an indication of whether or not the sample duration was sufficient to determine an 8 hour TWA. Since specific OSHA standards require a minimum of 7 hours of sampling, this column allows the selection of data to examine results under this criteria or any other duration criteria desired. In many instances it is just not possible to get a 7 hour sample, but sample information of shorter duration is essential for documentation and analysis. Frequent shorter duration sampling is also essential for epidemiologic surveillance purposes.

If there is no code after the duration, the sample is a full period single sample greater than 420 minutes. The Code C would indicate a full period (> 420 minutes) consecutive sample, and the code P would indicate a partial period sample including a partial period consecutive sample (< 420 minutes).

⑫ *Explanation*. This column allows the entry of a single code (below), in the first space which in itself can provide background information. In the remainder of the field other codes or clear text can be entered to further describe the circumstances of the sample. This enables the entry of exposure data for unusual circumstances that can be separated from data of normal activities. It also allows for the consideration of respiratory and hearing protection. This column can also be used to further identify the data to the extent the user requires. For example, a universally applied grid system might be used for area samples, and exact location information could be input at this point.

EXPLANATION CODES:

Blank = Normal Activity
I = Infrequent Activity (less than a few hours per week)
S = Spill Condition
U = Upset Condition
R = Respiratory Protection Used During Sampling
H = Hearing Protection Used During Sampling

⑬ *Employee number*. The purpose of this column is to provide the capability to get back to individuals where personal samples have been taken. Here, yearly summaries of personal exposure information can be automatically compiled. This mechanism can also be used to inform employees of their present and past experiences. It also provides an important interface with the medical module for providing epidemiologic background information.

output reports

The critical aspect in the development of this system is to demonstrate functionality through useful output reports.

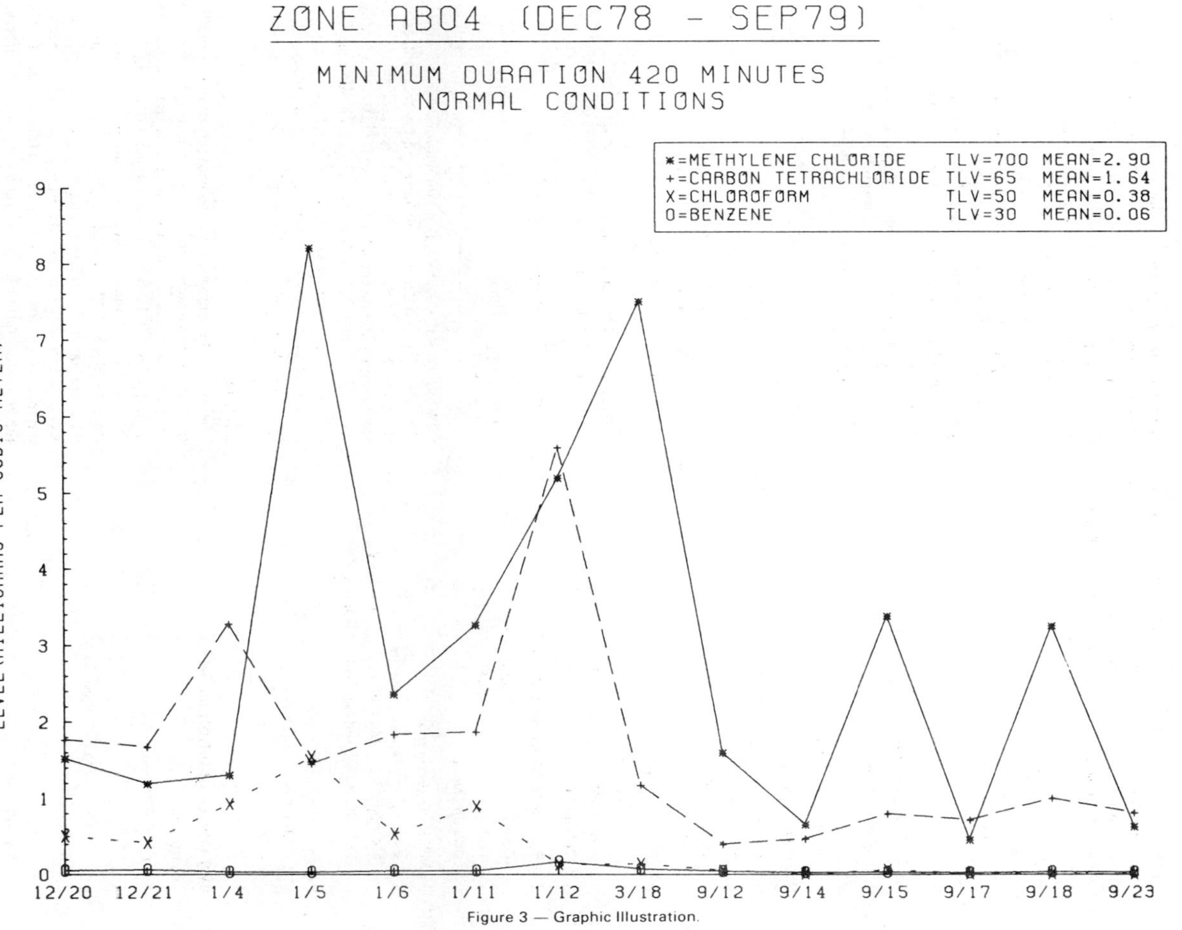

Figure 3 — Graphic Illustration.

This is facilitated by the data input format. The most basic and most flexible output is a selection program. Specific selection criteria in menu format are available to the user. The menu is presented through a sequence of screen displays.

    BY ZONE (Screen 1)

        Site Code: __________

        Zone Code(s): ______   ______   ______   ______

                       ______   ______   ______   ______

        Default: All Zones

    BY FACTOR (Screen 2)

        CAS Numbers: ______   ______   ______   ______

                     ______   ______   ______   ______

        Default: All Factors

    BY DATE (Screen 3)

        Period __________ Through __________

        Month to Date: __________

        Year to Date: __________

        Default: All

    BY LEVEL (Screen 4)

        < Amount __________ Unit ________

        > Amount __________ Unit ________

        Default: All

    OVER STANDARD (Screen 5)

        OSHA __________   Action Level __________

        TLV __________   STEL   __________

        Default: None

    BY METHOD (Screen 6)

        Sample   Analytical
        Method: ____   Method: ____   Laboratory: ____
        Default: All   Default: All   Default: All

    BY DURATION (Screen 7)

        ≤ Minutes __________

        ≥ Minutes __________

        Default: All

    BY SAMPLE CONDITION (Screen 8)

        Normal: ________

        Abnormal: ________

        Default: All

    BY TYPE OF SAMPLE (Screen 9)

        Personal: ________

        Area: ________

        Default: All

    BY EMPLOYEE NUMBER (Screen 10)

        Employee Number: ________

        Default: All

This selection process may be by one or more of the above mentioned criteria. Upon selecting the desired criteria, the user is provided a printed copy of the information requested that is in column format identical to the Zone Description (Figure 2). At the end of the report, a listing is given of the zones selected and the number of exposure data elements selected for each zone.

Other packaged statistical and graphic systems are used to analyze and present data reported from this system. These packages are not an integral part of this program at this time. The output programs from this CRT system are used to establish the data base.

Figure 3 illustrates a simple example of a data selection output in graphic form. In this example the selection criteria, which is printed prior to the report, included:

Zone	= AB04	
Factor	= 71432 (Benzene)	76663 (Chloroform)
	= 56235 (Carbon Tetra-chloride)	75092 (Methylene Chloride)
Date	= 12/01/78 through 9/30/79	
Level	= Default: All	
Standard	= Default: None	
Method	= Default: All	
Duration	= Minutes > 420	
Sample Condition	= Normal	
Type of Sample	= Personal	
Employee Number	= Default: All	

While complex statistical manipulations can be made, figure 3 demonstrates an elementary ability to graphically illustrate selected data through a packaged plot system. The total number of data entries found under the stated selection parameters, the arithmetic mean, and a graphic comparison of this data can be used as a basic report of exposure information to management and also as a method of notifying employees in a zone of their exposure data.

summary

The overriding emphasis in the development of this industrial hygiene exposure data computer system has been to provide a functional system that specifically addresses issues of data retrieval. In order to accomplish this, essential data elements were defined; then entry procedures and formats were developed. Efforts were made to simplify the procedures for data entry and to reduce the total volume of data. Key elements in the factor data provide the ability to generate specific reports. A simplified format for exposure data in the zone description facilitates ready access to the data as well as easy data interpretation.

The provision for direct benefit to the user, in conjunction with the ease of data entry, facilitates user acceptance and cooperation. The intent of the system is work reduction, and this must be demonstrable. Redundant record keeping requirements are minimized by providing sorted input information in printout form immediately after data entry. However, it must be emphasized that the intent of this system is not to totally eliminate requirements for manual record keeping.

The use of intelligent CRT terminals greatly facilitates the use of file checks, error messages and immediate editing. Therefore, data quality controls enhance the reliability of the information. The plain English conversation lead-through and prompting minimizes the amount of computer experience necessary to effectively utilize this system.

references

1. **Corn, M. and N.A. Esmen:** Workplace Exposure Zones for Classification of Employee Exposures to Physical and Chemical Agents. *Am. Ind. Hyg. Assoc. J. 40*:47-57 (1979).
2. **Socha, G.E., R.R. Langer, R.D. Olson and G.L. Story:** Computer Handling of Occupational Exposure Data. *Am. Ind. Hyg. Assoc. J. 40*:553-561 (1979).